STOCHASTIC GLOBAL OPTIMIZATION

Springer Optimization and Its Applications

VOLUME 1

Aims and Scope
Optimization has been expanding in all directions at an astonishing rate during the last few decades. New algorithmic and theoretical techniques have been developed, the diffusion into other disciplines has proceeded at a rapid pace, and our knowledge of all aspects of the field has grown even more profound. At the same time, one of the most striking trends in optimization is the constantly increasing emphasis on the interdisciplinary nature of the field. Optimization has been a basic tool in all areas of applied mathematics, engineering, medicine, economics and other sciences.

The series *Springer Optimization and Its Applications* publishes undergraduate and graduate textbooks, monographs and state-of-the-art expository works that focus on algorithms for solving optimization problems and also study applications involving such problems. Some of the topics covered include nonlinear optimization (convex and nonconvex), network flow problems, stochastic optimization, optimal control, discrete optimization, multi-objective programming, description of software packages, approximation techniques and heuristic approaches.

STOCHASTIC GLOBAL OPTIMIZATION

By

Anatoly Zhigljavsky
School of Mathematics
Cardiff University, UK

Antanas Žilinskas
Institute of Mathematics and Informatics
Vilnius, Lithuania

Anatoly Zhigljavsky
School of Mathematics
Cardiff University, UK

Antanas Žilinskas
Institute of Mathematics and Informatics
Vilnius, Lithuania

ISBN: 978-1-4419-4485-6 e-ISBN: 978-0-387-74740-8

Printed in the United States of America.

Printed on acid-free paper.

9 8 7 6 5 4 3 2 1

springer.com

Preface

This book aims to cover major methodological and theoretical developments in the field of stochastic global optimization. This field includes global random search and methods based on probabilistic assumptions about the objective function.

We discuss the basic ideas lying behind the main algorithmic schemes, formulate the most essential algorithms and outline the ways of their theoretical investigation. We try to be mathematically precise and sound but at the same time we do not often delve deep into the mathematical detail, referring instead to the corresponding literature. We often do not consider the most general assumptions, preferring instead simplicity of arguments. For example, we only consider continuous finite dimensional optimization despite the fact that some of the methods can easily be modified for discrete or infinite-dimensional optimization problems.

The authors' interests and the availability of good surveys on particular topics have influenced the choice of material in the book. For example, there are excellent surveys on simulated annealing (both on theoretical and implementation aspects of this method) and evolutionary algorithms (including genetic algorithms). We thus devote much less attention to these topics than they merit, concentrating instead on the issues which are not that well documented in literature. We also spend more time discussing the most recent ideas which have been proposed in the last few years.

We hope that the text of the book is accessible to a wide circle of readers and will be appreciated by those interested in theoretical aspects of global optimization as well as practitioners interested mostly in the methodology. The target audience includes graduate students and researchers in operations research, probability, statistics, engineering (especially mechanical, chemical and financial engineering). All those interested in applications of global optimization can also benefit from the book.

The structure of the book is as follows. In Chapter 1, we discuss general concepts and ideas of global optimization in general stochastic global optimization in particular. In Chapter 2, we describe basic global random search algorithms, study them from different view-points and discuss various probabilistic and statistical aspects associated with these algorithms. In Chapter 3, we discuss and study several more sophisticated global optimization techniques including random and semi-random coverings, random multistart, stratified sampling schemes, Markovian algorithms and finally the methods of generations. In Chapter 4, techniques based on the use of statistical models about the objective function are studied. The Introduction and Chapter 1 are written by both co-authors. Chapters 2 and 3 are written by A.Zhigljavsky, Chapter 4 is written by A.Žilinskas.

A.Zhigljavsky is grateful to his colleagues at Cardiff University (V.Savani, V.Reynish, E.Hamilton) who helped with typing and editing the manuscript and patiently tolerated his monologues on different aspects of global optimization. He is also grateful to his long-term friends and collaborators Luc Pronzato and Henry Wynn for stimulating discussions and to his former colleagues from St.Petersburg University – M.Chekmasov, V.Nevzorov, S.Ermakov, and especially to M.Kondratovich, V.Nekrutkin and A.Tikhomirov. Significant parts of Sects. 2.4, 2.5 and 3.3 are based on the joint work of A.Zhiglajvsky and M.Kondratovich; Sect. 3.4 is fully based on the results of V.Nekrutkin and A.Tikhomirov who very much helped with writing a summary of their results.

A.Žilinskas thanks the Institute of Mathematics and Informatics at Vilnuis for facilitating his work on the book, and J.Mockus for introducing him to the field of global optimization many years ago. The work by A.Žilinskas has been partly supported by the Lithuanian State Science and Studies Foundation. The material on one-dimensional algorithms included into Chapter 4 is based mainly on joint publications by A.Žilinskas and J.Calvin. Before starting work on the book, the authors invited Jim Calvin to become a co-author. Although he rejected our invitation in view of his involvement in other projects, we consider him a virtual co-author of the mentioned part of the book.

Both authors thank Rebecca Haycroft and Julius Žilinskas as well as the two referees for their careful reading of the manuscript and constructive remarks. Especially, the authors are very grateful to the editor of the series Panos Pardalos for his encouragement with this project.

Cardiff, Vilnuis

Anatoly Zhigljavsky
Antanas Žilinskas

Contents

Introduction

Global optimization is a fast growing area. Its importance is primarily related to the increasing needs of applications in engineering, computational chemistry, finance and medicine amongst many other fields.

The area of global optimization is well documented in publications. There is the *Journal of Global Optimization* fully devoted to the developments in this area. Other journals on optimization, such as *Journal of Optimization Theory and Applications* and *Mathematical Programming* regularly publish research papers on global optimization. During the last few years, monographs on global optimization regularly appear predominantly in the Kluwer/Springer series *Nonconvex Optimization and Its Applications* edited by Panos Pardalos. Several conferences each year are either fully devoted to global optimization or have sessions on this subject. For the state of the art in theory and methodology of global optimization we refer to two volumes of *Handbook of Global Optimization* [122], [180] and to the paper by Floudas et al, *Global Optimization in the 21st Century* [79].

The problem of global optimization is difficult. Although classical optimization theory can not be directly applied in the problems of global optimization, the traditional tools such as convex analysis are extensively used in constructing global optimization methods. This approach constitutes an essential part of deterministic global optimization. For example, remarkable achievements have been made in constructing minimization algorithms for concave functions in convex regions and also in the minimization of differences of convex functions.

Deterministic global optimization is a well developed mathematical theory which has many important applications. Recently, several monographs by Floudas [78], Horst and Tuy [124], Strongin and Sergeyev [233], Tuy [256], and a text-book by Horst, Pardalos and Thoai [123] on deterministic global optimization have been published. The current state of affairs in deterministic global optimization is presented, e.g. in the book by Floudas and Pardalos [84], and a special volume of the *Mathematical Programming* (Ser.B, v. 103, 2005). However, in a situation close to 'black box' optimization the deterministic models often do not adequately represent the available information about the objective function.

In the cases where the objective function is given as a 'black box' computer code, the optimization problem is especially difficult. Stochastic approaches can often deal with problems of this kind much easier and more efficiently than the deterministic algorithms. Other big advantages of stochastic methods are related to their relative simplicity, their suitability for the problems where the evaluations of the objective function are corrupted by noise, and their robustness with respect to the growth of dimension.

Many algorithms where randomness and/or statistical arguments are involved have been proposed heuristically. Some algorithms are based on analogies with natural processes. Well-known examples of such algorithms are evolutionary optimization and simulated annealing. Heuristic global optimization algorithms are very popular in applications. Results of their application are frequently published in various engineering journals. Papers on evolutionary global optimization can be found in books and journals on evolutionary computing. Simulated annealing, as an important method, has been intensively studied by many authors. However, these studies often consider simulated annealing only, without the broader context of global optimization.

Although there are many publications on stochastic global optimization algorithms and their applications, they are scattered throughout various sources. It is not easy to grasp the state of the art in the field. Therefore, the authors believed that there was a serious need for a book presenting the main ideas and methods in stochastic global optimization from a unified view-point. The authors also believe that they have made an honest attempt to achieve this aim.

The stochastic global optimization techniques are not represented in literature nearly as well as the deterministic approaches. The recent monographs by Mockus [165] and Zabinsky [267] mainly cover the results related to the authors' research. The monographs by the authors of this book [248, 271, 273, 276] represent the stochastic approach to global optimization as it was fifteen-twenty years ago. The authors' current aim is to summarize the current state of affairs in stochastic global optimization and present the recent progress. For completeness of presentation, the key material of the above mentioned monographs is also included into this book. The book also contains a fair amount of new results.

The theory of stochastic global optimization is the main topic of the book. Although the applications of corresponding algorithms are very important, we have restricted the discussion of applications to a few short examples only. We nevertheless believe that the monograph will be useful to a wide circle of readers whose main interest lies in the applications of global optimization. The target audience also includes specialists in operations research, probability, statistics, engineering and other fields.

Notation

$\mathbb{R}$	space of real numbers
$\mathbb{R}^d$	d-dimensional Eucledian space
A	feasible region (optimisation space); typically, A is a compact subset of $\mathbb{R}^d$ with non-zero volume
$f(\cdot)$	objective function given on A; this function is to be minimized
m	minimum of $f(\cdot)$ on A: $m = \min f = \min_{x \in A} f(x)$
x_*	any global minimizer of $f(\cdot)$; that is, x_* is any point such that $f(x_*) = m$
A_*	set of all global minimizers of $f(\cdot)$: $A_* = \{x_* \in A : f(x_*) = m\}$
$\mathcal{B}$	σ-algebra of Borel subsets of A
$\mathrm{vol}(Z)$	volume (d-dimensional Lebesgue measure) of $Z \in \mathcal{B}$
ρ	a metric on $\mathbb{R}^d$
ρ_2	Euclidean metric on $\mathbb{R}^d$
$\|\|\cdot\|\|$	Euclidean norm on $\mathbb{R}^d$
$\mathcal{F}$	set of all possible objective functions
$\mathrm{Lip}\,(A, L, \rho)$	class of functions satisfying the Lipschitz condition with known constant L in metric ρ: $\mathrm{Lip}\,(A, L, \rho) = \{f : \lvert f(x) - f(z)\rvert \le L\rho(x, z) \ \forall x, z \in A\}$
$B(x, \varepsilon)$	$= \{z \in A : \lVert z - x\rVert \le \varepsilon\}$, the ball (in Euclidean metric) in A of radius ε centred at x; more precisely, $B(x, \varepsilon)$ is the intersection of the set A with the ball in $\mathbb{R}^d$ of radius ε and centre at x
$B(\varepsilon)$	$= B(x_*, \varepsilon)$, the ball centered at the global minimizer x_*
$B(x, \varepsilon, \rho)$	$= \{z \in A : \rho(z, x) \le \varepsilon\}$, the ball (in metric ρ) in A of radius ε centred at x
$W(\delta)$	$= \{x \in A : f(x) \le m + \delta\}$
W_x	$= W(f(x) - m) = \{z \in A : f(z) \le f(x)\}$
$x_i (i = 1, ..., n)$	n points where the objective function $f(\cdot)$ has been evaluated
$y_i = f(x_i)$	result of the objective function evaluation at the point x_i
y_{on}	$= \min_{i=1 \dots n} y_i$, the smallest value of the objective function in n evaluations (*record value* or simply *record*)
x_{on}	the point x_i with smallest $i \le n$ such that $f(x_i) = y_{on}$ (*record point*)
$\mathrm{ess\,inf}\,\eta$	essential infimum of a random variable η: $\mathrm{ess\,inf}\,\eta = \inf\{a : \Pr\{\eta \ge a\} > 0\}$
c.d.f.	cumulative distribution function
$F^{-1}(s)$	$= \inf\{t : F(t) \ge s\}$, the inverse function of the c.d.f. $F(t)$
P	a probability distribution on A; more precisely, P is a probability distribution on the measurable space $(A, \mathcal{B})$
P_Z	the distribution on $Z \subseteq A$ defined by $P_Z(U) = P(U \cap Z)/P(Z)$ for all $U \in \mathcal{B}$, where P is a distribution on $(A, \mathcal{B})$ and $Z \in \mathcal{B}$
$\eta \stackrel{d}{=} \nu$	for random variables (vectors) η and ν means equality of their c.d.f.'s
$a_n \sim b_n,\ n \to \infty$	$\Longleftrightarrow$ the limit $\lim_{n \to \infty} a_n / b_n$ exists and equals 1; convergence in distribution is assumed if $\{a_n\}$ and $\{b_n\}$ are sequences of random variables
κ_n	$(1/n)$-quantile of the c.d.f. $F(\cdot)$: $\kappa_n = \inf\{u \mid F(u) \ge 1/n\}$
$\stackrel{\mathcal{D}}{\to}$	convergence in distribution; that is, $\xi_n \stackrel{\mathcal{D}}{\to} \xi \ (n \to \infty)$ for random variables ξ_n and ξ, if $\Pr(\xi_n \le x) \to \Pr(\xi \le x)$, $n \to \infty$, for all x such that $\Pr(\xi = x) = 0$
$\Phi(\cdot)$	the c.d.f. of the standard normal distribution: $\Phi(t) = \frac{1}{\sqrt{2\pi}} \int_{-\infty}^{t} e^{-u^2/2} du$
PRS	pure random search
i.i.d.r.v.	independent identically distributed random variables (vectors)
l.h.s. / r.h.s.	left-hand side / right-hand side

1

Basic Concepts and Ideas

1.1 The Scope of Global Optimization

In this section, we introduce the main concepts and discuss some difficulties arising in problems of global optimization.

1.1.1 General Minimization Problem

Statement of the problem

Let $f : A \to \mathbb{R}$ be a function defined on some set A. We shall call $f(\cdot)$ *the objective function* and A *the feasible region.*

Let $m = \min_{x \in A} f(x)$ be *the (global) minimum* of $f(\cdot)$ in A. Any point x_* in A such that $f(x_*) = m$, is called *a (global) minimizer* of $f(\cdot)$. We shall always assume that there exists at least one minimizer x_*.

By the problem of global minimization of the objective function $f(\cdot)$ we mean constructing a sequence of points $x_1, x_2, \ldots$ in A such that the sequence of values $y_{on} = \min_{i=1\ldots n} f(x_i)$ approaches the minimum m as n increases. Here y_{on} is the smallest value of the objective function in n observations; we shall often call the value y_{on} *the record value* or simply *the record.*

In addition to approximating the minimum m, one often needs to approximate at least one of the minimizers x_*. Let x_{on}, $n = 1, 2, ...$ be the sequence of points (called *record points*) associated with the record values y_{on}. That is, x_{on} is one of the points $x_i, i \leq n$, such that $f(x_i) = y_{on}$; we choose the point with the smallest index i if there are several points $x_i (i \leq n)$ with this property.

Under very general assumptions concerning $f(\cdot)$ and A, the construction of a sequence of points x_n, $n = 1, 2, ...$ such that

$$y_{on} \to m \text{ as } n \to \infty \tag{1.1}$$

implies that the associated sequence of record points x_{on} converges to the set $A_* = \{x_* \in A : f(x_*) = m\}$ of all minimizers of $f(\cdot)$; that is,

$$\rho(x_{on}, A_*) = \inf_{z \in A_*} \rho(x_{on}, z) \to 0 \quad \text{as} \quad n \to \infty, \tag{1.2}$$

where ρ is a metric on A. This is the only general conclusion we can deduce from the fact that the records y_{on} converge to m. Unless we assume that A_* contains only one point, the sequence of points x_{on} does not necessarily converge to a particular global minimizer x_*. Moreover, finding (approximating) all global minimizers of $f(\cdot)$ is typically a much more difficult (and often practically impossible) problem.

Feasible region

In continuous problems, the feasible region A is a subset of $\mathbb{R}^d$ ($d \geq 1$), typically with positive volume. There are different types of feasible regions appearing in optimization problems. However, the discrete (where A is a discrete set) and continuous cases represent the two main types of optimization problems concerning the classification of the problem with respect to the structure of the feasible region A. There are some similarities in the theory and practice of global optimization in both discrete and continuous problems. Traditionally, however, these two types of optimization problems are handled separately. In this book, we shall not try to bridge these two types of problems and shall deal with continuous problems only. In theoretical constructions, we do not need many additional assumptions concerning A (see, for example Sect. 2.1.1 of Chap. 2). In applications, however, feasible sets with complex structure may create serious difficulties in implementing the algorithms and interpreting the corresponding results.

Note that there is a certain duality between the complexity of the feasible set and the complexity of the objective function $f(\cdot)$. One may often reformulate the same problem in different ways by transferring the complexity of the problem from the objective function to the feasible region and vice versa. For example, when the feasible region is determined by complex constraints and as a consequence has a complex structure, it may be worthwhile reformulating the optimization problem using the so-called penalty function approach; this may significantly reduce the complexity of the feasible region at the expense of increased complexity of the objective function. The use of penalty functions approach is not always obligatory: for example, in Sect. 2.6.3 of Chap. 2 we show how to relate the surfaces determined by equality-type constraints to the subsets of $\mathbb{R}^d$ with positive volume.

It is a common practice to formulate the optimization problems in such a way that the feasible region A has a reasonably simple structure.

Objective function

We shall typically assume that the values of the objective function $f(\cdot)$ can be evaluated at any point of A without error (we shall allow for evaluation errors in Sects. 2.1.4 and 3.6 as well as in some general discussions). We need to impose some constraints on $f(\cdot)$ as the problem of minimizing an arbitrary function does not make much sense. For instance, there are no general minimization algorithms that would be able to converge to the minimum of the function

$$f(x) = \begin{cases} 0 & \text{if } x = \ln 2 \simeq 0.693, \\ 1 & \text{at all other points } x \text{ in } A = [0,1]. \end{cases}$$

An obvious attempt at making a reasonable assumption about $f(\cdot)$ would be to assume something like continuity or differentiability. However, the assumption of, say, differentiability of $f(\cdot)$ at all points of A is not the correct type of assumption to make. Firstly, it is too strong (very often, continuity of $f(x)$ for all $x \in A$ is not needed) and secondly, it is not sufficient to guarantee that the related global optimization problem can be resolved in a reasonable time.

Consider an example. Assume $A = [0,1]$ and let the objective function be

$$f_k(x) = \begin{cases} 1 - \frac{1}{2}\left(\sin \frac{5k\pi x}{4(k-1)}\right)^2 & \text{for } x \in \left[0, \frac{4(k-1)}{5k}\right], \\ 1 - \left(\sin \frac{5k\pi x}{4}\right)^2 & \text{for } x \in \left[\frac{4(k-1)}{5k}, \frac{4}{5}\right], \\ 1 - \frac{1}{2}\left(\sin 5\pi x\right)^2 & \text{for } x \in \left[\frac{4}{5}, 1\right], \end{cases} \tag{1.3}$$

where $k \geq 2$ is some integer. For illustration, the function $f_{12}(x)$ is depicted in Fig. 1.1.

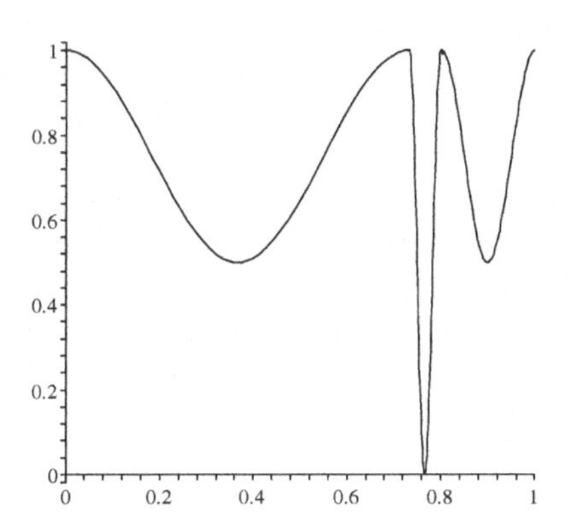

Fig. 1.1. Graph of the function (1.3) with $k = 12$.

For any $k \geq 2$, the function $f_k(\cdot)$ is continuously differentiable in $A = [0,1]$ and has three local minimizers. These local minimizers are:

$$z_1 = \frac{2(k-1)}{5k}, \quad z_2 = \frac{2(2k-1)}{5k}, \quad \text{and } z_3 = 0.9$$

with the point $x_* = z_2$ being the global minimizer.

The problem of finding x_* is difficult when k is large despite the fact that the function $f_k(\cdot)$ is continuously differentiable. Indeed, the *region of attraction* of x_* is $\left(\frac{4(k-1)}{5k}, \frac{4}{5}\right)$ (starting at any point in this region any monotonous local descent algorithm converges to $x_* = z_2$); the length of this interval is $4/(5k)$; this value can be uncomfortably small if k is large.

In different sections of the book we impose and discuss different constraints on the objective function. In addition to continuity, the most important of these constraints are certain regularity conditions of the objective function in the neighbourhood of the global minimizer.

1.1.2 Global Minimization Versus Local Minimization

Let $f : A \to \mathbb{R}$ be a function defined on A. A point $z \in A$ is called *a local minimizer* of $f(\cdot)$ if there exists a neighbourhood U_z of z such that

$$f(z) \leq f(x) \text{ for all } x \in U_z\,; \tag{1.4}$$

as the neighbourhood U_z of z one may always consider $\{x \in A : ||x - z|| < \varepsilon\}$, the intersection of A and an open ball of radius $\varepsilon > 0$ centered at z.

If z is a local minimizer of $f(\cdot)$, then the value $f(z)$ is called *a local minimum.* A local minimizer z becomes the global minimizer if the set U_z is replaced with the whole set A in (1.4). If the objective function $f(\cdot)$ has only one local minimizer in $\bar{A}$ (the closure of A), then this local minimizer is the global minimizer as well. The corresponding objective function (and sometimes the minimization problem also) is called *unimodal.* Otherwise, if either there is more than one local minimizer or the number of local minimizers is unknown, we shall say that the objective function is *multimodal.*

In applied optimization it is normally required to find the global minimum. Local optimization can be acceptable when theoretical or heuristic arguments show that the optimization problem is unimodal. However, many applied optimization problems correspond to a nearly 'black box' situation where the optimization problem is given by a computer code, and its theoretical investigation is severely restricted. Even if the underlying problem is unimodal, this property can seldom be proven. As an example, Fig. 1.2 displays a contour-plot of a 'banana function'

$$f(x,y) = (1-x)^2 + 10\left(y - x^2\right)^2, \tag{1.5}$$

and a plot of the one-dimensional cross-section $f(x,1)$. The function $f(x,y)$ is unimodal with minimum $m = 0$ achieved at the point $(1,1)$. However, the cross-section $f(x,1)$ (as well as many other cross-sections) is a bimodal function of x.

Moreover, it is frequently known from past experience that in similar problems the objective functions are multimodal, implying the need for global optimization algorithms. Sometimes multimodality of the objective function can

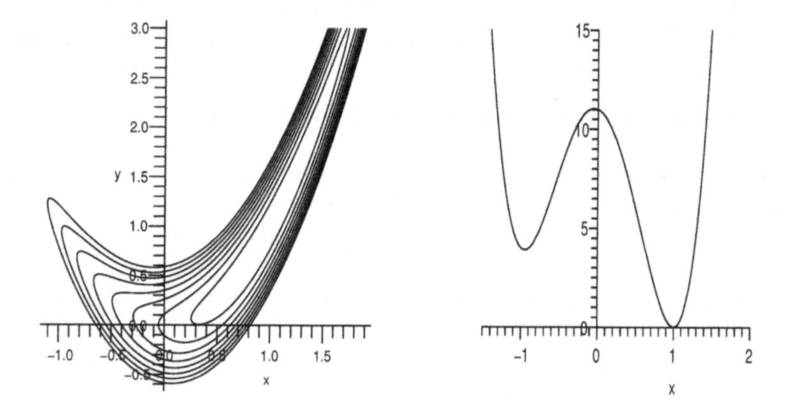

Fig. 1.2. Contour-plot of the banana-function (1.5) and its cross-section $f(x, 1)$

be established in a preliminary study. However, despite the fact that global optimization algorithms are of high practical importance, most available packages of optimization software contain local optimization routines only.

Some inconsistency between the demand and availability of global optimization software can be explained using the following arguments. Classical optimization methods are local. Further development of local optimization theory and algorithms is based on traditional mathematical disciplines aiding the development of efficient algorithms. For example, in the development of local optimization methods, numerous achievements in solving non-linear equations are often very helpful. Bearing in mind the availability of efficient local optimization algorithms, the researchers frequently simplify their theoretical models to maintain unimodality of their objective functions. If there are several local minima, local algorithms are often run from several starting points to find the best.

However, there are important applications where multimodality of the objective function can not be avoided, and in view of a large number of local minima these problems can not be efficiently solved using the multistart of local algorithms. A well known example is the problem of molecular conformation and protein folding [181].

Development of efficient global optimization methods is more difficult than that of the local optimization methods. Indeed, in global optimization there are no simple and widely acceptable models of the objective functions (such as the quadratic model in local optimization) and the diversity of multimodal functions is huge. Additionally, in global optimization there is no substitution for the idea of local descent, which is central in local optimization, and the rationality of search strategies for finding global minimum are difficult to justify. Therefore, it does not seem realistic to find a universal strategy of global search as well as a widely acceptable model of multimodal functions. This explains the wide variety of existing approaches towards global optimization.

1.1.3 Combining Locality and Globality of Search

If the objective function $f(\cdot)$ is multimodal, then the application of a local optimization algorithm will lead to finding a local minimum which may not be the global one. Thus, in order to increase the chances of finding the global minimum, one has to provide 'globality' of search.

Nevertheless, the local optimization techniques constitute an important part of global optimization methodology. This is related to the fact that a typical global optimization strategy is always a compromise between two competing objectives: globality and locality of search.

Globality of search means that the points x_i (where we evaluate the objective function $f(\cdot)$) are spread all over the feasible region A. If we assume that function evaluation $f(x_i)$ at a point x_i provides information about $f(\cdot)$ in some neighbourhood of x_i rather than at the point x_i only (we then need to assume something like the Lipschitz condition for f), then the globality of search would aim to cover the whole feasible region A with these neighbourhoods of x_i's.

There are global optimization algorithms that are fully based on the globality of search and completely ignore locality. One of the most well-known algorithms of this kind is the 'pure random search', where the points x_i's are random, independent and identically distributed with fixed probability distribution P, see Sect. 2 of Chap. 2. One can achieve a better coverage of A if the points x_i's are generated in some deterministic rather than random manner, see Sect. 1 of Chap. 3.

Adaptive coverings of A are often much more efficient than the non-adaptive ones, but the efficiency of any adaptive covering depends on how well the sequence of records $y_{oj} = \min_{1 \le i \le j} f(x_i)$ $(j = 1, 2, \ldots, n)$ approximates the minimum m.

In adaptive covering algorithms and in many other global optimization methods, whose efficiency depends on the closeness of the current records y_{oj} to the minimum m, it is always worthwhile to take the following general advice: *immediately after obtaining a new record point make several iterations of a local descent from this point.* An extreme would be to apply a local descent to each point we generate at the global stage. The corresponding algorithm is called 'multistart' and is studied in Sect. 2.6.2 of Chap. 3.

A reasonable intermediate rule is to alternate global steps with local ones, for instance, one each or k each. Here the global steps may correspond to generating random points in A and the local ones to performing one iteration of local descent from the k best available points (the points x_i's with the smallest function values). The corresponding algorithms are very simple and tend to work reasonably well in practice.

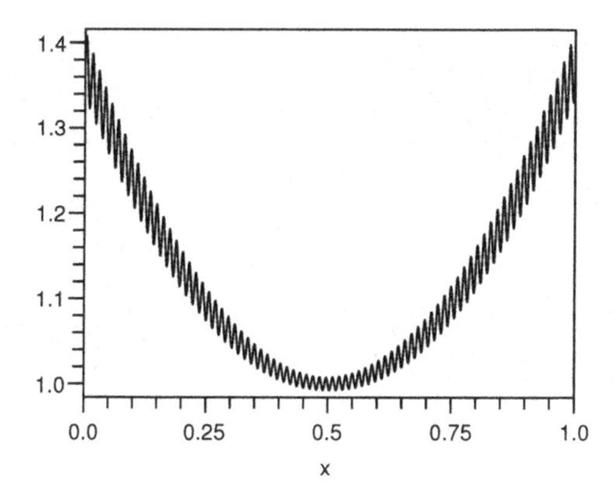

Fig. 1.3. Graph of a quadratic function with added high-frequency component

The second piece of general advice is: *every global optimization procedure must finish with a local descent from one or several of the best points x_i's.* This can be expressed by stating that the globality of algorithms serves to narrow the area of search for the minimizer x_* (hopefully reducing the area to the region of attraction of x_*), leaving the problem of finding the exact location of x_* to a local optimization technique.

Note that the neighbourhood of a minimizer x_* should not be confused with its region of attraction. Assume that the neighbourhood is $B(x_*, \varepsilon)$, where ε is a reasonably small positive number. Depending on the objective function, this neighbourhood can be both wider and narrower than the region of attraction. For instance, assume that the objective function $f(\cdot)$ is a sum of a smooth and slowly varying function $f_1(\cdot)$ (the Lipschitz constant of $f_1(\cdot)$ is small) and a small irregular function $f_2(\cdot)$ (the Lipschitz constant of $f_2(\cdot)$ can be large, but the width of $f_2(\cdot)$, which is $\sup f_2 - \inf f_2$, is much smaller than $\sup f_1 - \inf f_1$, the width of f_1); for examples of these functions see Fig. 1.3 and Fig. 1.4. In this case, what we are effectively dealing with is the function $f_1(\cdot)$ rather than $f(\cdot)$. If we know in advance that the objective function has this kind of structure (this is typical, for example, in finding optimal packings or coverings and in constructing optimum experimental designs with fixed numbers of points), then we can easily take it into account and construct very efficient optimization algorithms, see e.g. [152]. To a certain extent, therefore, the function of Fig. 1.3 is simpler to optimize than the function of Fig. 1.1; this is despite the fact that the former has 75 local minimizers on $A = [0, 1]$ and the latter has only three.

Of course, the best compromise between the globality and locality of search depends on specific features of the problem including the complexity of A, prior information about the number of (local) minima of $f(\cdot)$, sharpness of

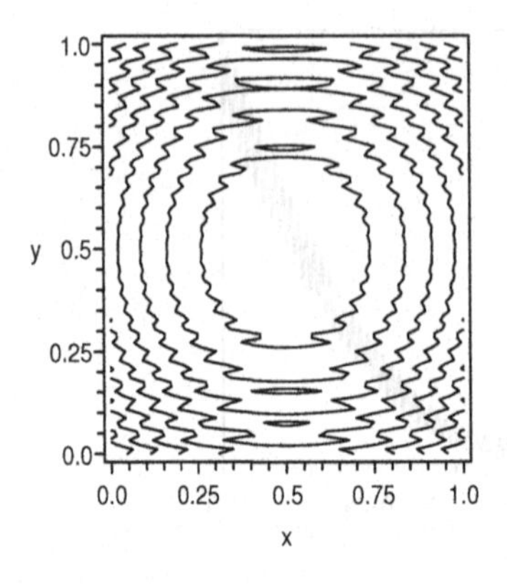

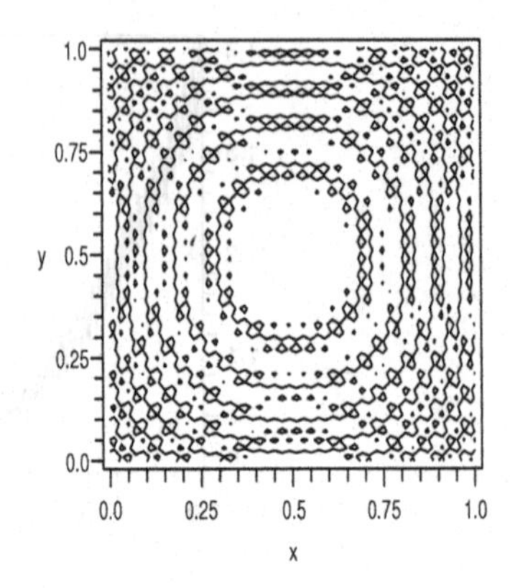

Fig. 1.4. Contour-plots of a quadratic function of two variables with added high frequency component in one variable and in both variables

$f(\cdot)$ in the neighbourhoods of the global minimizers and computational efforts required for evaluating the gradient of $f(\cdot)$.

1.1.4 Theory and Heuristics

Global optimization is very important for applications but it is one of the most difficult fields in computing in general. Thus, it is not surprising that many different approaches for constructing global minimization methods have been proposed. Some of the methods are called heuristic, even though it is not often easy to classify methods according to the dichotomy 'heuristic – not heuristic'. Also, a universal definition of 'not heuristic' methods does not exist. Scientific folklore suggests that which methods give acceptable solutions in reasonable time but lack underlying theoretical results are called heuristic methods. The practical importance of global optimization and the relative underdevelopment of its mathematical theory (compared to that for local optimization theory) initiated active research on heuristic methods, which was frequently made by experts in different subjects where difficult optimization problems occur. Some heuristically proposed algorithms subsequently have been investigated mathematically. A characteristic example is the simulated annealing method [207]. In contrast, clustering and topographic methods have been proven efficient in many applications, but there is little theory behind them [4, 248, 250, 251]; thus, they may be regarded as typical heuristic methods of stochastic global optimization. We want to emphasize that nowadays there is no negative connotation in the attribute 'heuristic'. Rephrasing the title of [266] we may say 'heuristic methods: once scorned, now highly respectable'.

Irrespective of whether a global minimization method has mathematical or heuristic origin, its main theoretical feature is convergence. Normally global optimization methods are targeted to broad classes of objective functions not assuming special properties like convexity, special analytical form etc. Therefore, we are also interested in necessary and sufficient conditions of convergence under weak assumptions concerning the optimization problem. It is known that without strong assumptions concerning an optimization problem, the only converging algorithms are those which generate everywhere dense sequences of observation points. Indeed, the following general result is valid, see [248, 292]:

Theorem 1.1. *Let the feasible region A be compact and the objective function be continuous in the neighbourhood of a global minimizer. Then a global minimization algorithm converges in the sense* $y_{\mathrm{o}n} \to m$ as $n \to \infty$ *iff the algorithm generates a sequence of points* x_i *which is everywhere dense in A.*

The theorem also remains valid for more narrow classes of objective functions, e.g. Lipschitz continuous functions. Stronger assumptions about the targeted problems should be made to ensure convergence without generating everywhere dense sequences of observation points; e.g. to construct a method whose observation points converge only to global minimizers; the assumptions concerning the optimization problem should allow exclusion of the subsets of the feasible region not containing global minimizers. This can be done, e.g. for a subclass of Lipschitz continuous function with known Lipschitz constant L. Further assumptions about class of objective functions should be valid to prove reasonable rates of convergence of global optimization algorithms.

This book is devoted to the theoretical aspects of stochastic global optimization, and we are mainly interested in general approaches to the analysis of these algorithms. The majority of global optimization methods with heuristic origins are very complicated, and special methods have been developed for analyzing different stochastic heuristics taking into account their special properties. Such special methods are not considered in this book. Nevertheless, some general results presented in the book can be applied for analyzing some heuristic global optimization methods. For example, evolutionary methods constitute an important class of heuristic global optimization methods; for the original ideas we refer to [119, 212]. A very large number of publications on applications as well as on theoretical aspects of evolutionary optimization have appeared recently. Special methods have been developed to investigate different classes of evolutionary optimization algorithms; see e.g. [8]. However, a general approach to random search can also be useful to investigate evolutionary algorithms as shown in Sect. 3.5.

1.2 Stochastic Methods

In global optimization, randomness can appear in several ways. The main three are: (i) the evaluations of the objective function are corrupted by random errors; (ii) the points x_i are chosen on the base of random rules, and (iii) the assumptions about the objective function are probabilistic. The way (ii) corresponds to global random search of Chapts. 2 & 3 while (iii) leads to the approach investigated in Chap. 4. Let us first consider the basic ideas of the stochastic methods in general. In several places of the book (see e.g. Sects. 2.1.4 and 3.5.2) we discuss the case (i); that is, the case when there are random errors in function evaluations.

1.2.1 Deterministic Versus Stochastic Methods in Global Optimization

Deterministic approaches to global optimization are outside the scope of this book. However, we will briefly discuss some important ideas of the deterministic methods to show common points with stochastic methods. The deterministic 'branch and bound' approach can be applied to construct methods of global minimization if a suitable lower bound for function values can be constructed for the considered class of objective functions. Let the feasible region $A \subset \mathbb{R}^d$ be subdivided into subsets A_i, $i = 1, ..., l$, and a convex underestimate (minorant) $\underline{f}_i(\cdot)$ for the objective function $f(\cdot)$ be constructed in each subset A_i. The subsets and underestimates are defined using observations made at previous steps of the algorithm: $y_j = f(x_j)$, $j = 1, ..., n$. The subsets with

$$\min_{x \in A_i} \underline{f}_i(x) \geq y_{on}$$

are excluded from further search. Choosing a subset for further subdivision and a point for the next function evaluation takes into account the minima of underestimates. Let us consider a simple case of one-dimensional minimization of Lipschitz continuous functions with known Lipschitz constant; that is, assume that $f \in \mathrm{Lip}(A, L, \rho)$. Assume that several values of the objective function are known; in Fig. 1.5, these values are represented by circles and the graph of the function is drawn by a solid line. Using the known Lipschitz constant L, we can construct the piecewise linear underestimate for $f(\cdot)$:

$$f(x) \geq \underline{f}_n(x) = \max_{j=1,\ldots,n} (y_j - L|x_j - x|), \tag{1.6}$$

which is shown in the figure by dashed line. The level of the record value is shown by dotted horizontal line. Using the underestimate and the record value, the prospective search region can be bounded: the subintervals of the interval [0, 1] where the graph of the underestimate is above the dotted line do not contain global minimizers, and therefore further search should be continued over the subintervals where underestimate is below the dotted line.

Different branching rules can be applied. In a well-known algorithm by Shubert-Pijavskij [123], the next observation is made at the point of minimum of the underestimate. We are not going to discuss this algorithm here, and refer for details to [98]. However, as Fig. 1.5 illustrates, the Lipschitz constant based underestimate is rarely tight. Improving the tightness of the underestimates

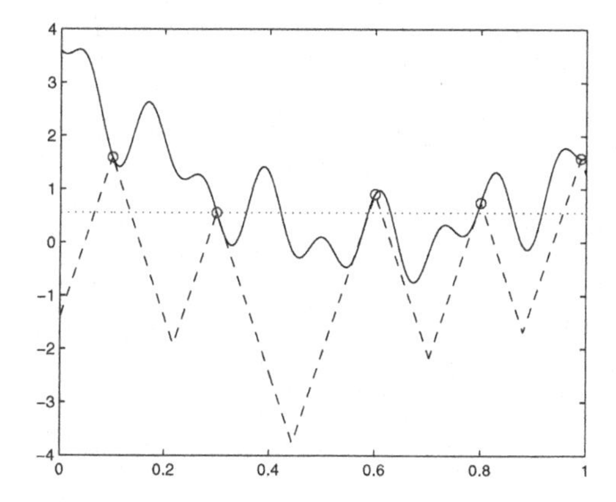

Fig. 1.5. Example of Lipschitz underestimate

is one of the most important topics in deterministic optimization theory, see [2, 79, 123, 124, 215, 225, 237, 256].

The idea of 'branching and bounding' can be extended in different ways to incorporate stochastic techniques. Branching of the feasible region A into a tree of subsets A_i $(i = 1, \dots, l)$ is done in a similar manner. In global random search algorithms, the test points $x_j \in A_i$ are random; therefore, statistical methods can be used for constructing estimates for $m_i = \inf_{x \in A_i} f(x)$ and testing hypothesis about the m_i's. These methods are described in Sects. 2.4, 2.5 and 3.2. By making the statistical inference about m_i we evaluate 'prospectiveness' of the subregions A_i's which enables to perform the 'bounding' of those subsets A_i that are considered as non-prospective for further search. As the statistical rather than deterministic bounds are used, a mistake in bounding (rejecting) the sets for further search must be allowed. The probability of this mistake can be controlled and kept as low as possible. The algorithms based on these ideas have been named 'branch and probability bound' methods; they are considered in Sect. 2.6.1.

One does not have to have random points to make statistical inference about m_i: in Chap. 4 we consider a number of probabilistic models for the objective function $f(\cdot)$ and the corresponding probabilistic bounds for m_i's. These bounds can also be used in the stochastic versions of the 'branch and bound' methods.

In addition to the typical ideas for the subject of global optimization, the general theory of optimal algorithms [254] can be applied to develop a mathematical theory of global optimization. Using such an approach, a class of multimodal function is postulated, and an optimal algorithm with respect to this functional class and a chosen precision criterion is defined.

A class of Lipschitz functions with known Lipschitz constant is the most popular class of functions of the traditional deterministic approach. The op-

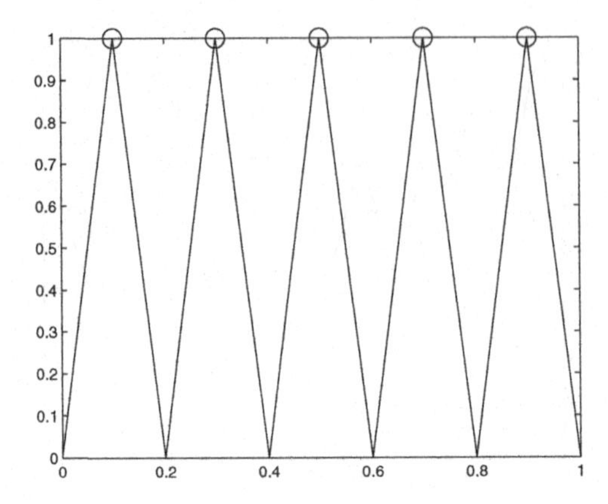

Fig. 1.6. Model function for minimax algorithm

timal algorithm with respect to the criterion of guaranteed minimal error is called minimax algorithm. Let us discuss passive (non-adaptive) algorithms first. In any passive algorithm, the trial points are chosen apriori, and this choice is not influenced by the function values obtained in the process of optimization. The trial points of passive minimax algorithms are defined as centres of spheres optimally covering the feasible region [234]. It may seem surprising, but passive minimax algorithms are the worst-case optimal also in the class of adaptive algorithms. Similar results are valid for the minimax algorithms in some other problems of computational mathematics; this may be formulated as 'in the worst-case situation the adaptation does not help'. Minimax algorithms in global optimization take into account only the worst case function, which for the class of Lipschitz functions with known Lipschitz constant is the underestimate $\underline{f}_n(\cdot)$ defined in (1.6) and illustrated in Fig. 1.6, where the circles denote the function values. For the detailed description of minimax algorithms based on deterministic models we refer to [234].

The worst case Lipschitz (saw-like) function seems rather unnatural from the applications point of view. But only this function is taken into account when constructing the worst-case optimal algorithm. If the trials are performed sequentially starting with the points of the corresponding passive algorithm then the actual objective function may be recognized as being not the worst case function. Taking into account this new information, the class of objective functions may be assumed to be narrower than initially, and the minimax algorithm may be improved. In such a case the model function is still assumed saw-like but with teeth of different sizes [234]. The suitably modified, so-called the best algorithm, becomes rather complicated; however, it again reduces to a passive algorithm in the worst case. Similar disadvantages of minimax algorithms are known in other fields of numerical analysis. We cite [236], page xv, for an opinion on the worst case analysis from a broader point of view of the theory of algorithms: 'Along these lines, however, the design of an algorithm is sometimes targeted at coping efficiently with unrealistic, even pathological inputs and the possibility is neglected that a simpler algorithm that works fast on average might perform just as well, or even better in practice'. Some deficiencies of minimax approach in designing optimal algorithms can be avoided using the approach based on statistical models and criterion of average optimality.

1.2.2 Methods Based on Statistical Models

To construct a mathematical method for solving a practical problem, one needs to formulate a mathematical model of the problems considered. The mathematical model should possess essential properties of the given class of practical problems but at the same time should not be too complicated for analysis. In local optimization, deterministic models are well suitable for construction and analysis of optimization methods. In global optimization, however, information on the class of objective functions frequently is very fragmented, not strict, and difficult to integrate into a consistent deterministic model. Therefore, statistical models seem attractive in these situations, similarly to such fields as financial markets, reliability of complex systems, wether forecasting etc. Involvement of statistical models requires reconsideration some basic concepts of the efficiency of the methods. In Chap. 4, the approach based on statistical models and average optimality criteria is considered in detail. In this subsection we will only highlight the main concepts of the global optimization approach based on statistical models of objective functions.

Assume a stochastic function is chosen for a model of objective functions. Under such an assumption objective functions can be interpreted as randomly chosen sample functions of the stochastic function considered. It is supposed that the stochastic function has properties similar to the properties of the class of aimed objective functions relevant to global optimization, e.g. that sample functions are continuous with probability 1. The efficiency of a method with respect to the stochastic function is understood as the average efficiency, e.g. the order of convergence of average error to zero. From a theoretical point of view, it is interesting to analyze the method which is optimal for the chosen statistical model. The analysis of average efficiency of optimization methods is similar to such analysis of methods for other computational problems, e.g. in [175, 201, 253, 254].

The criterion of average efficiency has a practical meaning if the method is intended to be applied many times to solve similar problems. It is a good criterion from the point of view of massive applications. However, an individual user may be interested in solving his single problem not related to any probabilities. Moreover, it may seem that involvement of probabilities for optimization of unique problems is not justified at all. However, we show in

Chap. 4 that a statistical model corresponds to consistent assumptions on uncertainty even in the case of solving unique optimization problems.

The global optimization methods based on statistical models are constructed so that they deal with the uncertainty present in a particular global optimization problem maintaining the rationality oriented to an average objective function. The convergence/efficiency of the algorithms based on statistical models can be analyzed with respect to underlying statistical models, e.g. the rate of convergence to zero of the average error. However, the convergence of the algorithms can also be analyzed with respect to a class of objective functions defined by standard assumptions (which are not related to the underlying model), e.g. for continuous functions with non degenerated global minimizers. Such analysis is especially important in comparing the algorithms based on statistical models with the algorithms based on deterministic models. We concentrate on the latter type of analysis. In some cases the theoretical results are supported by numerical experiments.

1.2.3 Basic Ideas of Global Random Search

Global random search algorithms are global optimization methods where random decisions are involved in the process of choosing the observation points.

A general global random search algorithm assumes that a sequence of random points $x_1, x_2, \ldots, x_n$ is generated where for each $j \geq 1$ the point x_j has some probability distribution P_j. For each $j \geq 2$, the distribution P_j may depend on the previous points $x_1, \ldots, x_{j-1}$ and on the results of the objective function's evaluations at these points (the function evaluations may not be noise-free). The number of points n, $1 \leq n \leq \infty$ (the stopping rule) can be either deterministic or random and may depend on the results of function evaluation at the points $x_1, \ldots, x_n$.

Global random search algorithms are very popular in both theory and practice. Their popularity is owed to several attractive features that many global random search algorithms share. These attractive features are:

- the structure of global random search algorithms is usually simple;
- these algorithms are often rather insensitive to the irregularity of the objective function's behaviour and the shape of the feasible region, to the presence of noise in the objective function evaluations, and even to the growth of dimensionality;
- it is very easy to construct algorithms guaranteeing theoretical convergence.

However, global random search algorithms have certain drawbacks. First, the practical efficiency of the algorithms often depends on a number of parameters, but the problem of the choice of these parameters frequently has little relevance to the theoretical results concerning the convergence of the algorithms. Secondly, a serious drawback of many global random search algorithms is the fact that the analysis on good parameter values is lacking or just impossible. Another serious drawback is slow convergence. Improving the convergence rate (or efficiency of the algorithms) is a problem that much research concerning the global random search technique is devoted to.

A very large number of specific global random search algorithms exist, but only a few main principles form their basis. These principles can be summarized as follows:

(i) random sampling of points at which $f(\cdot)$ is evaluated,
(ii) random covering of the space,
(iii) combination with a local optimisation technique,
(iv) use of cluster-analysis techniques to avoid clumping of points around a particular local minima,
(v) Markovian construction of algorithms,
(vi) more frequent selection of new trial points in the neighbourhood of 'good' previous points,
(vii) use of statistical inference, and
(viii) decrease of randomness in the selection rules for the trial points.

In constructing a particular global random search method, one usually incorporates a few of these principles. In Chaps. 2 & 3 we shall pay attention to all these principles. In Chap. 2 the main emphasis will be placed on principle (vii); that is, on the use of statistical procedures for improving efficiency of the global random search algorithms.

Let us briefly comment on the types of convergence in global random search methods. As in these methods the points x_i are random, the strongest possible convergence in (1.1) and (1.2) is convergence with probability one. In fact, in many cases we are only able to claim *convergence in probability* or even *convergence with probability* $\geq 1 - \gamma$ for some $\gamma > 0$. Talking about the rates of convergence involves much more subtle concepts. One of the implications is that theoretical comparison of methods of different natures (say, the methods based on the Lipschitz condition and random search methods) is difficult. First, the methods are based on different assumptions about the objective function and, even more importantly, the language of expressing optimality, convergence and convergence rate is different.

1.3 Testing, Software and Applications

1.3.1 Testing

Global optimization theory considers mathematical models of optimization of multimodal functions. Similarly to other mathematical theories, it is related to a practical problem as much as the theoretical assumptions adequately describe the practical problem. However, the correspondence of the theoretical assumptions to practical problems is not always obvious. Some classes of global optimization problems can be precisely described by mathematical formulas, e.g. minimization of concave quadratic functions subject to linear constrains.

It may seem that in such case of absolute certainty the results of the theoretical analysis of the general problem should be applicable to every particular case of the problem. However, the time to solve different particular cases by means of the same method can be very different. The worst case theoretical analysis gives rather pessimistic result: the problem considered is NP complete. In this case, testing can be complementary to the theoretical analysis assessing e.g. average efficiency of the algorithm for the sample of randomly generated test functions.

The example considered shows that application of the theoretical estimates to practical performance is not straightforward even for a class of problems defined by simple analytical formulas. The situation is more complicated for the class of functions defined by computer codes; frequently such problems are described as black/gray box optimization problems. In such situations only very general assumptions about the function can be made, and consequently only very general theoretical properties of an optimization method can be established. Although theoretical qualitative assessment of a method is interesting from scientific point of view, it is often not sufficient to predict efficiency of the algorithm in a practical application. Testing is supposed to aid potential users in a priori assessment of practical performance of tested algorithms. For example, a user can decide which of several known constrained global optimization and constraint satisfaction solvers to chose, taking into account the testing results reported in [169] where more than 1000 test functions have been minimized, and the estimates of reliability and speed are presented.

The concept of practical efficiency is polysemous; it is not easy to give its thorough definition although experienced developer of algorithms as well as concerned users agree about its main aspects. In spite of the similarity of goals, a methodology for testing global optimization algorithms may not be copied from the methodology for testing the local minimization algorithms. As opposed to the class of convex minimization problems, the diversity of non convex continuous (global) minimization problems is huge. For example, it includes the class of combinatorial optimization problems since the feasible set $\{0, 1\}$ can be expressed via a system of inequalities including continuous variables

$$\{0, 1\} = \{x : x \geq 0, 1 - x \geq 0, x(x - 1) \geq 0\}.$$

Because of the high complexity of the class of global optimization problems, guaranteed precise solutions of high dimensional problems in a reasonable time is impossible. Therefore an algorithm should be evaluated not only with respect to time of (possibly not successful) solution, but also with respect to the solution reliability (percentage of successfully solved problems in a representative subclass of problems). Concretization of 'successfully solved' and 'representative subclass' is an important part of the testing methodology. However, the results of experimental testing frequently are presented in research papers without proper methodological substantiation. Normally the results of 'competitive testing' (according to the definition of [121]) are presented. This means reporting the numbers of test functions' evaluations performed by the tested algorithms before stopping. However, such results are not always generalizable because of different termination conditions for different algorithms, unsubstantiated choice of algorithms' parameters, and some uncertainty in the results caused by the use of randomized subroutines. The title of the paper [121] 'Testing heuristics: we have it all wrong' emphasizes problems of testing of heuristics which are true also for testing of global optimization algorithms, irrespective of the attribute 'heuristic'.

The development of algorithms has not only a mathematical component but also an engineering component involving the experimental analysis of algorithms. We cite [157], page 489: 'Perhaps especially in the context of global optimization, where our ability to prove theorems trails far behind our need to understand and evaluate algorithmic solutions to hard problems, experimental research is needed to produce precise, reliable, robust, and generalizable results'. The experiments without proper methodology are not very useful in predicting performance of the algorithm. Moreover, such experimental results can be even misleading. There are some general principles of scientific experimentation with algorithms, e.g. the field of software engineering has general standards for software testing. The global optimization packages intended for commercial distribution are tested according to these standards. However, the software engineering standards ensure only the quality of implementation of an algorithm but not its practical efficiency. Two of the most important criteria in the competitive testing are reliability and efficiency. For a fair comparison, the trade-off between these criteria should be equally balanced in the algorithms considered, e.g. applying comparable termination conditions. In the book, we apply competitive testing only in few places where we do not doubt fairness of comparison.

A methodology for testing based on collective work of interested researchers has been proposed by Dixon and Szegö [64], where different authors have presented results of minimization of the same functions. The set of test functions of [64] currently seems rather outdated, but the volume [64] remarkably influenced concepts of testing. For the contemporary methodological principles of testing we refer to [156], [157]. The methodology related questions are also discussed in [12], [120], [121], [169], [191], [252], [248].

For the experimental testing of global optimization algorithms a set of test functions representing different application problems is needed. The choice of test functions is crucial in testing. The acceptance of a set of test functions by several respected authors may have strong influence on the development of algorithms, since the later authors aiming to publish their papers tend to tune their algorithms to the test functions used earlier. The choice of test functions not adequately representing real world problems may imply wrong conclusions, similarly to comparing caterpillar tractors according to their performance on a Formula One track. The main difficulty in selecting test functions is the lack of criteria describing classes of practical global optimization problems well. Some criteria, which seem intuitively acceptable, are difficult to measure, and they are not documented in publications about successfully solved applied problems. Therefore it is difficult to summarize consistently experimental results obtained by different authors using different test functions.

A broadly used set of test functions was proposed in the eighties in [64], where comparative testing of global optimization algorithms was initiated. Many multimodal test functions of the special structures are collected in [82], [83]. The test functions for Lipschitz one-dimensional and low dimensional optimization are presented in [71], [112].

The need for large families of test functions caused a development of special software for their generation, e.g. a generator of multimodal functions with desired properties has ben recently proposed by Gaviano, Kvasov, Lera and Sergeev [88]. For generators of other classes of test functions we refer to [155], [160], [209]. The sample functions of a random function are well suited to evaluate the average performance of the global optimization algorithms; of course, the properties of a random function should be similar to the properties of the targeted practical problems.

The majority of statistical global optimization algorithms considered in this book are experimental ones. The performance of these algorithms is not always well understood. The areas of their rational applications are still not well defined. Since testing of statistical methods for global optimization should correspond to the actual research/application aims, its methodology can not be directly copied from similar but more mature fields, e.g. from local optimization or from solution of systems of nonlinear equations. Among the goals of experimental testing we will emphasize the following two: to demonstrate the theoretically predicted behavior of the algorithm, and to understand its properties which are difficult to analyze theoretically. This kind of testing (called 'scientific testing' in [121]) is interesting mainly to the researchers in optimization. In the present book we apply such methodology to analyze convergence rate of an one-dimensional global optimization algorithm in Sect. 4.4.3.10.

1.3.2 Software

Many implementations of global optimization algorithms are available via the Internet. Some of them are free. Recently, global optimization software (GO solvers) have been included in commercially distributed packages.

One of the most widely used software packages is Microsoft Excel. Algorithms for optimization problems are included into add-in Excel Solver. The global optimization library consists of deterministic and statistical algorithms. Algorithms based on interval arithmetic and on branch and bound technique represent deterministic global optimization. The statistical part consists of multistart, genetic and evolutionary algorithms.

MATLAB is extended with global optimization algorithms included into TOMLAB library. Simple statistical algorithms (multistart and adaptive random search) are included as well as a deterministic algorithm based on the branch and bound approach. Two subroutines based on response surface methods are intended for expensive objective functions.

An advanced mathematics and computer algebra software package for symbolic and numeric computations MAPLE recently has been supplemented with multistart, adaptive random search and branch and bound global optimization subroutines.

The Branch And Reduce Optimization Navigator (BARON) is a computational system for solving nonconvex optimization problems. BARON combines enhanced branch and bound concepts with efficient techniques for reducing search space, e.g. constraint propagation, interval analysis, and duality. BARON has been proven to be a very efficient solver in many practical applications. Nowadays it can be purchased as implementations in GAMS and AIMMS modelling languages.

1.3.3 Applications

Applications of global optimization methods are very broad. A search in scientific data bases with the key words 'global optimization' finds thousands of references. The majority of papers found are devoted to applications. We are not going to discuss here particular applications. For the details of applications of global optimization we refer to the recently published exhaustive reviews by Biegler and Grossman [18], [104], Floudas et all [79], and Pinter [190], [191]. Important applications of global optimization in engineering, e.g. [4], [81], [85], [179], [237], [270], [299], and molecular conformation, e.g. [80], [94], [126], [148], [178], [187], may be especially noted. The other fields of applications where statistical methods can be efficient are optimal design [102], [248], training of artificial neural networks [26], [177], [223], and statistical inference [264].

To demonstrate the problems related to analysis, classification and solution of real world global optimization problems, we will discuss the problem of multidimensional scaling (MDS). This technique is aimed mapping a set of abstract objects with given mutual proximity to a space of low dimensionality, called the embedding space [25], [54]. An image of the objects considered is a set of points in the embedding space which is two-dimensional in case MDS images are aimed for visualization. The mutual dissimilarity of the objects is visualized by means of distances between the points. The image of the objects can be analyzed heuristically thus the heuristic human abilities are extended to abstract spaces. For example, in a special case the objects can be points in multidimensional Euclidean space where human heuristic abilities are very poor. On the contrary, such abilities can be efficiently applied to the analysis of two dimensional images. Therefore, by means of analysis of images in two dimensional embedding space, human heuristic abilities are extended to multidimensional spaces.

Mathematically MDS can be defined as a method of non-linear mapping of an abstract space of objects into the embedding space preserving the structure of proximity of the objects. The implementation of the mapping supposes minimization of a mapping error. As the minimization problem is multimodal, MDS is related to global optimization. Similar problems occur in the so called distance geometry, e.g. of determination of structures of complex molecules.

Before starting a discussion on global optimization problem occurring in MDS we will present an illustration of application of MDS to bio-medical data. Visualization of bio-medical data is important for the development of new diagnostics for different diseases. For example, in diagnosing sleep diseases and disorders the sleep structure is important; it can be defined by seven numerical parameters. These parameters are measured by means of the polysomnography (continuous recording of electrooculogram, electroencephalogram, and electromyogram throughout the night). A patient is characterized by a record of time series of seven dimensional vectors composed of the sleep structure parameters. Since the polysomnography is very expensive its prescription should be well justified, e.g. using subjective self assessment data. From the record a medical doctor experienced in sleep problems assesses sleep quality as 'good', 'average', and 'bad'. In parallel, the sleep quality is self assessed by the patient subjectively using a psychometric questionnaire.

The Institute of Psychophysiology and Rehabilitation of Kaunas University of Medicine (in Palanga, Lithuania) has collected a data base of objective and subjective sleep quality of a large population of patients. Their visualization is helpful in analyzing the correspondence between data on subjective and objective sleep quality [297]. For example, in Fig. 1.7 a two dimensional image of data on objective sleep quality of the population of patients considered is presented; i.e. the image of a set of 1500 seven dimensional vectors. In the picture '+' denotes the patients with good sleep quality, '·' denotes patients with average sleep quality, and '∇' denotes patients with bad sleep quality. For medical experts the figures are presented on computer screen denoting different classes in different colors. We are not going to discuss medical interpretations of the image but one structural conclusion is obvious without a medical interpretation: the set of patients with good sleep quality is disjointed and clearly separated from two other classes. The classes of average and bad sleep quality are relatively compact and are not well separated.

We have presented this example to show that MDS has interesting and important applications. Many other applications are described in [25], [54].

Several mathematical formulations of MDS problems are possible. We consider here only one of them, namely, metric MDS. Let the dissimilarities of n objects be defined by the matrix (δ_{ij}), where $i, j = 1, ..., n$. We search in the two-dimensional Euclidean space for the points x_i whose inter-point distances fit the dissimilarities. Most frequently the so called *STRESS* criterion is used to assess the precision of mapping from the space of objects to the two-dimensional embedding space:

$$STRESS = \sum_i \sum_j w_{ij}(\delta_{ij} - d(x_i, x_j))^2, \tag{1.7}$$

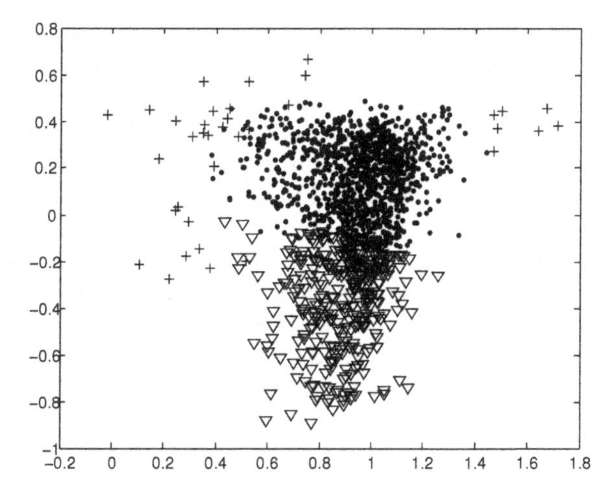

Fig. 1.7. Two dimensional visualization of data on sleep quality

where w_{ij} denotes the weight of error of dissimilarity between the i-th and the j-th objects, and $d(\cdot)$ denotes a distance in the embedding space; normally $d(\cdot)$ is the Euclidean distance but other distances are also considered, e.g. the city block distance. To implement the MDS method (1.7) should be minimized. We discuss the minimization difficulties later, but mention here that (1.7) is not everywhere differentiable. The other frequently used criterion *SSTRESS* is a differentiable function of variables x_i

$$SSTRESS = \sum_i \sum_j w_{ij}(\delta_{ij}^2 - d(x_i, x_j)^2)^2, \tag{1.8}$$

where $d(\cdot)$ is the Euclidean distance.

Both functions (1.7) and (1.8) are not unimodal [255], [302]. The minimization difficulty is also caused by the high dimensionality of MDS problems. For example, the dimensionality of the minimization problems with the data on sleep quality is 3000. Besides dimensionality, the difficulty of a global minimization problem depends on the number of local minimizers, and on the extent of the region attraction of the global minimizer. The worst case estimate of the number of local minimizers is exponential with respect to the dimensionality of the minimization problem. However, there are only a few documented investigations of example/practical problems in this respect. For example, in [155] a MDS problem with 10 objects was considered, and 9 different local minima indicated; the hypervolume of the regions of attraction of the global minimizers was about 4% of the feasible region. In [103] a problem with 1098 local minimizers is referenced, however in [136] it is shown that many of these points are not actually local minimizers, since the Newton method makes progress from many of these points. An experimentally evaluated number of local minimizers is not always reliable, e.g. solutions, found by a local descent method terminating where the norm of gradient is small, can be not only local minimizers but also points of flatness and saddle points.

The discussion on the properties of minimization problems related to MDS shows that assessment of the complexity of practical global optimization problems is difficult even in the case of a problem defined by simple analytical formulae. Therefore the classification of multimodal optimization problems is also difficult.

The known properties of a MDS problem are not sufficient to specify the most suitable global optimization method. Therefore different methods have been tried. In the case where $d(\cdot)$ is the Euclidean distance, gradient descent methods are applicable for local minimization of both functions (1.7) and (1.8). The criterion *SSTRESS* is obviously everywhere differentiable. Although *STRESS* is not differentiable on a small subset, gradient descent trajectories avoid this set as shown in [294]. Therefore for global optimization of (1.7) and (1.8) it seems rational to construct an algorithm combining gradient descent (taking into account the second order information on the objective function) with a random global search. As shown in [153], [154] a version of the evolutionary algorithm combined with a quasi Newton descent is quite efficient for solving global minimization problems in MDS. Such a method was also used to obtain the images presented in Fig. 1.7 and Fig. 1.8; the first minimization problem was 3000 dimensional, the second was 2200 dimensional. In these examples the results are satisfactory from an applications point of view.

Nevertheless, properties of MDS related optimization problems depend on data. Therefore further research is needed to justify the application of different global optimization methods to different classes of MDS problems, and it seems reasonable to continue experiments with hybrid methods combining evolutionary global search and gradient based local descent [154], [296]. Since the MDS related minimization problems exhibit many properties of real world problems they can be useful as test functions [155].

MDS is not only an interesting problem of global optimization but can also be applied as a tool for creating interactive global optimization methods. For example, in [248] MDS was applied to analyze globality/locality properties of global optimization algorithms. By means of visualization of a set of observation points various properties of an optimization problem can be grasped. In the paper [298] a minimization problem with an implicitly defined very small feasible region is considered. It is important to understand the shape of the feasible region and its location inside a known larger set (five dimensional unit hypercube). The volume of the feasible region is less than 0.01% of the volume of hypercube. In Fig. 1.8 the two dimensional image of the 1000 uniformly random generated trial points in the hypercube is presented (small points in the figure) together with the images of the cube vertices (thick points on the boarder of the figure), and the images of one hundred points belonging to the feasible region (a cluster of thick points in the figure). Analyzing the shape of the cluster of points representing the feasible region, and taking into account the small hypervolume of the feasible region, we may conclude that

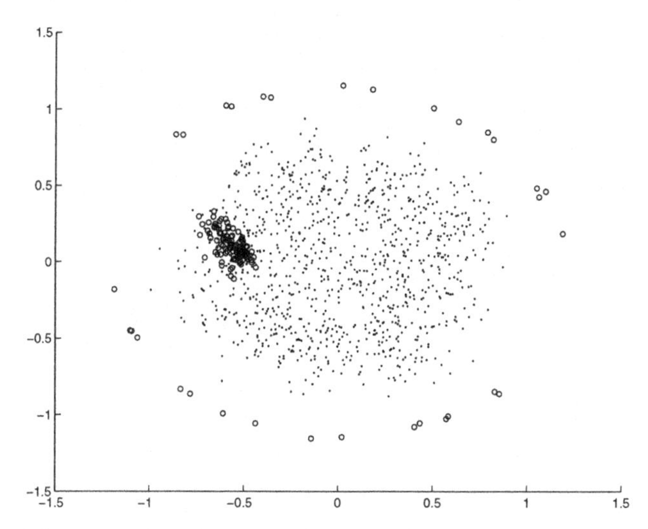

Fig. 1.8. Two dimensional image of trial points in the optimization problem with a small implicitly defined feasible region.

the feasible region is not disjoint, it has different extent in different directions, and is probably convex.

2

Global Random Search: Fundamentals and Statistical Inference

2.1 Introduction to Global Random Search

In this section, we formulate and discuss basic assumptions concerning the feasible region and the objective function, introduce a general scheme of global random search algorithms and provide a general result establishing convergence of global random search algorithms.

2.1.1 Main Assumptions

Consider a general minimization problem

$$f(x) \to \min_{x \in A}$$

with objective function $f(\cdot)$ and feasible region A. We shall always assume that the minimum value $m = \min_{x \in A} f(x)$ is attained in A. In general, $f(\cdot)$ may have more than one minimizer x_*.

Let us formulate other common assumptions about A and $f(\cdot)$. Most of these assumptions will be assumed true throughout this chapter and the next.

Assumptions concerning the feasible region A:

C1: A is a bounded closed subset of $\mathbb{R}^d$ $(d \geq 1)$;

C2: $\mathrm{vol}(A) > 0$, where 'vol$(\cdot)$' denotes the volume (or d-dimensional Lebesgue measure) of a set;

C3: A is a finite union of the sets defined by a finite number of the inequality-type constraints $g_i(x) \leq 0$, where the functions $g_i(\cdot)$ defining the constraints are continuously differentiable (however, we do not need to know the explicit forms of these functions);

C4: there exist constants $c > 0$ and $\varepsilon_0 > 0$ such that for at least one minimizer x_* and all ε, $0 < \varepsilon < \varepsilon_0$, we have $\mathrm{vol}(B(x_*, \varepsilon)) \geq c\varepsilon^d$; that is, at least a uniformly constant proportion of a ball in $\mathbb{R}^d$ with centre at x_* and small radius must intersect A;

C5: the structure of A is simple enough for distribution sampling algorithms on A and some of its subsets, to be of acceptable complexity.

These conditions are satisfied for an extremely wide class of practically interesting sets A. We do not require, in particular, for A to be a cube; moreover, neither convexity nor connectivity for A are generally required.

Conditions C1 and C2 are very simple and natural (note that Condition C1 implies $\text{vol}(A) < \infty$ so that $0 < \text{vol}(A) < \infty$). Conditions C3 and C4 are needed to avoid difficulties at the boundaries of A. Thus, Condition C3 prevents fractal boundaries and Condition C4 helps avoid the configurations where random search algorithms would almost certainly fail. An example of such a configuration is shown in Figure 2.1. For such a configuration, simple random search algorithms (which are not using local descents) will not be able to approach the minimizer x_* as the path to this point is 'too narrow'.

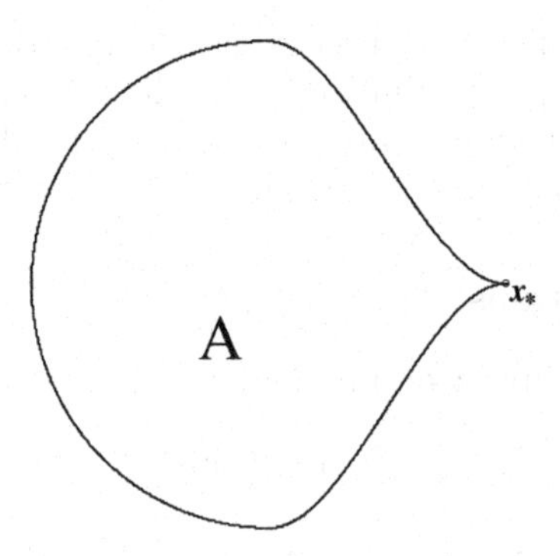

Fig. 2.1. An example of a disallowed combination of the set A and the minimizer x_*.

Of course, Condition C4 can be substituted with the following simpler but somewhat stronger condition on A:

C4′: there exist constants $c > 0$ and $\varepsilon_0 > 0$ such that $\text{vol}(B(x, \varepsilon)) \geq c\varepsilon^d$ for all $x \in A$ and all ε, $0 < \varepsilon < \varepsilon_0$.

Assumptions concerning the objective function $f(\cdot)$*:*

C6: $f(\cdot)$ can be evaluated at any point of A without error (note, however, that we allow evaluation errors in Sects. 2.1.4, 3.5 and in some general discussions);

C7: the number of minimizers x_* is finite.

As a condition additional to C7, we shall sometimes assume that the minimizer x_* is unique.

Rather than demanding continuity of $f(x)$ for all $x \in A$, we shall demand the following two weaker conditions:

C8: function $f(\cdot)$ is bounded and piece-wise continuous on A;
C9: there exists $\delta_0 > 0$ such that for all $0 < \delta \leq \delta_0$ the sets

$$W(\delta) = \{x \in A \colon f(x) \leq m + \delta\}$$

are closed and $f(x)$ is continuous for all $x \in W(\delta_0)$.

Note that if the objective function $f(\cdot)$ is continuous for all $x \in A$ then Condition C9 holds for all δ_0 (and, in view of Condition C1, Condition C8 also holds).

The sets $W(\delta)$ and their behaviour as $\delta \to 0$ (an example)

The boundaries of the sets $W(\delta)$ are the level sets of $f(\cdot)$:

$$\partial W(\delta) = \{x \in A : \; f(x) = m + \delta\} = f^{-1}(m + \delta)\,.$$

These level sets can be easily visualized when $d \leq 2$; see e.g. Fig. 1.2 and Fig. 2.2, where the contour-plot of the function

$$g(x, y) = f_{3,3}(x) + f_{5,5}(y) + f_{3,3}(x) f_{5,5}(y), \quad (x, y) \in [0,1] \times [0,1]\,, \tag{2.1}$$

is provided; the function $f_{(k,l)}(\cdot)$ is defined below in (2.2).

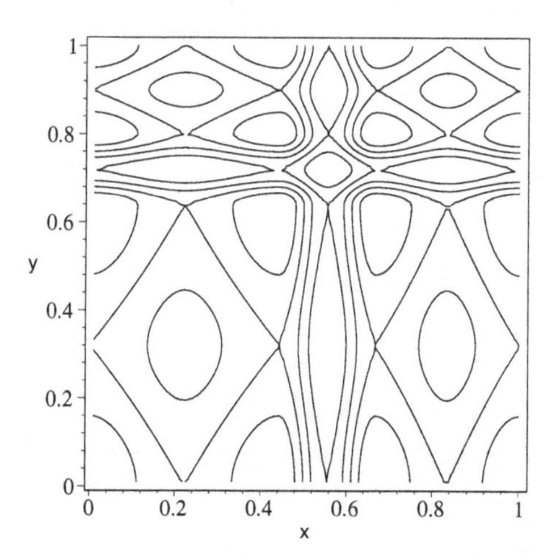

Fig. 2.2. Contour-plot of the function $g(x, y)$ defined in (2.1); the minimum value of this function is equal to 0 and is achieved at the global minimizer $(x_*, y_*) = (\frac{5}{9}, \frac{18}{25})$.

Let $A = [0,1]$ and let $k \geq 2$ and $l \geq 2$ be some integers. The function $f_{(k,l)}(\cdot)$ is defined as

$$f_{(k,l)}(x) = \begin{cases} 1 - \frac{1}{2}\left(\sin\frac{lk\pi\, x}{(k-1)(l-1)}\right)^2 & \text{for } x \in \left[0, \frac{(k-1)(l-1)}{kl}\right] \\ 1 - \left(\sin\frac{lk\pi\, x}{l-1}\right)^2 & \text{for } x \in \left[\frac{(k-1)(l-1)}{kl}, \frac{l-1}{l}\right] \\ 1 - \frac{1}{2}\left(\sin l\pi\, x\right)^2 & \text{for } x \in \left[\frac{l-1}{l}, 1\right]. \end{cases} \quad (2.2)$$

As illustrations, the functions $f_{(12,5)}(x)$ and $f_{(3,5)}(x)$ are depicted in Figs. 1.1 and 3.2(A), respectively.

For all integers $k, l \geq 2$, the functions $f_{(k,l)}(x)$ are continuously differentiable in $A = [0, 1]$ and have three local minima. These local minima are achieved at the points:

$$x_{(1)} = \frac{(k-1)(l-1)}{2k\,l}, \quad x_{(2)} = \frac{(2k-1)(l-1)}{2k\,l}, \quad \text{and } x_{(3)} = 1 - \frac{1}{2l}$$

with the point $x_* = x_{(2)}$ being the global minimizer. The values of the function $f_{(k,l)}(\cdot)$ at these points are

$$f_{(k,l)}(x_{(1)}) = f_{(k,l)}(x_{(3)}) = \frac{1}{2}, \quad f_{(k,l)}(x_{(2)}) = 0.$$

Despite the fact that the functions $f_{(k,l)}(x)$ are continuously differentiable, the problem of finding x_* is very difficult when k is large. Indeed, the complexity of the problem of global optimization is very much related to the rate of decrease of the ratio $\mathrm{vol}(W(\delta))/\mathrm{vol}(A)$ as δ decreases. Thus, for large k, the complexity of the function $f_{(k,l)}(x)$ defined in (2.2) is expressed in terms of the values of $\mathrm{vol}(W(\delta))$ which are small even for moderately large values of δ; for instance,

$$\mathrm{vol}(W(0.5)) = \frac{l-1}{2kl} < \frac{1}{2k}.$$

In fact, for $f(\cdot) = f_{(k,l)}(\cdot)$, we can easily compute $\mathrm{vol}(W(\delta))$ for all δ:

$$\mathrm{vol}(W(\delta)) = \begin{cases} 0 & \text{if } \delta \leq 0 \\ \frac{l-1}{kl}\left(1 - \frac{2}{\pi}\arcsin\sqrt{1-\delta}\right) & \text{if } 0 \leq \delta \leq \frac{1}{2}, \\ 1 - \frac{2}{\pi kl}\left((l-1)\arcsin\sqrt{1-\delta} + (kl-l+1)\arcsin\sqrt{2-2\delta}\right) & \text{if } \frac{1}{2} < \delta \leq 1, \\ 1 & \text{if } \delta \geq 1. \end{cases}$$

For general $f(\cdot)$, the rate of convergence of $\mathrm{vol}(W(\delta))/\mathrm{vol}(A)$ to zero as $\delta \to 0$ is studied in Sect. 2.5.3. In particular, it is shown in that section that if Conditions C7, C8 and C9 hold and if additionally for each global minimizer x_* the objective function $f(\cdot)$ is locally quadratic in the neighbourhood of x_*, then there exists a constant $c > 0$ such that

$$\mathrm{vol}(W(\delta)) = c\delta^{d/2}(1 + \mathrm{o}(1)) \quad \text{as } \delta \to 0. \quad (2.3)$$

This means that the rate of convergence of $\mathrm{vol}(W(\delta))$ to zero as $\delta \to 0$ is the same for a very broad class of objective functions. Of course, the complexity of the function $f(\cdot)$ is also related to the value of the constant c in (2.3) and the range of values of δ, where the asymptotic relation (2.3) can be applied.

In a particular case of $f(\cdot) = f_{(k,l)}(\cdot)$, we have

$$\mathrm{vol}(W(\delta)) = \frac{2(l-1)\sqrt{\delta}}{kl\pi}\left(1 + \frac{\delta}{6} + O\left(\delta^2\right)\right), \quad \delta \to 0 .$$

2.1.2 Formal Scheme of Global Random Search Algorithms

In a general global random search algorithm, a sequence of random points $x_1, x_2, \ldots, x_n$ is generated where for each j, $1 \le j \le n$, the point x_j has some probability distribution P_j. For each $j \ge 2$, the distribution P_j may depend on previous points $x_1, \ldots, x_{j-1}$ and the results of the objective function evaluations at these points (the function evaluations may not be noise-free). The number of points n, $1 \le n \le \infty$ (the stopping rule) can be either deterministic or random and may depend on the results of function evaluation at the points $x_1, \ldots, x_n$. For convenience, we shall refer to this general scheme as Algorithm 2.1.

Algorithm 2.1.

1. *Generate a random point x_1 according to a probability distribution P_1 on A; evaluate the objective function at x_1; set iteration number $j = 1$.*
2. *Using the points $x_1, \ldots, x_j$ and the results of the objective function evaluation at these points, check whether $j = n$; that is, check an appropriate stopping condition. If this condition holds, terminate the algorithm.*
3. *Alternatively, generate a random point x_{j+1} according to some probability distribution P_{j+1} and evaluate the objective function at x_{j+1}.*
4. *Substitute $j+1$ for j and return to step 2.*

In the algorithm which is often called 'pure random search' all the distributions P_j are the same (that is, $P_j = P$ for all j) and the points x_j are independent. In Markovian algorithms the distribution P_{j+1} depends only on the previous point x_j and its function value $f(x_j)$. There is also a wide class of global random search algorithms where the distributions are not updated at each iteration but instead after a certain number of points have been generated. We can formally write down this scheme as follows.

Algorithm 2.2.

1. *Choose a probability distribution P_1 on the n_1–fold product set $A \times \ldots \times A$, where $n_1 \ge 1$ is a given integer. Set iteration number $j = 1$.*
2. *Obtain n_j points $x_1^{(j)}, \ldots, x_{n_j}^{(j)}$ in A by sampling from the distribution P_j. Evaluate the objective function $f(\cdot)$ at these points.*
3. *Check a stopping criterion.*

4. *Using the points* $x_{l(i)}^{(i)}$ $(l(i)=1,\ldots,n_i;\ i=1,\ldots,j)$ *and the objective function values at these points, construct a probability distribution* P_{j+1} *on the* n_{j+1}*–fold product set* $A\times\ldots\times A$*, where* n_{j+1} *is some integer that may depend on the search information.*
5. *Substitute* $j+1$ *for* j *and return to Step 2.*

Of course, if $n_j = 1$ (for all j) in Algorithm 2.2 then it becomes Algorithm 2.1. On the other hand, Algorithm 2.1 allows more freedom in defining the distributions of points where $f(\cdot)$ is evaluated and therefore can seem to be more general than Algorithm 2.2. Thus, the difference between Algorithms 2.1 and 2.2 is purely formal; sometimes one form is more convenient and in other cases the other form is more natural.

There are two important issues to deal with while constructing global random search algorithms (in either form, Algorithm 2.1 or 2.2):

(i) choosing the stopping rule n, and
(ii) choosing the way of constructing the distributions P_j.

Consider issue (i). Commonly, a fixed number of points is generated (that is, the total number of points n is fixed). A more sophisticated approach would be to estimate the closeness of the current record value of the objective function $f(\cdot)$ to its minimum value $m = \min f$. This can be done in different ways. Any of the deterministic approaches (based, for example, on the use of Lipschitz constant estimates) can be applied. An enormous advantage of many global random search algorithms is related to the fact that because of the randomness of the points where $f(\cdot)$ is evaluated, probabilistic and statistical considerations can be applied to infer about the closeness of the current record value of $f(\cdot)$ to the minimum m; many of these considerations can be used in defining the stopping rule. A large part of the present chapter is devoted to these probabilistic and statistical considerations.

Issue (ii) concerns the construction of the distributions P_j (here by 'construction' we do not mean 'giving an analytic formula' but rather 'formulating an algorithm for sampling from the distribution'). This is the issue of how we use prior information about $f(\cdot)$ and the information we obtain in the process of the search, as well as how we compromise between the globality and locality of our search. The former problem (of extracting and using information about f) is complex and versatile; significant parts of this chapter and the next deal with it. The latter problem is potentially simpler, it was briefly considered in Sect. 1.1.3.

2.1.3 Convergence of Global Random Search Algorithms

In the early stages of development of global random search theory (in the nineteen seventies and eighties), a number of papers were published establishing sufficient conditions for convergence (in probability and with probability one) of random search algorithms; see, for example, [63, 188, 229]. The main idea in most of these, and in many other results on convergence of global random search algorithms, is the classical, in probability theory, 'zero-one law', see e.g. [226]. The following simple theorem stated and proved in [273], Sect. 3.2, illustrates this technique in a very general setup.

Let us consider a general global random search algorithm in the form of Algorithm 2.1, where the point x_j has some distribution P_j which may depend on previous points $x_1, \ldots, x_{j-1}$ and the results of the objective function evaluation at these points.

Theorem 2.1. *Let the objective function $f(\cdot)$ satisfy Condition C7, x_* be a global minimizer of $f(\cdot)$ and let $f(\cdot)$ be continuous in the vicinity of x_*. Assume that*

$$\sum_{j=1}^{\infty} q_j(\varepsilon) = \infty \tag{2.4}$$

for any $\varepsilon > 0$, where

$$q_j(\varepsilon) = \inf P_j(B(x_*, \varepsilon)), \tag{2.5}$$

with $B(x_*, \varepsilon) = \{x \in A \colon ||x - x_*|| \leq \varepsilon\}$; the infimum in (2.5) is taken over all possible previous points and the results of the objective function evaluations at them. Then, for any $\delta > 0$, the sequence of points x_j with distributions P_j falls infinitely often into the set $W(\delta) = \{x \in A \colon f(x) - m \leq \delta\}$, with probability one.

Proof is given in Sect. 2.7; it is a simplified version of the proof given in [273].

Note that Theorem 2.1 holds in the general case where evaluations of the objective function $f(\cdot)$ can be noisy (and the noise is not necessarily random). If the function evaluations are noise-free, then the conditions of the theorem ensure that the sequence $\{x_n\}$ converges to the set $A_* = \{\arg\min f\}$ of global minimizers with probability one; similarly, the sequence of records $y_{on} = \min_{j \leq n} f(x_j)$ converges to $m = \min f$ with probability one.

If for a particular sequence $\{x_j\}$ we have

$$\sum_{j=1}^{\infty} P_j(B(x_*, \varepsilon)) < \infty\,,$$

then the Borel-Cantelli lemma (see e.g. [226]) implies that the points $x_1, x_2, \ldots$ fall into $B(x_*, \varepsilon)$ only a finite number of times, with probability one. Moreover, looking at the family of functions (2.2), we conclude that (2.4) cannot be improved upon for a wide enough class $\mathcal{F}$ of objective functions. That is, if (2.4) is not satisfied then there exists $f \in \mathcal{F}$ such that for any n, none of the points $x_1, \ldots, x_n$ fall into $B(x_*, \varepsilon)$ with any fixed probability γ, $0 < \gamma < 1$.

Since the location of x_* is not known a priori, the following simple sufficient condition for (2.4) can be used:

$$\sum_{j=1}^{\infty} \inf P_j(B(x,\varepsilon)) = \infty \tag{2.6}$$

for all $x \in A$ and $\varepsilon > 0$.

In practice, a very popular rule for selecting probability measures P_j's is

$$P_{j+1} = \alpha_{j+1} P + (1 - \alpha_{j+1}) Q_j \,, \tag{2.7}$$

where $0 \le \alpha_{j+1} \le 1$, P is the uniform distribution on A (extension to other probability distributions P is straightforward) and Q_j is an arbitrary probability measure on A which may depend on the results of the evaluation of the objective function at the points $x_1, \ldots, x_j$. For example, sampling from Q_j may correspond to performing several iterations of a local descent from the current record point x_{oj}.

Sampling from the distribution (2.7) corresponds to taking a uniformly distributed random point in A with probability α_{j+1} and sampling from Q_j with probability $1 - \alpha_{j+1}$.

If the probability measures P_j in Algorithm 2.1 are chosen according to (2.7), then a simple and rather weak condition

$$\sum_{j=1}^{\infty} \alpha_j = \infty \tag{2.8}$$

is sufficient for (2.4) and (2.6) to hold.

The rate of convergence of the global random search algorithms, represented in the form of Algorithm 2.1 with distributions P_j chosen according to (2.7) and (2.8), is discussed at the end of Sect. 2.2.2.

2.1.4 Random Errors in Observations

Many global random search algorithms can easily be modified so that they can be used in the case where there are random errors in the observations of the objective function values. To give an example, several versions of the 'simulated annealing' algorithm considered in Sect. 3.3.2, have been devised for optimizing objective functions corrupted by noise, even before the simulated annealing algorithms became widely known. The corresponding algorithms are often called global (or multiextremal) stochastic approximation algorithms, see [260, 277] and [273], Sects. 3.3.3 and 3.3.4. The theoretical study of these algorithms is often related to the study of stochastic differential equations and in particular to the study of diffusion processes, see e.g. [142].

Providing the globality of search (see Sect. 1.1.3) is simple whether or not there are errors in observations. What is not that simple is recognizing the neighbourhood of the global minimizer and making local steps (as it is difficult to estimate gradients of the objective function). However, many statistical and heuristic arguments can be employed for monitoring the arrival at the neighbourhood of the global minimizer (see e.g. Sect. 4.1 in [273]).

Rather than further developing this topic (which is not particularly challenging), we briefly consider a different problem related to the fact that there are random errors in observations of $f(\cdot)$. This is the problem of estimating the values of the objective function and its gradients in the case where the distribution of noise is known. We assume that the objective function is specified as the expectation

$$f(x) = E_x g(x,Y) = \int g(x,y)\phi_x(y)dy, \tag{2.9}$$

where $g(x,y)$ is a known function and $\phi_x(\cdot)$ is the density of the random variable Y; note that the random variable $Y = Y_x$ and the density $\phi_x(\cdot)$ may depend on x.

Assuming that the integral in (2.9) cannot be evaluated analytically, a natural way of approximating it is to use the following Monte Carlo estimator

$$f(x) \cong \frac{1}{n}\sum_{i=1}^{n} g(x, Y_x^{(i)}) \tag{2.10}$$

where $\{Y_x^{(1)}, \dots, Y_x^{(n)}\}$ is a sample from a distribution with density $\phi_x(\cdot)$.

Assume that there exists a density $\pi(\cdot)$ such that $\phi_x(y) = 0$ whenever $\pi(y) = 0$ so that the ratio

$$w_x(y) = \phi_x(y)/\pi(y)$$

is well defined. Then, to estimate $f(x)$, we can use the method known as the *importance sampling:*

$$f(x) \cong \frac{1}{n}\sum_{i=1}^{n} g(x, Y^{(i)})w_x(Y^{(i)}) \tag{2.11}$$

where $\{Y^{(1)}, \dots, Y^{(n)}\}$ is a sample from a distribution with density $\pi(\cdot)$. Note that this sample does not have to be independent or even random; it can be, for instance, a stratified sample, a MCMC sample or even a quasi-random sample, see Sects. 3.1, 3.2.1 and 3.3.2.

The main advantage of using (2.11) over (2.10) is the fact that we can use the same sample $\{Y^{(1)}, \dots, Y^{(n)}\}$ for estimating values of $f(x)$ for all required values of the argument x. Moreover, using the same sample we can approximate the components of the gradient

$$\nabla f(x) = \left(\partial f(x)/\partial x_{(1)}, \dots, \partial f(x)/\partial x_{(d)}\right)\,, \quad x = (x_{(1)}, \dots, x_{(d)})\,.$$

Indeed, the j-th derivative $\partial f(x)/\partial x_{(j)}$ can be written as

$$\frac{\partial f(x)}{\partial x_{(j)}} = \int \left(w_x(y) \frac{\partial g(x,y)}{\partial x_{(j)}} + g(x,y) \frac{\partial w_x(y)}{\partial x_{(j)}} \right) \pi(y) dy$$

and approximated by

$$\frac{\partial f(x)}{\partial x_{(j)}} \cong \frac{1}{n} \sum_{i=1}^{n} \left(w_x(Y^{(i)}) \frac{\partial g(x, Y^{(i)})}{\partial x_{(j)}} + g(x, Y^{(i)}) \frac{\partial w_x(Y^{(i)})}{\partial x_{(j)}} \right), \quad (2.12)$$

where $\{Y^{(1)}, \ldots, Y^{(n)}\}$ is the same sample as above. Similarly one can approximate higher-order derivatives of $f(\cdot)$.

The method based on (2.10) is often called the *many-samples method* as it requires a new sample $Y_x^{(1)}, \ldots, Y_x^{(n)}$ for every function evaluation. The method based on (2.11) and (2.12) is called the *single-sample method* as it only uses one sample to estimate all required function values and its derivatives. The single-sample method has numerous advantages over the more traditional many-samples method, see [92] for references and more discussion.

2.2 Pure Random and Pure Adaptive Search Algorithms

Pure random search (PRS for short) is the simplest global random search algorithm. It consists of taking a sample of n independent random points x_j $(j = 1, \ldots, n)$ in A and evaluating the objective function $f(\cdot)$ at these points. Studying this algorithm is relatively simple. However, knowing the properties of this algorithm is very important as PRS is a component of many other random search algorithms. Additionally, PRS is often a bench-mark for comparing properties of other global optimization algorithms (not necessarily random search ones).

In this section, we also consider a version of PRS which is called 'pure adaptive search' and generalize it to the 'pure adaptive search of order k'.

2.2.1 Pure Random Search and the Associated c.d.f.

The algorithm

PRS is an algorithm where n random points x_j $(j = 1, \ldots, n)$ are generated and the objective function $f(\cdot)$ at these points is evaluated. The points x_j are i.i.d.r.v. in A with common distribution P. Here n is a stopping rule which is not necessarily a fixed number (typically, however, it is a fixed number); P is some given probability measure on A, not necessarily uniform (although the case where P is uniform is the main special case).

The probability distribution P should be simple enough to sample from and must not be much different from the uniform measure on A (otherwise PRS may lose the property of being a global optimization algorithm). It is often enough to assume that the distribution P is equivalent to the uniform distribution on A, see Condition C10 below.

Of course, PRS can be represented in the form of Algorithm 2.1, with $P_j = P$ for all $j = 1, \ldots, n$ and independent points $x_1, \ldots, x_n$.

The most common estimators (of course, they can only be used where the evaluations of $f(\cdot)$ are noise-free) of the minimum $m = \min f$ and the minimizer $x_* = \arg\min f$ are respectively the record values $y_{on} = \min_{1 \le j \le n} f(x_j)$ and the corresponding record points x_{on} which satisfy $f(x_{on}) = y_{on}$. We shall see below that the estimator y_{on} of m can often be significantly improved.

The c.d.f. of major importance

As a result of the application of PRS we obtain an independent sample $X_n = \{x_1, \ldots, x_n\}$ from a distribution P on A. Additionally, we obtain an independent sample $Y_n = \{y_1, \ldots, y_n\}$ of the objective function values at these points. The elements $y_j = f(x_j)$ of the sample Y_n are i.i.d.r.v. with the c.d.f.

$$F(t) = \Pr\{x \in A \,:\, f(x) \le t\} = \int_{f(x) \le t} P(dx)\,. \tag{2.13}$$

Fig. 2.3 displays the c.d.f. (2.13) for the case where the distribution P is unform on $A = [0, 1]$ and the objective function $f(x) = f_{(k,l)}(x)$ is as defined in (2.2) with $l = 5$ and $k = 2, 5$ and 20.

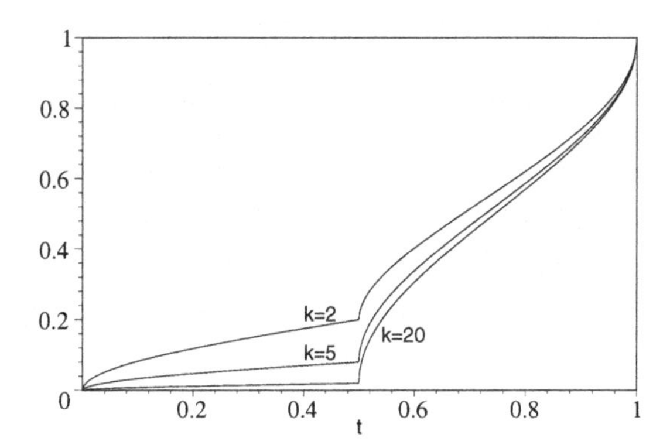

Fig. 2.3. Graphs of the c.d.f. (2.13) when P is uniform on $[0, 1]$ and $f(x) = f_{(k,l)}(x)$ is as defined in (2.2) with $l = 5$ and $k = 2, 5, 20$.

Assumption about the distribution P

Additional to the assumptions formulated in Sect. 2.1.1 we shall need an assumption about the probability distribution P. We shall assume that P is equivalent to the standard Lebesque measure on A; that is, we assume the following condition:

C10: the probability distribution P has a density $p(x)$ such that $c_{(1)} \le p(x) \le c_{(2)}$ for all $x \in A$ and some positive constants $c_{(1)}, c_{(2)}$.

Condition C10, along with condition C4 of Sect. 2.1.1, implies that for every minimizer x_* we have

$$P(B(x_*,\varepsilon)) \ge c\varepsilon^d \quad \text{for some } c>0 \text{ and all } 0<\varepsilon\le 1; \tag{2.14}$$

here, as usual, $B(x_*,\varepsilon) = \{z \in A : ||x_* - z|| \le \varepsilon\}$. As a consequence, we obtain, in particular, that the elements of the sample X belong to the vicinity of x_* with positive probability.

General properties of the c.d.f. (2.13)

The c.d.f. (2.13) is of major importance in studying PRS as well as some associated global random search algorithms. This is related to the fact that

$$F(t) = P(W(t-m)) \quad \text{for all } t \ge m\,, \tag{2.15}$$

where $W(\delta) = \{x \in A\colon f(x) \le m + \delta\}$, $\delta \ge 0$. Therefore, for $t \ge m$, $F(t)$ has the interpretation of the probability that a random point x_i distributed according to P falls into the set $W(t-m)$.

If the probability measure P is uniform on A, then the representation (2.15) can be written in the form

$$F(m+\delta) = \mathrm{vol}(W(\delta))/\mathrm{vol}(A)\,. \tag{2.16}$$

The importance of the ratio in the r.h.s. of (2.16) has already been discussed at the end of Sect. 2.1.1. Furthermore, as we shall see below, the behaviour of $F(m+\delta)$ for small $\delta > 0$ is a very important characteristic of the efficiency of PRS and, more generally, of the complexity of the objective function $f(\cdot)$.

Since the set $W(\delta)$ is empty for $\delta < 0$, we have $F(t) = 0$ for $t < m$. In view of Conditions C7 and C10 we also have $F(m) = 0$. On the other hand, the inequality (2.14) and Condition C9 imply that $F(t) > 0$ for all $t > m$. Moreover, Conditions C1–C4 and C7–C9 imply that the c.d.f. $F(t)$ is continuous at $t = m$.

Certain properties of the c.d.f. $F(\cdot)$, see Sect. 2.3, are different depending on whether this c.d.f. is continuous or not. In addition to Conditions C8 and C10, to guarantee the continuity of $F(\cdot)$ defined in (2.15) we have to assume that $\mathrm{vol}(f^{-1}(t)) = 0$ for all t, where

$$f^{-1}(t) = \{x \in A : f(x) = t\}\,.$$

That is, $F(\cdot)$ is continuous if the volume of every level set of $f(\cdot)$ is zero.

Let η denote a random variable with c.d.f. $F(\cdot)$. The fact that $F(m) = 0$ and $F(t) > 0$ for all $t > m$ is equivalent to the statement that the essential infimum of η is equal to m:

$$F(m) = 0 \text{ and } F(t) > 0 \text{ for all } t > m \iff \operatorname{ess\,inf} \eta = m\,.$$

Finally, let $M = \sup_{x \in A} f(x)$. Condition C8 implies that $M < \infty$. For the c.d.f. $F(\cdot)$, this means that $F(M) = 1$ and correspondingly, $F(t) = 1$ for all $t \geq M = \sup f$. The value of M is never important; however, it is sometimes important that the random variable η is concentrated on a bounded interval.

Poisson process representation

Let us follow [44] and give a representation of PRS through a Poisson process. Assume that vol(A)=1 and the distribution P is uniform on A. Let x_0 be an internal point of A (for instance, x_0 is one of the global minimizers of f). Define a sequence of point processes N_n on $\mathcal{B}$ by

$$N_n(B) = \sum_{j=1}^{n} \mathbf{1}_B(n^{1/d}(x_j - x_0))\,, \quad B \in \mathcal{B}\,,$$

where $\mathbf{1}_B(\cdot)$ is the indicator function

$$\mathbf{1}_U(z) = \begin{cases} 1 & \text{if } z \in U \\ 0 & \text{otherwise.} \end{cases}$$

That is, for a fixed measurable set B, $N_n(B)$ is defined as the number of points among $x_1, \ldots, x_n$ that belong to the set $x_0 + n^{-1/d}B$. The sequence of point processes N_n converges in distribution (as $n \to \infty$) to N, a Poisson point process with intensity 1 defined on A. For this process,

$$\Pr\{N(B) = k\} = \frac{[\text{vol}(B)]^k}{k!} \exp(-\text{vol}(B)), \;\; k \geq 0, \;\; B \in \mathcal{B};$$

additionally, for disjoint $B_1, \ldots, B_i \in \mathcal{B}$, the values $N(B_j)$ $(j = 1, \ldots, i)$ are independent random variables.

Therefore, a suitably normalized point process of observations near x_0 looks like a standard Poisson point process. This does not give us new results about the rate of convergence of PRS but permits us to look at the algorithm from a different prospective.

2.2.2 Rate of Convergence of Pure Random Search

Let us consider a PRS where x_j $(j = 1, \ldots, n)$ are i.i.d.r.v. distributed according to P and let the stopping rule n be a fixed number. In this section, our aim is to study the rate of convergence of PRS. We assume that all conditions of Sect. 2.1 concerning the feasible region A and the objective function $f(\cdot)$ are satisfied.

Rate of convergence to a neighbourhood of a global minimizer

Let $\varepsilon > 0$ be fixed, $x_* = \arg\min f$ be a global minimizer of $f(\cdot)$ and let our objective be hitting the set

$$B = B(x_*, \varepsilon, \rho) = \{x \in A\colon \rho(x, x_*) \leq \varepsilon\}$$

with one or more of the points x_j ($j = 1, \ldots, n$). Let us regard the event 'a point x_j hits the set B' as success and the alternative event as a failure. Then PRS generates a sequence of independent Bernoulli trials with a success probability $P(B)$; Conditions C4 and C5 of Sect. 2.1 imply that $P(B) > 0$ for all $\varepsilon > 0$.

A sequence of independent Bernoulli trials is perhaps the most celebrated sequence in the probability theory. Below, we use some well-known results concerning this sequence to obtain results concerning the rate of convergence of PRS.

For fixed j, we have

$$\Pr\{x_j \in B\} = P(B)\,. \tag{2.17}$$

Therefore,

$$\Pr\{x_j \notin B\} = 1 - P(B), \qquad \text{for all } j\,.$$

In view of the independence of x_j,

$$\Pr\{x_1 \notin B, \ldots, x_n \notin B\} = (1 - P(B))^n$$

and therefore

$$\Pr\{x_j \in B \text{ for at least one } j,\ 1 \leq j \leq n\} = 1 - (1 - P(B))^n\,. \tag{2.18}$$

Since $P(B) > 0$, this probability tends to one as $n \to \infty$.

Let τ_B be a random moment of first hitting the set B. Then the average number of PRS iterations required for reaching B is

$$E\tau_B = \frac{1}{P(B)}\,.$$

Typically, $P(B)$ is very small even if ε is not small (see below) and the rate of convergence of the probability (2.18) to one is very slow. Additionally, if $P(B)$ is small then $E\tau_B$ is large.

Taking $n \approx 1/P(B)$ is not enough to guarantee that B is reached with high probability. Indeed, for small $x > 0$ we have

$$(1 - x)^{\frac{1}{x}} \cong e^{-1} \cong 0.36788$$

and therefore for $n = \lceil 1/P(B) \rceil$

$$1-(1-P(B))^n \cong 0.63212 \qquad \text{as } P(B) \to 0\,.$$

To achieve a probability of 0.95 for the r.h.s. of (2.18) we need to almost triple this value:

$$1-(1-P(B))^n \cong 1-\frac{1}{e^3} \cong 0.950213 \;\text{ for } n = \lceil 3/P(B) \rceil \;\text{ as } P(B) \to 0\,.$$

Furthermore, let us assume that we are required to reach the set B with probability at least $1-\gamma$ for some $0<\gamma<1$. This gives us the following inequality for n:

$$1-(1-P(B))^n \geq 1-\gamma\,.$$

Solving it we obtain

$$n \geq n(\gamma) = \frac{\ln \gamma}{\ln(1-P(B))}\,. \tag{2.19}$$

Since we assume that $P(B)$ is small, $\ln(1-P(B)) \cong -P(B)$, and we can replace (2.19) with

$$n \geq -\frac{\ln \gamma}{P(B)}\,; \tag{2.20}$$

that is, we need to make at least $\lceil -\ln\gamma/P(B) \rceil$ evaluations in PRS to reach the set B with probability $1-\gamma$.

Note that

$$\Pr\{x_j \in B(x_*,\varepsilon,\rho) \text{ for at least one } j, 1\leq j \leq n\} = \Pr\left\{ \min_{1\leq j\leq n} \rho(x_j,x_*) \leq \varepsilon \right\}$$

and therefore the discussion above can be considered as a discussion about the rate of convergence in probability of the sequence

$$\min_{1\leq j \leq n} \rho(x_j,x_*)$$

to zero, as $n \to \infty$.

Rate of convergence with respect to function values

If we want to study the rate of convergence with respect to the function values, that is, of

$$y_{on} - m = \min_{1\leq j\leq n} |f(x_j)-m| \qquad \text{as } n \to \infty\,,$$

then in the above study we have to replace the set $B = B(x_*,\varepsilon,\rho)$, with the set

$$W(\delta) = \{x \in A \colon f(x) - m \leq \delta\}$$

with some $\delta > 0$. In particular, we have $E\tau_{W(\delta)} = 1/P(W(\delta))$,

$$\Pr\{y_{on} - m \leq \delta\} = 1 - (1 - P(W(\delta)))^n \to 1 \text{ as } n \to \infty,$$

and in order to reach the set $W(\delta)$ with probability $1-\gamma$, we need to perform approximately $-\ln\gamma/P(W(\delta))$ iterations of PRS.

In view of (2.16), these formulae can be expressed in terms of the c.d.f. $F(\cdot)$. We have, in particular,

$$\Pr\{y_{on} - m \leq \delta\} = 1 - (1 - F(m+\delta))^n, \quad E\tau_{W(\delta)} = 1/F(m+\delta),$$

and to reach $W(\delta)$ with probability $1-\gamma$, we need to perform approximately $-\ln\gamma/F(m+\delta)$ iterations of PRS.

Particular case of the uniform distribution

Consider an important particular case, where the distribution P is uniform on A and ρ is the Euclidean metric (that is, $\rho = \rho_2$). Then for every $Z \in \mathcal{B}$ (this means that Z is any measurable subset of A) we have

$$P(Z) = \text{vol}(Z)/\text{vol}(A)$$

and for $B = B(x_*, \varepsilon)$ we have

$$P(B) = \frac{\text{vol}(B)}{\text{vol}(A)} \leq \frac{\pi^{\frac{d}{2}}\varepsilon^d}{\Gamma\left(\frac{d}{2}+1\right)\cdot\text{vol}(A)}, \tag{2.21}$$

where $\Gamma(\cdot)$ is the Gamma-function. If x_* is an interior point of A and ε is small enough so that the ball $\left\{x \in \mathbb{R}^d \colon \rho_2(x, x_*) \leq \varepsilon\right\}$ is fully inside A, then the inequality in (2.21) becomes an equality.

The formulae (2.19) and (2.20) then say that if we want to reach the set

$$B = B(x_*, \varepsilon) = \{x \in A \colon \rho_2(x, x_*) \leq \varepsilon\}$$

with probability at least $1-\gamma$, then we need to perform at least

$$n_* = \left\lceil \frac{\ln\gamma}{\ln\left(1 - \pi^{\frac{d}{2}}\varepsilon^d/(\Gamma(\frac{d}{2}+1)\cdot\text{vol}(A))\right)} \right\rceil \simeq \left\lceil -\ln\gamma\cdot\frac{\Gamma\left(\frac{d}{2}+1\right)}{\pi^{\frac{d}{2}}\varepsilon^d}\cdot\text{vol}(A)\right\rceil \tag{2.22}$$

iterations of PRS. Table 2.1 and Fig. 2.4 illustrate the dependence of $n_* = n_*(\gamma, \varepsilon, d)$ on γ, ε and d.

Note that in the majority of cases considered in Table 2.1, the approximation for n_* given in the r.h.s. of (2.22) over-estimates the true value of n_* by 1. Taking into account the fact that the values of n_* are typically very large, we can conclude that the r.h.s. of (2.22) gives a very good approximation for n_*.

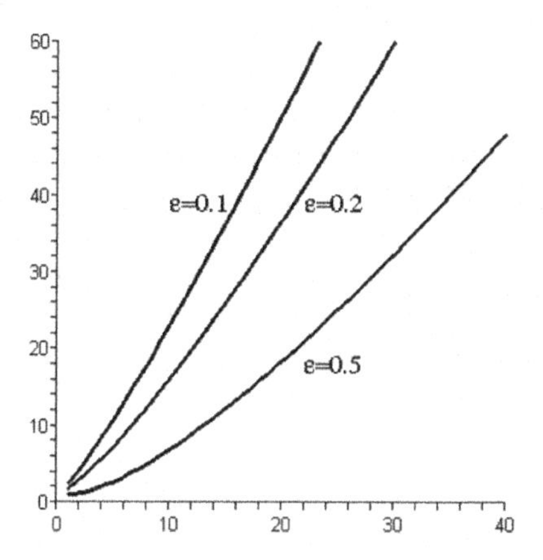

Fig. 2.4. Values of $\ln n_*$ as a function of d; here $\mathrm{vol}(A)=1$, $\gamma=0.1$, $\varepsilon=0.5, 0.2, 0.1$ and n_* is as defined in (2.22).

d	$\gamma=0.1$			$\gamma=0.05$		
	$\varepsilon=0.5$	$\varepsilon=0.2$	$\varepsilon=0.1$	$\varepsilon=0.5$	$\varepsilon=0.2$	$\varepsilon=0.1$
1	0	5	11	0	6	14
2	2	18	73	2	23	94
3	4	68	549	5	88	714
4	7	291	4665	9	378	6070
5	13	1366	43743	17	1788	56911
7	62	38073	$4.9\cdot 10^{6}$	80	49534	$6.3\cdot 10^{6}$
10	924	$8.8\cdot 10^{6}$	$9.0\cdot 10^{9}$	1202	$1.1\cdot 10^{7}$	$1.2\cdot 10^{10}$
20	$9.4\cdot 10^{7}$	$8.5\cdot 10^{15}$	$8.9\cdot 10^{21}$	$1.2\cdot 10^{8}$	$1.1\cdot 10^{16}$	$1.2\cdot 10^{22}$
50	$1.5\cdot 10^{28}$	$1.2\cdot 10^{48}$	$1.3\cdot 10^{63}$	$1.9\cdot 10^{28}$	$1.5\cdot 10^{48}$	$1.7\cdot 10^{63}$
100	$1.2\cdot 10^{70}$	$7.7\cdot 10^{109}$	$9.7\cdot 10^{139}$	$1.6\cdot 10^{70}$	$1.0\cdot 10^{110}$	$1.3\cdot 10^{140}$

Table 2.1. Values of $n_* = n_*(\gamma, \varepsilon, d)$, see (2.22), for $\mathrm{vol}(A)=1$, $\gamma=0.1$ and 0.05, $\varepsilon=0.5, 0.2$ and 0.1, for various d.

We can see that the dependence of n_* on γ is not crucial; on the other hand, $n_* = n_*(\gamma, \varepsilon, d)$ increases exponentially as the dimension d increases. As a matter of fact, the Stirling approximation gives for fixed $0<\gamma<1$ and $\varepsilon>0$:

$$\ln n_*(\gamma,\varepsilon,d) = \frac{d+1}{2}\ln(d) - d\ln(\sqrt{2\pi e}\varepsilon) + \ln\left[\sqrt{\pi}\mathrm{vol}(A)\,(-\ln\gamma)\right] + O\left(\frac{1}{d}\right)$$

as $d\to\infty$, and this approximation is extremely good even for small d.

If one is interested in the asymptotic behaviour of the value of $n_*(\gamma,\varepsilon,d)$ when d is fixed and the required precision ε tends to 0, then (2.22) implies

$$n_* = O\left(\frac{1}{\varepsilon^d}\right) \quad \text{as } \varepsilon \to 0\,. \tag{2.23}$$

Since we are not specifying the constant in (2.23), this formula holds not only for the case where $n_* = n_*(\gamma, \varepsilon, d)$ with $\rho = \rho_2$ and $P = P_0$ is the uniform distribution, but also for arbitrary $\rho = \rho_p$ $(1 \le p \le \infty)$, for any probability measure P equivalent to the uniform measure P_0 on A (see Condition C10 above). The same formula is true in the case when n_* has the meaning of the average number of iterations required to reach the ball $B(x^*, \varepsilon, \rho)$. If the objective function satisfies Conditions C7–C9 of Sect. 2.1, then we can replace the ball $B(x^*, \varepsilon, \rho)$ with the set $W(\varepsilon)$; the formula (2.23) will still hold.

Multivariate spacings

Let us mention a relevant result of S.Janson [127] on multivariate spacings. In general, the maximum (multivariate) spacing of a set of points $x_1, \ldots, x_n$ with respect to a convex set $B \subset \mathbb{R}^d$ is defined as the largest possible subset $x + rB$ of A which does not contain any of the points $x_j (j = 1, \ldots, n)$. Let $\mathrm{vol}(A) = 1$, B be either a cube or a Euclidean ball in $\mathbb{R}^d$ of unit volume: $\mathrm{vol}(B) = 1$; let also $x_1, \ldots, x_n$ be i.i.d.r.v. with the uniform distribution on A. Set

$$\triangle_n = \sup\{t\colon \text{there exists } x \in \mathbb{R}^d \text{ such that } x + tB \subset A \setminus \{x_1, \ldots, x_n\}\} \tag{2.24}$$

and define the volume of the maximum spacing as $V_n = (\triangle_n)^d$, which is the volume of the largest ball (or cube of fixed orientation) that is contained in A and avoids all n points $x_1, \ldots, x_n$. Then we have

$$\lim_{n \to \infty} \frac{nV_n - \ln n}{\ln \ln n} = d - 1 \quad \text{with probability 1} \tag{2.25}$$

(this result generalizes the result of P. Deheuvels [62]; see also [72]). Moreover, the sequence of random variables

$$nV_n - \ln n - (d-1) \ln \ln n + \beta_d$$

converges (as $n \to \infty$) in distribution to the r.v. with c.d.f. $\exp(-e^{-u})$, $u > 0$, where $\beta_d = 0$ if A is a cube and

$$\beta_d = \ln \Gamma(d{+}1) - (d{-}1) \left[\frac{1}{2} \ln \pi + \ln \Gamma\left(\frac{d}{2}{+}1\right) - \ln \Gamma\left(\frac{d{+}1}{2}\right)\right] \tag{2.26}$$

in the case when A is a ball. Since $\beta_d \ge 0$, the spherical spacings are a little bit smaller than the cubical ones. For large d, we can use the approximation

$$\beta_d = \frac{d}{2} \ln \frac{2d}{\pi} - d + \ln(\pi d) - \frac{1}{4} + \mathrm{O}\left(\frac{1}{d}\right), \quad d \to \infty\,,$$

for the quantity β_d defined in (2.26). This approximation is very accurate, especially if d is not very small (say, $d \ge 5$).

Extension to general global random search algorithms

The main results on the rate of convergence of PRS can be extended to a much wider class of global random search algorithms.

Consider the general Algorithm 2.1 of Sect. 2.1.2 and assume that the probability measures P_j are chosen according to (2.7), where the probability measure P satisfies Condition C10 of Sect. 2.2.1. Assume also that the measures Q_j are arbitrary and the condition (2.8) guaranteeing the convergence of the algorithm is met. Let us generalize the arguments that led us to the estimates of the convergence rate of PRS into this more general situation.

As a replacement of (2.17), for all $j \geq 1$ we have

$$\Pr\{x_j \in B\} \geq \alpha_j P(B)\,. \tag{2.27}$$

Arguments similar to those used in deriving (2.18) imply the inequality

$$\Pr\{x_j \in B \text{ for at least one } j,\ 1 \leq j \leq n\} \geq 1 - \prod_{j=1}^{n} (1 - \alpha_j P(B))\,. \tag{2.28}$$

In view of the condition (2.8) and the fact that $P(B) > 0$, the r.h.s. of (2.28) tends to one as $n \to \infty$.

Assume now that $P(B)$ is small and define $n(\gamma)$ as the smallest integer such that the following inequality is satisfied:

$$\sum_{j=1}^{n(\gamma)} \alpha_j \geq -\frac{\ln \gamma}{P(B)}\,.$$

Similarly to (2.20) we deduce that one has to perform at least $n(\gamma)$ iterations of Algorithm 2.1 to guarantee that at least one of the points x_j reaches the set B with probability $\geq 1 - \gamma$. Of course, $n(\gamma)$ is smallest if all $\alpha_j = 1$, that is when Algorithm 2.1 is PRS.

Extension of the main results concerning the rate of convergence with respect to function values and specialization to the case where P is the uniform distribution on A can be similarly made.

Slow rate of convergence may imply that the convergence is not practically achievable

Paying much attention to local search reduces the values of α_j's in (2.27). It may be tempting to perform many local searches leaving α_j's very small (for example, by setting $\alpha_j = 1/j$), just to guarantee the global convergence of the algorithm. Let us check what happens with the rate of convergence in the case when $\alpha_j = 1/j$. Since for large n we have $\sum_{j=1}^{n} 1/j \simeq \ln(n)$, from (2.20) we obtain

$$n(\gamma) \simeq \exp\{(-\ln \gamma)/P(B)\}\,.$$

Assuming that vol(A)=1 and that the distribution P is the uniform this, roughly speaking, implies that to compute the number of required iterations we need to exponentiate the numbers presented in Table 2.1. For instance, for very reasonable parameters $\varepsilon = 0.1$ and $d = 5$, we would need about 10^{19000} iterations of Algorithm 2.1 to guarantee that at least one of the points x_j will reach the ball $B(x_*, \varepsilon)$ with probability ≥ 0.9 (note that the total number of atoms in the universe is estimated to be smaller than 10^{81}).

The discussion above is very similar to the discussion provided by G.H.Hardy in Appendix III of his book [113]. Its consequence is that the fact of convergence of some global optimization algorithms is only a theoretical fiddle and does not mean anything in practice.

2.2.3 Pure Adaptive Search and Related Methods

In recent years there has been a great deal of activity (see e.g. papers [182, 265, 268, 269], the monograph [267] by Z.Zabinsky and references therein) related to the so-called 'pure adaptive search'. Unlike PRS, where the points x_j are independent and distributed in A with the same distribution P, at iteration $j+1$ of the pure adaptive search one chooses a random point x_{j+1} within the set

$$S_j = \{x \in A \colon f(x) < y_{oj}\}, \tag{2.29}$$

where $y_{oj} = \min\{f(x_1), \ldots, f(x_j)\}$ is the current record (in fact, every new point in the pure adaptive search is a new record point so that $x_j = x_{oj}$ and $y_j = y_{oj}$ for all $j \geq 1$). More precisely, x_1 has the probability distribution P and for each $j \geq 1$, x_{j+1} is a random point with the distribution P_{j+1} defined for all Borel sets $U \subset A$ by

$$P_{j+1}(U) = \frac{P(U \cap S_j)}{P(S_j)}, \tag{2.30}$$

where P is the original distribution and S_j is defined in (2.29). If P is the uniform distribution on A, then P_{j+1} is the uniform distribution on S_j. Of course, the points $x_1, x_2, \ldots$ generated in the pure adaptive search are dependent (unlike in PRS), see Sect. 2.3.3 for details.

If we replace the strict inequality $<$ in the definition of the sets S_j with $\leq$, then these sets are exactly the sets

$$W_{x_j} = \{x \in A : f(x) \leq f(x_j)\};$$

that is, $\overline{S}_j = W_{x_j}$, where $\overline{Z}$ denotes the closure of a set Z. The corresponding method (using the sets W_{x_j} in place of S_j) is called 'weak pure adaptive search'.

The study of the sequence of function values $f(x_1), f(x_2), \ldots$ in the pure adaptive search is equivalent to the study of the record values in PRS. This is the subject of Sect. 2.3.3; that section provides, therefore, a detailed investigation of the properties of the pure adaptive search. Note that in previous literature on the pure adaptive search this kind of investigation was lacking.

Of course, the sequence $f(x_j)$ converges to m much faster for the pure adaptive search than for PRS. The major obstacle preventing the application of the pure adaptive search to the practice is the fact that it is very hard to find points in the sets (2.29). To some extent, the problem of finding points in the sets (2.29) is one of the major objectives of all global optimization strategies. In particular, the set covering methods of Sect. 3.1 can help in removing the subregions of A that have no intersection with the sets (2.29) and thus simplify the problem of generating random points in these sets.

There are several papers fully devoted to the problem of generating random points in the sets (2.29), see e.g. [28, 194, 269] and Chapt. 5 in [267]; the corresponding methods either resemble or are fully based on the celebrated Markov Chain Monte Carlo methods. However, the problem is too difficult and cannot be resolved adequately. In general, there is no algorithmically effective way of generating (independent) random points from the sets (2.29) apart from using PRS in the first place (perhaps, with the bounding of certain subsets of A) and waiting for a new record value of the objective function (which is equivalent to obtaining a new point x_j in the pure adaptive search). If this is the way of performing the pure adaptive search then:

(a) the average waiting time of a new record is infinite for all $j > 1$, see (2.63);
(b) by discarding the k-th record values in PRS ($k > 1$) we lose an enormous amount of information contained in the evaluations made during PRS.

Taking these points into account we can state that generally, despite the fast convergence, the pure adaptive search only has theoretical interest as it is either impractical or much less efficient that PRS.

A similar conclusion can be drawn about different modifications of the pure adaptive search. These modifications include:

(i) weak pure adaptive search defined above;
(ii) 'hesitant random search' (see e.g. [28, 30]), where for all $j > 1$ the next point x_{j+1} is random and has distribution P_{j+1} with probability α_{j+1} and any other distribution on A with probability $1-\alpha_{j+1}$; here α_{j+1} $(0 \leq \alpha_{j+1} \leq 1)$ may depend on $f(x_j)$ and the probability measures P_{j+1} are as defined in (2.30);
(iii) 'backtracking adaptive search', see [29, 265], where for all $j > 1$ the new point x_{j+1} is sampled from the sets:

$$\begin{cases} \{x \in A\colon f(x) < y_{oj}\} = S_j & \text{with probability } \alpha_{j+1} \\ \{x \in A\colon f(x) = y_{oj}\} & \text{with probability } \beta_{j+1} \\ \{x \in A\colon f(x) > y_{oj}\} & \text{with probability } 1-\alpha_{j+1}-\beta_{j+1} \end{cases}$$

for some α_{j+1} and β_{j+1} which may depend on $f(x_j)$.

2.2.4 Pure Adaptive Search of Order k

Aiming to resolve the problems (a) and (b) of the pure adaptive search, we can suggest the following extension of this algorithm (similar extensions can be suggested for its modifications (i)-(iii) above) which improves both pure random search and pure adaptive search.

Algorithm 2.3 (Pure adaptive search of order k).

1. *Choose points $x_1, \ldots, x_k$ by independent random sampling from the uniform distribution on A. Compute the objective function values $y_i = f(x_i)$ $(i = 1, \ldots, k)$. Set iteration number $j = k$.*
2. *For given $j \geq k$, we have points $x_1, \ldots, x_j$ in A and values of the objective function at these points. Let $y_j^{(k)}$ be the k-th record value corresponding to the sample $\{y_i = f(x_i),\ i = 1, \ldots, j\}$. Define the set*

$$S_j^{(k)} = \{x \in A\colon f(x) < y_j^{(k)}\}. \tag{2.31}$$

3. *Choose x_{j+1} as a uniform random point from the set $S_j^{(k)}$ and evaluate the objective function value at x_{j+1}.*
4. *Substitute $j+1$ for j and return to step 2.*

For simplicity, Algorithm 2.3 is formulated under the assumption that the underlying distribution of points in A is uniform. It can be easily generalized to the case of a general distribution: in this case, $x_1, \ldots, x_k$ are distributed in A according to a distribution P and x_{j+1} has the distribution $P_{j+1}^{(k)}$ defined for all Borel sets $U \subset A$ by $P_{j+1}^{(k)}(U) = P(U \cap S_j^{(k)})/P(S_j^{(k)})$; this formula is an extension of (2.30). Thus, the pure adaptive search of order 1 is just the pure adaptive search of Sect. 2.2.3.

Similarly to the case $k = 1$, we can define 'weak pure adaptive search of order k' by replacing the strict inequality $<$ in the definition of the sets $S_j^{(k)}$ with $\leq$. Analogously, we can define 'hesitant adaptive search of order k', 'backtracking adaptive search of order k' and other versions of the pure adaptive search.

Let, at iteration $j \geq k$ of Algorithm 2.3, $Y_j^{(k)} = \{y_j^{(1)}, \ldots, y_j^{(k)}\}$ be the set of k record values and $X_j^{(k)} = \{x_1^{(j)}, \ldots, x_j^{(k)}\}$ be the corresponding set of k record points. We have $y_j^{(1)} \leq \ldots \leq y_j^{(k)}$ and these values are the k smallest values of the objective function computed so far. At iteration $j+1$, the value $y_j^{(k)}$ is never in the set $Y_{j+1}^{(k)}$ (as $f(x_{j+1}) < y_j^{(k)}$) but the value $y_j^{(k-1)}$ always belongs to the new set of records: $y_j^{(k-1)} \in Y_{j+1}^{(k)}$ (the value $y_j^{(k-1)}$ can be either $y_{j+1}^{(k)}$ or $y_{j+1}^{(k-1)}$). Similarly, $x_j^{(k)} \notin X_{j+1}^{(k)}$ and $x_j^{(k-1)} \in X_{j+1}^{(k)}$. Thus, the pure adaptive search of order k (that is, Algorithm 2.3) is probabilistically equivalent to performing PRS and keeping k records and record points (rather than just one record value and one record point in the original pure adaptive search).

The two main advantages of choosing $k > 1$ over $k = 1$ are:

(a) the set (2.31) is bigger than (2.29) and it is therefore easier to find random points belonging to the set (2.31). In particular, if at an iteration $j > k$ of Algorithm 2.3 we perform random sampling from A and wait for a point to arrive in the set (2.31), then the average waiting time is infinite when $k = 1$ and finite when $k > 1$, see (2.63) and (2.64), respectively;
(b) the set of records $Y_k^{(j)}$ contains much greater information about m than the set $Y_1^{(j)}$ consisting of the single record y_{oj} (see Sect. 2.4 on how to use this information).

Note also that if at each iteration of Algorithm 2.3, in order to obtain random points in the set (2.31) we sample points from A at random, then we can use the theory of k-th records (see Sect. 2.3) to devise the stopping rules (as this theory predicts the number of independent random points we need to obtain to improve the set of records $Y_k^{(j)}$). Fig.nnnn illustrates typical sequential updating of the set of records $Y_k^{(j)}$ obtained by performing random sampling of points from A. In this figure, the trajectories of three records $y_j^{(k)}$ $(k = 1, 2, 3)$ are plotted as we sequentially sample random points from A (the sample size n increases from 50 to 10000).

2.3 Order Statistics and Record Values: Probabilistic Aspects

Let $F(\cdot)$ be some c.d.f. and η be a random variable on $\mathbb{R}$ with this c.d.f. Our main particular case will be the c.d.f. (2.13) but the results of this section can be applied to many other c.d.f. as well. In this section, we shall not use the specific form of the c.d.f. (2.13) but we shall use the following two properties of this c.d.f.:

(i) the c.d.f. $F(\cdot)$ and the corresponding r.v. η have finite lower bound $m = \text{ess inf}\, \eta > -\infty$ so that $F(t) = 0$ for $t < m$ and $F(t) > 0$ for $t > m$,
(ii) the c.d.f. $F(\cdot)$ is continuous at some vicinity of m.

We shall sometimes use stronger assumptions:

(i′) the c.d.f. $F(\cdot)$ has bounded support $[m, M]$ with $-\infty < m < M < \infty$ implying, additionally to (i), $F(t) = 1$ for $t \leq M$,
(ii′) the c.d.f. $F(\cdot)$ is continuous.

In this section, always bearing in mind applications to the theory and methodology of global random search, we formulate and discuss numerous results of the theory of extreme order statistics and the associated theory of records.

2.3.1 Order Statistics: Non-Asymptotic Properties

Below, we collect several useful facts from the non-asymptotic theory of extreme value statistics. For more information about the theory we refer to the classical book by H.A. David [57] and to its extension [58].

Exact distributions and moments

Let $\eta_1, \eta_2, \ldots$ be i.i.d.r.v. with common c.d.f. $F(\cdot)$. If we rearrange the first n random variables $\eta_1, \ldots \eta_n$ so that $\eta_{1,n} \leq \eta_{2,n} \leq \cdots \leq \eta_{n,n}$, then the resulting variables are called order statistics corresponding to $\eta_1, \ldots \eta_n$. Two extreme order statistics are $\eta_{1,n}$ and $\eta_{n,n}$, the minimum and maximum order statistics respectively. Their c.d.f.'s are:

$$F_{1,n}(t) = \Pr\{\eta_{1,n} \leq t\} = 1 - (1 - F(t))^n \tag{2.32}$$

and

$$F_{n,n}(t) = \Pr\{\eta_{n,n} \leq t\} = (F(t))^n \, .$$

The c.d.f. of $\eta_{k,n}$ with $1 \leq k \leq n$ can also be easily computed:

$$F_{k,n}(t) = \Pr\{\eta_{k,n} \leq t\} = \sum_{m=k}^{n} \binom{n}{m} (F(t))^m (1 - F(t))^{n-m}$$

$$= \int_0^{F(t)} \frac{n!}{(k-1)!(n-k)!} u^{k-1}(1-u)^{n-k} du, \qquad -\infty < t < \infty \, . \tag{2.33}$$

The joint c.d.f of $\eta_{i,n}$ and $\eta_{j,n}$ $(1 \leq i < j \leq n)$ is given by $(-\infty < u < v < \infty)$:

$$\Pr\{\eta_{i,n} \leq u, \eta_{j,n} \leq v\} =$$

$$\sum_{s=j}^{n} \sum_{r=i}^{n} \frac{n!}{r!(s-r)!(n-s)!} (F(u))^r (F(v) - F(u))^{s-r} (1 - F(v))^{n-s} \, . \tag{2.34}$$

If η has density $p(t) = F'(t)$, then (2.34) implies the following expression for the joint density of $\eta_{i,n}$ and $\eta_{j,n}$ $(1 \leq i < j \leq n)$:
$p_{(i,j)}(u, v) =$

$$\frac{n!}{(i-1)!(j-i-1)!(n-j)!} (F(u))^{j-1} (F(v) - F(u))^{j-i-1} (1 - F(v))^{n-j} p(u)p(v),$$

where $u \leq v$. The joint distributions of several order statistics can also be written down, if needed.

The expression for the β-th moment of $\eta_{k,n}$ easily follows from (2.33):

$$EX_{k,n}^{\beta} = \int_{-\infty}^{\infty} t^{\beta} dF_{k,n}(t)$$

$$= \frac{n!}{(k-1)!(n-k)!} \int_{-\infty}^{\infty} t^{\beta}(F(t))^{k-1}(1-F(t))^{n-k} dF(t) .$$

We shall also need the following expression for the joint moment $E\eta_{i,n}\eta_{j,n}$ with $1 \leq i < j \leq n$:

$$E\eta_{i,n}\eta_{j,n} = \frac{n!}{(i-1)!(j-i-1)!(n-j)!} \times$$

$$\int_{-\infty}^{\infty} \int_{x}^{\infty} xy(F(x))^{k-1}(F(y) - F(x))^{j-i-1}(1 - F(y))^{n-j} dF(x)dF(y) ;$$

this expression is a direct consequence of (2.34).

Two useful representations

The following representation for the order statistics has proven to be extremely useful:

$$\eta_{k,n} \stackrel{d}{=} F^{-1}\left(\exp\left\{-\left(\frac{\nu_1}{n} + \frac{\nu_2}{n-1} + \cdots + \frac{\nu_k}{n-k+1}\right)\right\}\right) , \tag{2.35}$$

where $\nu_1, \nu_2, \ldots, \nu_k$ are i.i.d.r.v. with exponential density e^{-t}, $t \geq 0$; the formula (2.35) is called the Rényi representation, and was derived in [196] (The inverse function $F^{-1}(s)$ is defined here as $F^{-1}(s) = \inf\{t : F(t) \geq s\}$ and the equality $\stackrel{d}{=}$ means that the distributions of the random variables (vectors) in the l.h.s. and r.h.s. of the equation are the same.)

When studying the joint distributions of order statistics, the following representation is often used:

$$(\eta_{1,n}, \ldots, \eta_{n,n}) \stackrel{d}{=} (F^{-1}(U_{1,n}), \ldots, F^{-1}(U_{n,n})) \tag{2.36}$$

where $U_{1,n} \leq \cdots \leq U_{n,n}$ are the order statistics corresponding to the n i.i.d.r.v. with the uniform distribution on $[0, 1]$.

Order statistics as a Markov chain

Of course, the order statistics $\eta_{k,n}$ are dependent random variables (we assume that n is fixed and k varies). One of their important properties is that if the original i.i.d.r.v. $\eta_1, \eta_2, \ldots$ have a continuous distribution (that is, η_j's have a common density $F'(t)$), then the order statistics $\eta_{k,n}$ form a Markov chain. The (forwards and backwards) transition probabilities of the Markov chain are:

$$\Pr\{\eta_{k,n} \leq t \mid \eta_{k+1,n} = v\} = \left(\frac{F(t)}{F(v)}\right)^k , \quad t \leq v; \tag{2.37}$$

$$\Pr\{\eta_{k+1,n} \leq t \mid \eta_{k,n} = v\} = 1 - \left(\frac{1 - F(t)}{1 - F(v)}\right)^{n-k} , \quad t \geq v. \tag{2.38}$$

If we make the substitution $\eta \to -\eta$, then (2.38) will become (2.37) and vice versa.

Using the representations (2.35) and (2.36) we can express $\eta_{k+1,n}$ through $\eta_{k,n}$ as follows:

$$\eta_{k+1,n} \stackrel{d}{=} F^{-1}\left(\exp\left\{\ln F\left(\eta_{k,n}\right) - \frac{\nu_{k+1}}{n-k}\right\}\right); \tag{2.39}$$

here we assume that the c.d.f. $F(\cdot)$ is continuous, $1 \le k < n$ and ν_{k+1} is as in (2.35). Since ν_{k+1} is independent of $\eta_{k,n}$, the representation (2.39) also implies the fact that the sequence of order statistics $\{\eta_{k,n}\}$ forms a Markov chain.

If the original distribution is discrete with at least three support points, then the order statistics do not form a Markov chain.

2.3.2 Extreme Order Statistics: Asymptotic Properties

In this section, we collect classical facts from the asymptotic theory of extreme value statistics. These facts will play the key role in deriving statistical inference procedures in Sects. 2.4 and 2.5.

More information about the asymptotic theory of extreme value statistics and its numerous applications can be found in [11, 13, 68, 86, 105] and in many other books. No proofs of the classical results are given below; these proofs can easily be found in literature.

Let $\eta_1, \eta_2, \dots$ be i.i.d.r.v. with common c.d.f. $F(\cdot)$ and $\eta_{1,n} \le \dots \le \eta_{n,n}$ be the order statistics corresponding to the first n random variables $\eta_1, \dots, \eta_n$. We are interested in the limiting behaviour, as $n \to \infty$, of the minimal order statistic $\eta_{1,n}$. Also, for fixed k and $n \to \infty$, we shall look at the asymptotic distributions of the k-th smallest order statistics $\eta_{k,n}$. As we are only interested in applying the theory to global random search problems, we always assume the properties (i) and (ii) stated in the beginning of Sect. 2.3 and sometimes we additionally assume one of the stronger properties (i′) or (ii′).

Note that the classical theory of extremes is usually formulated in terms of the maximum order statistics but we formulate all statements for the minimal order statistics.

Asymptotic distribution of the minimum order statistic

Consider first the asymptotic distribution of the sequence of minimum order statistics $\eta_{1,n}$, as $n \to \infty$. In the case $m = \operatorname{ess\,inf} > -\infty$ (where η has c.d.f. $F(t)$), there are two possible limiting distributions. However, in global random search applications, where $F(\cdot)$ has the form (2.13), only one asymptotic distribution arises; specifically, the Weibull distribution with the c.d.f.

$$\Psi_\alpha(z) = \begin{cases} 0 & \text{for } z < 0 \\ 1 - \exp\{-z^\alpha\} & \text{for } z \ge 0\,. \end{cases} \tag{2.40}$$

This c.d.f. only has one parameter, α, which is called the 'tail index'. The mean of the Weibull distribution with tail index α is $\Gamma(1+1/\alpha)$; the density corresponding to the c.d.f. (2.40) is

$$\psi_\alpha(t) = (\Psi_\alpha(t))' = \alpha t^{\alpha-1} \exp\{-t^\alpha\}\,, \quad t > 0\,. \tag{2.41}$$

Figure 2.5 displays the density $\psi_\alpha(t)$ for $\alpha = 2, 3$ and 8.

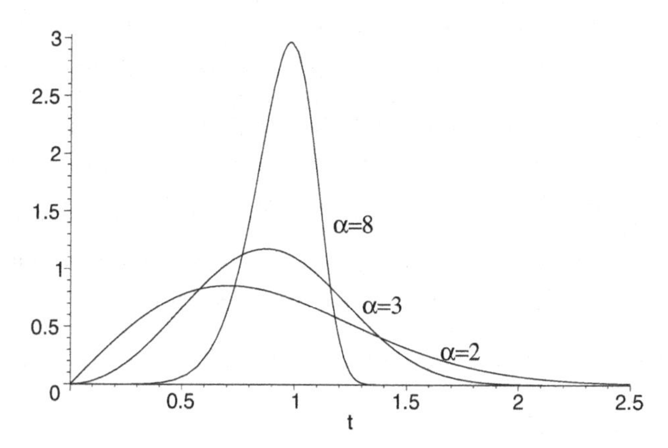

Fig. 2.5. The density $\psi_\alpha(t)$ for $\alpha = 2, 3$ and 8.

Let κ_n be the $(1/n)$-quantile of a c.d.f. $F(\cdot)$; that is, $\kappa_n = \inf\{u|F(u) \geq 1/n\}$. Note that since we assume that the c.d.f. $F(\cdot)$ is continuous in the vicinity of m, for n large enough we have $F(\kappa_n) = 1/n$. The following classical result from the theory of extreme order statistics is of primary importance to us.

Theorem 2.2. *Assume* $\operatorname{ess\,inf} \eta = m > -\infty$, *where* η *has c.d.f.* $F(t)$, *and the function*

$$V(v) = F\left(m + \frac{1}{v}\right), \qquad v > 0,$$

regularly varies at infinity with some exponent $(-\alpha)$, $0 < \alpha < \infty$; *that is,*

$$\lim_{v\to\infty} \frac{V(tv)}{V(v)} = t^{-\alpha}, \qquad \textit{for all } t > 0\,. \tag{2.42}$$

Then

$$\lim_{n\to\infty} F_{1,n}(m + (\kappa_n - m)z) = \Psi_\alpha(z)\,, \tag{2.43}$$

where $F_{1,n}$ *is the c.d.f.* (2.32), *the c.d.f.* $\Psi_\alpha(z)$ *is defined in* (2.40) *and* κ_n *is the* $(1/n)$-*quantile of* $F(\cdot)$.

The asymptotic relation (2.43) means that the distribution of the sequence of random variables $(\eta_{1,n} - m)/(\kappa_n - m)$ converges (as $n \to \infty$) to the random variable with c.d.f. $\Psi_\alpha(z)$.

The family of c.d.f.'s $\Psi_\alpha(z)$, along with its limiting case

$$\Psi_\infty(z) = \lim_{\alpha\to\infty} \Psi_\alpha(1 + z/\alpha) = 1 - \exp\{-\exp(z)\},\ z>0,$$

are the only non-degenerate limits of the c.d.f.s of the sequences $(\eta_{1,n}-a_n)/b_n$, where $\{a_n\}$ and $\{b_n\}$ are arbitrary sequences of positive numbers.

If there exist numerical sequences $\{a_n\}$ and $\{b_n\}$ such that the c.d.f.'s of $(\eta_{1,n}-a_n)/b_n$ converge to Ψ_α, then we say that $F(\cdot)$ belongs to the domain of attraction of $\Psi_\alpha(\cdot)$ and express this as $F \in D(\Psi_\alpha)$. The conditions stated in Theorem 2.2 are necessary and sufficient for $F \in D(\Psi_\alpha)$. There are two conditions: $m = \operatorname{ess\,sup} \eta < \infty$ and the condition (2.42). The first one is always valid in global random search applications. The condition (2.42) demands more attention. For example, it is never valid in discrete optimization problems since the c.d.f. $F(\cdot)$ has to be continuous in the vicinity of $m = \operatorname{ess\,inf} \eta$. In fact, for a c.d.f. with a jump at its lower end-point no non-degenerate asymptotic distribution for $\eta_{1,n}$ exists, whatever the normalization (that is, sequences $\{a_n\}$ and $\{b_n\}$).

The condition (2.42) can be written as

$$F(t) = c_0(t - m)^\alpha + o((t - m)^\alpha) \quad \text{as } t \downarrow m\,, \tag{2.44}$$

where c_0 is a function of $v = 1/(t - m)$, slowly varying at infinity as $v \to \infty$. Of course, any positive constant is a slowly varying function, but the actual range of eligible functions c_0 is much wider.

The following sufficient condition (the so-called von Mises condition) for (2.42) and (2.43) is often used: if $F(t)$ has a positive derivative $F'(t)$ for all $t \in (m, m + \varepsilon)$ for some $\varepsilon > 0$ and

$$\lim_{t\downarrow m} \frac{(t - m)F'(t)}{F(t)} = \alpha\,,$$

then (2.42) holds.

The following condition is stronger that the condition (2.44) and is often used for justifying properties of the maximum likelihood estimators:

$$F(t) = c_0(t - m)^\alpha \left(1 + O((t - m)^\beta)\right) \quad \text{as } t \downarrow m \tag{2.45}$$

for some positive constants c_0, α and β.

The quantity $\kappa_n - m$, where $m = \operatorname{ess\,inf} \eta$ and κ_n is the $(1/n)$-quantile of $F(\cdot)$ enters many formulae below and therefore its asymptotic behaviour is very important. Fortunately, the asymptotic behaviour of $\kappa_n - m$ is clear. Indeed, provided that (2.44) holds with some c_0, we have

$$\frac{1}{n} = F(\kappa_n) \sim c_0\,(\kappa_n - m)^\alpha \quad \text{as } n \to \infty$$

implying

$$(\kappa_n - m) \sim (c_0 n)^{-1/\alpha} \qquad \text{as } n \to \infty\,. \tag{2.46}$$

Extensions to k-th order statistics

There is a one-to-one correspondence between the convergence of the smallest order statistics $\eta_{1,n}$ and of the k-th smallest statistics $\eta_{k,n}$. Assume that $m = \operatorname{ess\,inf} \eta > -\infty$, $n \to \infty$ and let k be fixed. Then it is easy to prove that $F \in D(\Psi_\alpha)$ if and only if the sequence of random variables $(\eta_{k,n} - m)/(\kappa_n - m)$ converges in distribution to the random variable with c.d.f.

$$\begin{aligned}\Psi_\alpha^{(k)}(t) &= 1 - (1 - \Psi_\alpha(t)) \sum_{m=0}^{k-1} \frac{(-\ln(1-\Psi_\alpha(t)))^m}{m!} \\ &= 1 - \exp(-t^\alpha) \sum_{m=0}^{k-1} \frac{t^{\alpha m}}{m!}, \qquad t > 0\,. \end{aligned} \tag{2.47}$$

The corresponding density is

$$\psi_\alpha^{(k)}(t) = \left(\Psi_\alpha^{(k)}(t)\right)' = \frac{\alpha}{(k-1)!} t^{\alpha k - 1} \exp\{-t^\alpha\}\,, \quad t > 0\,. \tag{2.48}$$

The following statement is a generalisation of this fact and reveals the joint asymptotic distribution of the k smallest order statistics: if $m > -\infty$, $F \in D(\Psi_\alpha)$, $n \to \infty$, then for any fixed k the asymptotic distribution of the random vector

$$\left(\frac{\eta_{1,n} - m}{m - \kappa_n}, \frac{\eta_{2,n} - m}{m - \kappa_n}, \ldots, \frac{\eta_{k,n} - m}{m - \kappa_n}\right) \tag{2.49}$$

converges to the distribution with density

$$\psi_\alpha(t_1, \ldots, t_k) = \alpha^k (t_1 \ldots t_k)^{(\alpha-1)} \exp(-t_k^\alpha), \quad 0 < t_1 < \cdots < t_k < \infty\,. \tag{2.50}$$

The density (2.50) is the density of the random vector

$$\left(\nu_1^{1/\alpha}, (\nu_1 + \nu_2)^{1/\alpha}, \ldots, (\nu_1 + \cdots + \nu_k)^{1/\alpha}\right), \tag{2.51}$$

where $\nu_1, \ldots, \nu_k$ are i.i.d.r.v. with exponential density e^{-t}, $t > 0$.

As an important particular case, we find that the joint asymptotic density of the random vector

$$\left(\frac{\eta_{1,n} - m}{\kappa_n - m}, \frac{\eta_{k,n} - m}{\kappa_n - m}\right)$$

coincides with the joint density of the vector $(\nu_1^{1/\alpha}, (\nu_1 + \ldots + \nu_k)^{1/\alpha})$.

The following corollary of this result will be the basic tool in constructing confidence intervals for m.

Proposition 2.1. *If the conditions of Theorem 2.2 hold, then for any fixed integer $k \geq 2$ and $n \to \infty$ the sequence of random variables*

$$D_{n,k} = \frac{\eta_{1,n} - m}{\eta_{k,n} - m}$$

converges in distribution to a random variable with c.d.f.

$$F_k(u) = 1 - \left(1 - \left(\frac{u}{1+u}\right)^{\alpha}\right)^{k-1}, \quad u \geq 0. \tag{2.52}$$

The proof of this statement is given in Sect. 2.7; it is a simplified and corrected version of the proof of Lemma 7.1.4 in [273].

In the following proposition we use the asymptotic distributions (2.47) and (2.50) to derive the asymptotic formulae for the moments of the random variables $(\eta_{k,n} - m)$ and the first joint moment $\mathrm{E}(\eta_{j,n} - m)(\eta_{k,n} - m)$.

Proposition 2.2. *Let $m = \operatorname{ess\,inf} \eta > -\infty$ and $F \in D(\Psi_\alpha)$ with $\alpha > 1$. Assume that k is either fixed or k tends to infinity as $n \to \infty$ so that $k^2/n \to 0$, $n \to \infty$. Then*

$$E(\eta_{k,n} - m)^{\beta} \sim (\kappa_n - m)^{\beta} \frac{\Gamma(k + \beta/\alpha)}{\Gamma(k)} \quad \text{as } n \to \infty \tag{2.53}$$

for any $\beta > 0$ and

$$E(\eta_{j,n} - m)(\eta_{k,n} - m) \sim (\kappa_n - m)^2 \lambda_{jk} \quad \text{as } n \to \infty, \tag{2.54}$$

where $k \geq j$ and

$$\lambda_{jk} = \frac{\Gamma(k + 2/\alpha)\, \Gamma(j + 1/\alpha)}{\Gamma(k + 1/\alpha)\, \Gamma(j)}\,. \tag{2.55}$$

We give a proof of this statement in Sect. 2.7; this proof is easier than the one given in [273], Sect. 7.1.2.

General results on the rate of convergence of the normalised minima to the extreme value distribution (see e.g. [70] and §2.10 in [86]) imply that in the case considered in Theorem 2.2 this rate is $\mathrm{O}(1/n)$ as $n \to \infty$ (note that [60] and [197], Chapt. 2 contain more sophisticated results on the rate of convergence to the extreme value distribution). This fact along with the asymptotic relation (2.46) imply that for $\alpha \leq 1$ we have

$$\mathrm{E}(\eta_{k,n} - m)^{\beta} = \mathrm{O}(1/n^{\beta})$$

rather than (2.53). Similarly, we have to have $\alpha > 1$ for (2.54) to hold. The reasons why the condition $k^2/n \to 0$ as $n \to \infty$ must be satisfied are explained in [273], p. 245-246.

2.3.3 Record Values and Record Moments

In this section, we survey the theory of record values and record moments. The importance of this topic in global random search is related, first of all, to its link with pure adaptive search and its modifications, see Sects. 2.2.3 and 2.2.4. For all missing proofs and more information on record values and record moments we refer to [6, 171].

Definitions

Let $\eta_1, \eta_2, \ldots$ be a sequence of random variables. Define the sequences of related random variables $L(n)$ and $\eta(n)$ as follows: $L(1) = 1$, $\eta(1) = \eta_1$,

$$L(n+1) = \min\{j > L(n) : \eta_j < \eta_{L(n)}\},\ \eta(n) = \eta_{L(n)},\ n = 1, 2, \ldots; \quad (2.56)$$

$L(n)$ are called (lower) record moments corresponding to the sequence $\eta_1, \eta_2, \ldots$ and $\eta(n)$ are the associated (lower) record values.

If we change the inequality sign in (2.56) to be $>$, then we obtain the upper record order moments and upper record values. By changing η_j to $1/\eta_j$ or $(-\eta_j)$ for $j = 1, 2, \ldots$ we correspond the upper record moments to the lower ones. We will only consider the lower moments and values and omit the word 'lower'.

In addition to $L(n)$ and $\eta(n)$, we shall also use the following random variables: $\mathcal{N}(n)$ is the number of record values among $\eta_1, \ldots, \eta_n$ (note that $\mathcal{N}(L(n)) = n$) and $\Delta(n) = L(n) - L(n-1)$, the waiting time between $(n-1)$-th and n-th record moments.

Properties of record moments

Assume that $\eta_1, \eta_2, \ldots$ are i.i.d.r.v. with common continuous c.d.f. $F(\cdot)$. First, consider the non-asymptotic properties of the record moments $L(n)$.

P1: The distribution of $L(n)$ does not depend on $F(\cdot)$.

P2: The sequence of random variables $L(n)$ forms a Markov chain with the starting point $L(1) = 1$ and transition probabilities

$$\Pr\{L(n) = j \mid L(n-1) = i\} = \frac{i}{j(j-1)} \quad \text{for } j > i \geq n-1,\ \ n = 2, 3, \ldots$$

P3: The joint distribution of the record moments is

$$\Pr\{L(2) = i_2, \ldots, L(n) = i_n\} = \frac{1}{(i_2-1)\ldots(i_n-1)i_n} \text{ with } 1 < i_2 < \cdots < i_n.$$

Property P3 implies

$$\Pr\{L(2) = j\} = \frac{1}{j(j-1)}, \qquad j > 1; \quad (2.57)$$

$$\Pr\{L(n) = j\} = \frac{|S_{j-1}^{n-1}|}{j!}, \qquad j \geq n > 1, \quad (2.58)$$

where S_a^b are the Stirling numbers of the first kind.

Property P3 follows from P2, and properties P1 and P2 are simple consequences of the relation between $L(n)$ and $\mathcal{N}(n)$,

$$\Pr\{\mathcal{N}(n) < n\} = \Pr\{L(n) > n\}, \tag{2.59}$$

and of the representation

$$\mathcal{N}(n) \stackrel{d}{=} \zeta_1 + \cdots + \zeta_n \quad (n = 1, 2, \ldots), \tag{2.60}$$

where ζ_j are independent r.v. with $\Pr\{\zeta_j = 1\} = 1/j$ and $\Pr\{\zeta_j = 0\} = 1 - 1/j$. For each j, the random variable ζ_j can be interpreted as the indicator of the event that η_j is the new record value, which is the event $\eta_j < \min\{\eta_1, \ldots \eta_{j-1}\}$.

The sequence of record times $L(n)$ has another useful representation:

$$L(1) = 1, \quad L(n+1) \stackrel{d}{=} \left\lceil \frac{L(n)}{U_n} \right\rceil \quad (n = 1, 2, \ldots),$$

where $U_1, U_2, \ldots$ are i.i.d.r.v. uniformly distributed on [0,1].

For any integer $x > 1$, we have

$$\Pr\left\{ \frac{L(n+1)}{L(n)} > x \right\} = \frac{1}{x}; \tag{2.61}$$

if x is not an integer, then (2.61) holds asymptotically, as $n \to \infty$.

The representations (2.59) and (2.60) enable the application of classical techniques to obtain the law of large numbers, the central limit theorem and the law of iterated logarithm for the random variables $L(n)$ and $\mathcal{N}(n)$. In particular, as $n \to \infty$ we have

$$\Pr\left\{ \lim_{n\to\infty} \frac{1}{n} \ln L(n) = 1 \right\} = 1,$$

$$\lim_{n\to\infty} \Pr\left\{ (\ln L(n) - n) \le t\sqrt{n} \right\} = \Phi(t)$$

where $\Phi(t)$ is the c.d.f. of the standard normal distribution:

$$\Phi(t) = \frac{1}{\sqrt{2\pi}} \int_{-\infty}^{t} e^{-u^2/2} du. \tag{2.62}$$

Consider now the moments of $L(n)$. Using (2.57) and (2.58) we obtain:

$$EL(n) = \infty \text{ for all } n = 2, 3, \ldots \tag{2.63}$$

(note that $L(1) = 1$). Moreover, for any $n \ge 2$, the average waiting time $\mathrm{E}[L(n) - L(n-1)]$ of a new record is infinite. This is a very unsatisfactory result for the theory of global random search: it says that on average one has to make infinitely many iterations of PRS to get any improvement over the current best value of the objective function.

The distributions of the inter-record times $\Delta(n+1) = L(n+1) - L(n)$ can be easily computed; they are:

$$\Pr\{\Delta(n+1)=j\} = F(\eta(n)) \cdot (1-F(\eta(n)))^{j-1}, \quad j=1,2,\ldots, \quad n=1,2,\ldots$$

That is, the distribution of $\Delta(n+1)$ only depends on the n-th record value $\eta(n)$ and is in fact geometric with parameter of success $F(\eta(n))$.

The logarithmic moments of $L(n)$ can asymptotically be expressed as follows:

$$\mathrm{E}\ln L(n) = n - \gamma + \mathrm{O}\left(\frac{n^2}{2^n}\right), \qquad n \to \infty,$$

$$\mathrm{var}(\ln L(n)) = n - \frac{\pi^2}{6} + \mathrm{O}\left(\frac{n^3}{2^n}\right), \qquad n \to \infty,$$

where $\gamma = 0.5772\ldots$ is the Euler's constant.

The number of records in a given sequence

Consider again the sequence of random variables $\mathcal{N}(n)$, the number of records among the random variables $\eta_1, \ldots, \eta_n$:

$$\mathcal{N}(1) = 1, \qquad \mathcal{N}(n) = 1 + \sum_{j=2}^{n} I_{[\eta_j < \min\{\eta_1, \ldots, \eta_{j-1}\}]}\,.$$

In accordance with (2.60) and the fact that

$$\mathrm{E}\zeta_j = \tfrac{1}{j} \quad \text{and} \quad \mathrm{var}(\zeta_k) = \tfrac{1}{j} - \tfrac{1}{j^2}\,,$$

for any continuous c.d.f. $F(\cdot)$ we obtain

$$\mathrm{E}\mathcal{N}(n) = \sum_{j=1}^{n} \frac{1}{j} \quad \text{and} \quad \mathrm{var}(\mathcal{N}(n)) = \sum_{j=1}^{n} \left(\frac{1}{j} - \frac{1}{j^2}\right).$$

This implies that both $\mathrm{E}\mathcal{N}(n)$ and $\mathrm{var}(\mathcal{N}(n))$ are of order $\ln n$; the approximation $\mathrm{E}\mathcal{N}(n) \cong \ln n + \gamma$ with $\gamma = 0.5772\ldots$ (the Euler's constant) is very accurate. Additionally, taking into account the asymptotic normality of $(\mathcal{N}(n) - \ln n)/\sqrt{\ln n}$ (the normalized sequence of $\mathcal{N}(n)$), we can use Table 2.2 for making good guesses (using, say, the '3σ-rule') about the number of records $\mathcal{N}(n)$ for given n. This table shows the expected number of records $\mathrm{E}\mathcal{N}(n)$ in a sequence of i.i.d.r.v. $\eta_1, \ldots, \eta_n$, along with the standard deviation of $\mathcal{N}(n)$, for some values of n. One can see that as n increases, the number of records grows very slowly.

n	10	10^2	10^3	10^4	10^5	10^6	10^7	10^8	10^9
$E\mathcal{N}(n)$	2.9	5.2	7.5	9.8	12.1	14.4	16.7	19.0	21.3
$\sqrt{\mathrm{var}(\mathcal{N}(n))}$	1.2	1.9	2.4	2.8	3.2	3.6	3.9	4.2	4.4

Table 2.2. Expected number of records $E\mathcal{N}(n)$ among n i.i.d.r.v. $\eta_1, \ldots, \eta_n$, along with the standard deviation of $\mathcal{N}(n)$.

Record values

Let $\eta(1), \eta(2), \ldots$ be the sequence of record values in the sequence of i.i.d.r.v. $\eta_1, \eta_2, \ldots$ Assume that the c.d.f. $F(t)$ of η_j is continuous with density $p(t)=F'(t)$. Under these assumptions, it is easy to see that the joint density of $\eta(1), \ldots, \eta(n)$ is

$$p(x_1, \ldots, x_n) = \frac{p(x_1)}{F(x_1)} \cdots \frac{p(x_{n-1})}{F(x_{n-1})} \cdot p(x_n), \qquad x_1 \le \cdots \le x_n \,.$$

This implies, in particular, that the sequence $\eta(1), \eta(2), \ldots$ is a Markov chain with transition probabilities

$$\Pr\left\{\eta(n+1) \le t \mid \eta(n) = u\right\} = \frac{F(t)}{F(u)}\,, \qquad m \le t \le u \,.$$

For each $n \ge 1$, the c.d.f. of the record value $\eta(n)$ is

$$\Pr\{\eta(n) < t\} = 1 - F(t) \sum_{j=0}^{n-1} \frac{(-\ln(F(t)))^j}{j!} \,.$$

Consider now the asymptotic distribution of the record values $\eta(n)$ corresponding to the sequence of i.i.d.r.v. $\eta_1, \eta_2, \ldots$ as $n \to \infty$, assuming that $m = \operatorname{ess\,inf} \eta_i > -\infty$ and $F \in D(\Psi_\alpha)$; that is, the c.d.f. $F(\cdot)$ belongs to the domain of attraction of the c.d.f. $\Psi_\alpha(\cdot)$ defined in (2.40). Theorem 2.2 implies that the condition $F \in D(\Psi_\alpha)$ yields that the sequence of random variables $(\eta(n) - m)/(\kappa_{L(n)} - m)$ has the asymptotic distribution with the c.d.f. $\Psi_\alpha(\cdot)$; here $L(n)$ is the sequence of record moments, and $\kappa_{L(n)}$ is $(1/L(n))$-quantile of $F(\cdot)$.

Since $L(n)$ is a random variable, $\kappa_{L(n)}$ is a random variable too, which is not very satisfactory. There is, however, another limiting law for the properly normalized record values $(\eta(n) - a(n))/b(n)$ with non-random coefficients $a(n)$ and $b(n)$. Specifically, the conditions $m > -\infty$ and $F \in D(\Psi_\alpha)$ imply that the sequence of random variables $(\eta(n) - m)/(\kappa_{\exp(n/2)} - m)$ converges in distribution to the r.v. with c.d.f.

$$\tilde{\Psi}_\alpha(z) = \begin{cases} 0, & z \le 0 \\ \Phi(-\alpha \ln(z))\,, & z > 0\,; \end{cases}$$

here $\Phi(\cdot)$ is as in (2.62).

Extension to k-th records

To perform statistical inference in global random search algorithms we need several minimal order statistics, rather than just one of them. Similarly, we can use the k-th record moments and the k-th record values. The record moments and the record values considered above are the first record moment and the first record value, respectively (in this case $k = 1$). Assume now the general case $k \geq 1$ and start with the so-called 'Gumbel's method of exceedances', see [105]. This method is aimed to answer the question: 'how many values among future observations exceed past records?'. Specifically, let the c.d.f. $F(\cdot)$ be continuous, $\eta_{1,n} \leq \cdots \leq \eta_{n,n}$ be the order statistics as usual and denote by $S_r^k(n)$ the number of exceedances of $\eta_{k,n}$ among the next r observations $\eta_{n+1}, \ldots, \eta_{n+r}$; that is,

$$S_r^k(n) = \sum_{i=1}^{r} I_{\{\eta_{n+i} < \eta_{k,n}\}} \,.$$

It is an easy consequence of (2.33) that the random variable $S_r^k(n)$ has the hypergeometric distribution with

$$\Pr\left\{S_r^k(n) = j\right\} = \frac{\binom{r+n-k-j}{n-k}\binom{j+k-1}{k-1}}{\binom{r+n}{n}}, \qquad j = 0, 1, \ldots, r \,.$$

In particular, the mean number of exceedances is equal to

$$\mathrm{E}S_r^k(n) = \frac{rk}{n+1} \,.$$

Let us now consider ways of generalizing other results discussed earlier in this section from the case $k = 1$ to the general case $k \geq 1$. We start with definitions.

For each $n \geq k$, we rearrange the random variables $\eta_1, \ldots, \eta_n$ so that

$$\eta_{1,n} \leq \eta_{2,n} \leq \cdots \leq \eta_{n,n} \,.$$

The random variable $\eta_{k,n}$ is the k-th order statistic. The sequence of k-th order statistics is

$$\eta_{k,k} \geq \eta_{k,k+1} \geq \ldots \geq \eta_{k,n} \geq \ldots$$

Let us select the indices n such that there is strict inequality in this sequence:

$$\ldots \geq \eta_{k,n-1} > \eta_{k,n} \geq \ldots$$

This gives us the sequence of k-th record moments $L^{(k)}(n)$. Formally, the sequence $L^{(k)}(n)$ can be defined as follows: $L^{(k)}(0) = 0$, $L^{(k)}(1) = k$ and

$$L^{(k)}(n+1) = \min\left\{j > L^{(k)}(n) \text{ such that } \eta_j < \eta_{k,j-1}\right\}, \quad n=1,2,\ldots$$

The sequence of random variables $\eta^{(k)}_{(n)} = \eta_{k,L^{(k)}(n)}$, $n = 1,2,\ldots$, is the sequence of k-th record values; the differences $\Delta^{(k)}(n) = L^{(k)}(n) - L^{(k)}(n-1)$ are k-th record waiting times; $\mathcal{N}^{(k)}(n)$ is the number of k-th record values among $\eta_1,\ldots,\eta_n$. Of course, $L^{(1)}(n) = L(n)$, $\eta^{(1)}_{(n)} = \eta_{(n)}$, $\Delta^{(1)}(n) = \Delta(n)$ and $\mathcal{N}^{(1)}(n) = \mathcal{N}(n)$.

For each n, the k-th record waiting times have all moments $E\left(\Delta^{(k)}(n)\right)^{\beta}$ of order $0 < \beta < k$; the mean is

$$E\Delta^{(k)}(n) = (k/(k-1))^{n-1}, \; k \geq 2. \tag{2.64}$$

Properties of the k-th record moments $L^{(k)}(n)$ are similar to the properties of the ordinary record moments $L(n)$. In particular, for any fixed $k \geq 1$ and any continuous c.d.f. $F(\cdot)$, the sequence of $L^{(k)}(n)$ forms a Markov chain; many limit theorems for $L^{(k)}(n)$ are direct extensions of the related theorems for $L(n)$, see [171], Lectures 18–20.

A very useful tool in studying the k-th record sequences is the so-called 'Ignatov's Theorem' which says that the processes of the k-th records are independent and identically distributed copies of the same random sequence (here the process of the k-th records is defined as a sequence of time moments when the current observation of a sequence of i.i.d.r.v. has rank k), see e.g. [96, 125, 230] and [197], Sect. 4.6. The underlying c.d.f. $F(\cdot)$ of the i.i.d.r.v. is almost arbitrary; in particular, it does not have to be continuous. The Ignatov's Theorem implies that the sequence of moments when the current observation has rank k is the same for any k, in particular, for $k = 1$ where this sequence is the sequence of record moments $\{L(n)\}$ discussed above. The second extremely informative part of this theorem is the independence of the k-th record processes for all $k = 1, 2, \ldots$. Different implications of this theorem are discussed in literature; see, for example, [31].

2.4 Statistical Inference About m: Known Value of the Tail Index

In this section, we consider statistical inference about m =ess inf η (in random search applications $m = \min f$) based on the asymptotic theory of extreme order statistics described in Sect. 2.3.2. These statistical procedures will be using only the k smallest order statistics

$$\eta_{1,n} \leq \eta_{2,n} \leq \cdots \leq \eta_{k,n}$$

corresponding to the independent sample $\{\eta_1,\ldots,\eta_n\}$ of the values of a random variable η with c.d.f. $F(t)$. The sample size n is assumed large (formally,

$n \to \infty$) and k is assumed small relative to n (see Sect. 2.4.3 concerning the choice of k).

We assume throughout this section that the conditions of Theorem 2.2 hold for the c.d.f. $F(\cdot)$ (in random search application this c.d.f. is defined by (2.13)) and the value of the tail index α is known. We shall also assume that $\alpha > 1$ (results of Sect. 2.5.3 show that this is the main case of interest in global optimization). As we are interested in the applications of the methodology in global random search, we always assume that the assumptions (i′) and (ii) of Sect. 2.3 are met.

Note that the statistical inference about m and the behaviour of the c.d.f. $F(\cdot)$ in the vicinity of m are much simpler when the value of α is known. Fortunately, in problems of global random search this case can be considered as typical in view of the results of Sect. 2.5.3, where the direct link between the form (2.13) of the c.d.f. $F(\cdot)$ and the value of the tail index α is considered.

A detailed consideration of the theory of asymptotic statistical inference about the bounds of random variables in the case of known α is given in [273], Chap. 7; there has not been much progress in this area since 1991, the time of publication of [273]. Hence, in this section, we only discuss the results that can be directly applied to global random search algorithms, those outlined in Sect. 2.6.1.

2.4.1 Estimation of m

The maximum likelihood estimator

The maximum likelihood estimators of m have been introduced and investigated in [109]. These estimators are constructed under the assumption that $\alpha \geq 2$ and that the distribution of the sample is the asymptotic one, which is the Weibull distribution with c.d.f. (2.40).

For fixed n and k, set

$$\beta_j(\hat{m}) = (\eta_{k,n} - \eta_{j,n})/(\eta_{j,n} - \hat{m}), \;\; j < k. \tag{2.65}$$

Differentiating the logarithm of the likelihood function, see (2.89) below, with respect to m and c_0 and equating the derivatives to zero, we obtain the following likelihood equation for $\hat{m}$:

$$(\alpha - 1)\sum_{j=1}^{k-1} \beta_j(\hat{m}) = k\,. \tag{2.66}$$

Hence, when α is known, the maximum likelihood estimator $\hat{m}$ of m is the solution of the equation (2.66) under the condition $\hat{m} \leq \eta_{1,n}$; if there is no solution of this equation satisfying the inequality $\hat{m} \leq \eta_{1,n}$, then $\hat{m}$ is defined as $\eta_{1,n}$.

If the conditions (2.45), $\alpha \geq 2$, $k \to \infty$, $k/n \to 0$ (as $n \to \infty$) are satisfied, then the maximum likelihood estimators of m are asymptotically normal and asymptotically efficient in the class of asymptotically normal estimators and their mean square error $E(\hat{m}-m)^2$ is asymptotically

$$E(\hat{m}-m)^2 \sim \begin{cases} (1-\frac{2}{\alpha})(\kappa_n - m)^2\, k^{-1+2/\alpha} & \text{for } \alpha > 2, \\ (\kappa_n - m)^2 \ln k & \text{for } \alpha = 2\,. \end{cases} \tag{2.67}$$

As usual, κ_n is the $(1/n)$-quantile of the c.d.f. $F(\cdot)$.

Linear estimators

Linear estimators of m are simpler than the maximum likelihood ones. However, the best linear estimators possess similar asymptotic properties.

Introduce the following notation:

$$a=(a_1,\ldots,a_k)' \in \mathbb{R}^k, \;\; \mathbf{1}=(1,1,\ldots,1)' \in \mathbb{R}^k,$$

$$b_i = \Gamma(i+1/\alpha)\,/\,\Gamma(i), \;\; b=(b_1,\ldots b_k)' \in \mathbb{R}^k,$$

$$\lambda_{ji}=\lambda_{ij}=\frac{\Gamma(i{+}2/\alpha)\,\Gamma(j{+}1/\alpha)}{\Gamma(i{+}1/\alpha)\,\Gamma(j)} \;\; \text{for } i \geq j, \;\; \Lambda = \|\lambda_{ij}\|_{i,j=1}^{k}\,; \tag{2.68}$$

here $\Gamma(\cdot)$ is the gamma–function.

A general linear estimator of m can be written as

$$\hat{m}_{n,k}(a) = \sum_{i=1}^{k} a_i \eta_{i,n}\,, \tag{2.69}$$

where $a=(a_1,\ldots,a_k)'$ is the vector of coefficients.

Using (2.53) with $\beta = 1$, for any linear estimator $\hat{m}_{n,k}(a)$ of the form (2.69) we obtain:

$$\mathrm{E}\hat{m}_{n,k}(a)=\sum_{i=1}^{k} a_i \mathrm{E}\eta_{i,n} = m\sum_{i=1}^{k} a_i - (\kappa_n - m)a'b + o(\kappa_n - m), \;\; n \to \infty. \tag{2.70}$$

Since $\kappa_n - m \to 0$ as $n \to \infty$, see (2.46), and the variances of all $\eta_{i,n}$ are finite (this is true, in particular, if the c.d.f. $F(\cdot)$ has bounded support, see assumption (i′) of Sect. 2.3), the estimator $\hat{m}_{n,k}(a)$ is a consistent estimator of m if and only if

$$a'\mathbf{1} = \sum_{i=1}^{k} a_i = 1\,. \tag{2.71}$$

The additional condition

$$a'b = 0 \quad \left(\Longleftrightarrow \;\; \sum_{i=1}^{k} a_i b_i = 0 \right) \tag{2.72}$$

guarantees that for $\alpha > 1$ the corresponding estimator $\hat{m}_{n,k}(a)$ has a bias of the order $o(\kappa_n - m) = o(n^{-1/\alpha})$, as $n \to \infty$, rather than $O(n^{-1/\alpha})$ for a general consistent linear estimator.

For a general consistent estimator $\hat{m}_{n,k}(a)$, we obtain from (2.70):

$$E\hat{m}_{n,k}(a) - m \sim (\kappa_n - m)\, a'b, \quad n \to \infty. \tag{2.73}$$

The mean square error of a general consistent estimator $\hat{m}_{n,k}(a)$ is obtained by applying (2.54):

$$E(\hat{m}_{n,k}(a) - m)^2 \sim (\kappa_n - m)^2\, a'\Lambda a, \quad n \to \infty. \tag{2.74}$$

Examples of linear estimators

For the simplest and most commonly used estimator of m, where only the minimal order statistic is used, $\hat{m}_{n,k}(a^{(0)}) = \eta_{1,n}$, where $a^{(0)}$, the vector of coefficients, is $a^{(0)} = (1, 0, \ldots, 0)'$. For this estimator we easily obtain

$$E(\hat{m}_{n,k}(a^{(0)}) - m)^2 \sim (\kappa_n - m)^2\, \Gamma(1 + 2/\alpha), \quad n \to \infty. \tag{2.75}$$

This is, however, a rather poor estimator, see (2.82) and Fig. 2.6.

The r.h.s. of (2.74) is a natural optimality criterion for selecting the vector a. The optimal consistent estimator $\hat{m}_{n,k}(a^*)$, we shall call it *the optimal linear estimator*, is determined by the vector of coefficients

$$a^* = \arg \min_{a: a'\mathbf{1}=1} a'\Lambda a = \frac{\Lambda^{-1}\mathbf{1}}{\mathbf{1}'\Lambda^{-1}\mathbf{1}}. \tag{2.76}$$

The estimator $\hat{m}_{n,k}(a^*)$ has been suggested in [52], where the form (2.76) for the vector of coefficients was obtained.

Solving the quadratic programming problem in (2.76) is straightforward. In the process of doing that, we obtain

$$\min_{a: a'\mathbf{1}=1} a'\Lambda a = (a^*)'\Lambda a^* = 1/\mathbf{1}'\Lambda^{-1}\mathbf{1}. \tag{2.77}$$

Lemma 7.3.4 in [273] gives us the following expression for the r.h.s. of (2.77):

$$\mathbf{1}'\Lambda^{-1}\mathbf{1} = \begin{cases} \frac{1}{\alpha-2}\left(\frac{\alpha\Gamma(k+1)}{\Gamma(k+2/\alpha)} - \frac{2}{\Gamma(1+2/\alpha)}\right) & \text{for } \alpha \neq 2, \\ \sum_{i=1}^{k} 1/i & \text{for } \alpha = 2; \end{cases} \tag{2.78}$$

this expression is valid for all $\alpha > 0$ and $k = 1, 2, \ldots$

The components a_i^* $(i = 1, \ldots, k)$ of the vector a^* can be evaluated explicitely: $a_i^* = u_i/\mathbf{1}'\Lambda^{-1}\mathbf{1}$ for $i = 1, \ldots, k$ with

$$\begin{aligned} u_1 &= (\alpha + 1)\,/\,\Gamma(1 + 2/\alpha), \\ u_i &= (\alpha - 1)\,\Gamma(i)/\Gamma(i + 2/\alpha) \qquad \text{for } i = 2, \ldots, k-1, \\ u_k &= -(\alpha k - \alpha + 1)\Gamma(k)\,/\,\Gamma(k + 2/\alpha). \end{aligned}$$

Deriving this expression for the coefficients of the vector a^* is far from trivial, see [273], Sect. 7.3.3.

The asymptotic properties (when both n and k are large) of the optimal linear estimators coincide with the properties of the maximum likelihood estimators and hold under the same regularity conditions (we again refer to [273], Sect. 7.3.3). In particular, the optimal linear estimators $\hat{m}_{n,k}(a^*)$ of m are asymptotically normal (as $n \to \infty$, $k \to \infty$, $k/n \to 0$) and their mean square error $E(\hat{m}_{n,k}(a^*)-m)^2$ asymptotically behaves like the r.h.s. of (2.67).

Consider two other linear estimators which have similar asymptotic properties (as $n \to \infty$, $k \to \infty$ and $k/n \to 0$).

The first one is the estimator $\hat{m}_{n,k}(a^+)$ which is optimal in the class of linear estimators satisfying the consistency condition (2.71) and the additional condition (2.72); it is determined by the vector

$$a^+ = \arg \min_{\substack{a:\, a'\mathbf{1}=1,\\ a'b=0}} a'\Lambda a = \frac{\Lambda^{-1}\mathbf{1} - (b'\Lambda^{-1}\mathbf{1})\Lambda^{-1}b/(b'\Lambda^{-1}b)}{\mathbf{1}\Lambda^{-1}\mathbf{1} - (b'\Lambda^{-1}\mathbf{1})^2/(b'\Lambda^{-1}b)} \tag{2.79}$$

(the solution to the above quadratic minimization problem is easily found using Lagrange multipliers). For the estimator $\hat{m}_{n,k}(a^+)$, the additional condition (2.72) guarantees a faster rate of decrease of the bias $E\hat{m}_{n,k}(a) - m$, as $n \to \infty$.

The Csörgő–Mason estimator $\hat{m}_{n,k}(a^{CM})$ (proposed in [55]) is determined by the vector a^{CM} with components

$$a_i = \begin{cases} v_i & \text{for } \alpha > 2, \quad i = 1, \ldots, k-1 \\ v_k + 2 - \alpha & \text{for } \alpha > 2, \quad i = k \\ 2/\ln(k) & \text{for } \alpha = 2, \quad i = 1 \\ \ln(1+1/i)/\ln(k) & \text{for } \alpha = 2, \quad i = 2, \ldots, k-1 \\ \left(\ln(1+1/k) - 2\right)/\ln(k) & \text{for } \alpha = 2, \quad i = k \end{cases}$$

with

$$v_j = (\alpha - 1)k^{2/\alpha - 1}\left(j^{1-2/\alpha} - (j-1)^{1-2/\alpha}\right).$$

The finite–sample behaviours of the optimal unbiased consistent estimator $\hat{m}_{n,k}(a^+)$ and the Csörgő–Mason estimator $\hat{m}_{n,k}(a^{CM})$ are slightly worse than that of the optimal consistent estimator $\hat{m}_{n,k}(a^*)$.

For practical use, a very simple estimator

$$\hat{m}_{n,k}(a^U) = (1 + C_k)\eta_{1,n} - C_k \eta_{k,n} \tag{2.80}$$

with $a^U = (1+C_k, 0, \ldots, 0, -C_k)'$ may be recommended, where $C_k = b_1/(b_k - b_1)$ is found from the condition $a'b = 0$. (An estimator resembling (2.80) was proposed in [257].)

For large values of α, which is an important case in global optimization practice,

$$\Gamma(k+1/\alpha) - \Gamma(k) \sim \frac{1}{\alpha}\Gamma'(k) \quad \text{as } \alpha \to \infty$$

and therefore

$$C_k \sim \frac{\Gamma(1) + \frac{1}{\alpha}\Gamma'(1)}{1 + \frac{1}{\alpha}\psi(k+1) - \Gamma(1) - \frac{1}{\alpha}\Gamma(1)} = \frac{\alpha - \gamma}{\psi(k+1) + \gamma}, \qquad \alpha \to \infty,$$

where $\psi(\cdot) = \Gamma'(\cdot)/\Gamma(\cdot)$ is the psi-function and $\gamma \cong 0.5772$ is the Euler constant.

Asymptotic efficiency of the estimators

If k is fixed, then the asymptotic efficiency of any consistent linear estimator $\hat{m}_{n,k}(a)$ can naturally be defined as

$$\text{eff}(\hat{m}_{n,k}(a)) = \left(\min_{c \in \mathbb{R}^k:\, c'\mathbf{1}=1} c'\Lambda c\right) / a'\Lambda a\,.$$

Obviously, $0 \le \text{eff}(\hat{m}_{n,k}(a)) \le 1$ for any $a \in \mathbb{R}^k$ satisfying the consistency condition (2.71). In view of (2.77) we obtain

$$\text{eff}(\hat{m}_{n,k}(a)) = \frac{1}{\mathbf{1}'\Lambda^{-1}\mathbf{1} \cdot a'\Lambda a}, \tag{2.81}$$

where $\mathbf{1}'\Lambda^{-1}\mathbf{1}$ can be computed using the expression (2.78).

In particular, the asymptotic efficiency of the simplest estimator $\hat{m}_{n,k}(a^{(0)}) = \eta_{1,n}$ is

$$\text{eff}(\hat{m}_{n,k}(a^{(0)})) = \frac{1}{\mathbf{1}'\Lambda^{-1}\mathbf{1} \cdot \Gamma(1+2/\alpha)}. \tag{2.82}$$

This result easily follows from (2.75) and (2.85). For large k and α, the asymptotic efficiency of the estimator $\hat{m}_{n,k}(a^{(0)})$ is low, see Fig. 2.6.

Asymptotic efficiency of the estimator $\hat{m}_{n,k}(a^U)$ is higher (especially for small k) than that of the simplest estimator $\hat{m}_{n,k}(a^{(0)}) = \eta_{1,n}$. Fig. 2.7 displays this efficiency for $k = 3, 5$ and 20 and varying α (for $k = 2$ the asymptotic efficiency of $\hat{m}_{n,k}(a^U)$ is equal to 1).

Note that the asymptotic efficiency of the optimal estimator $\hat{m}_{n,k}(a^*)$ can be low if an incorrect value of α is used to construct this estimator. This issue is considered in Sect. 2.5.2.

Finite-sample efficiency (simulation results)

Let us make a comparison of the efficiency for the maximum likelihood and linear estimators of m given finite samples of size n drawn from the Weibull distribution with tail index α. Considering the Weibull distribution means assuming that the original sample size n is large enough for the asymptotic distribution for the minimal statistics to be reached.

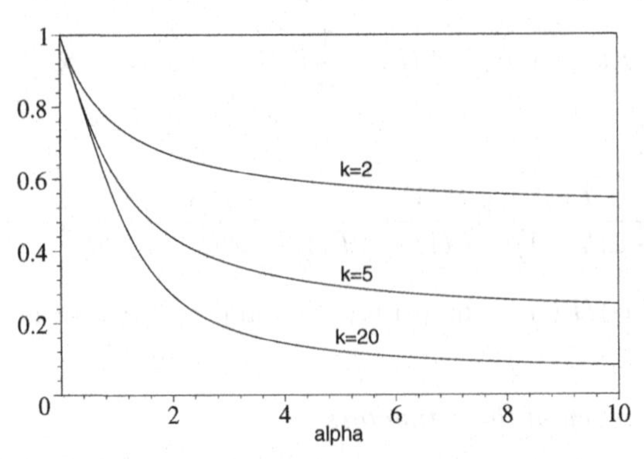

Fig. 2.6. Asymptotic efficiency of the simplest estimator $\hat{m}_{n,k}(a^{(0)}) = \eta_{1,n}$, see (2.82), for $k = 2, 5$ and 20 and varying α.

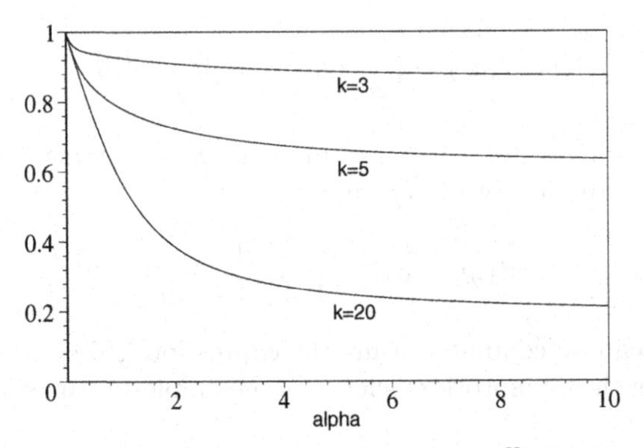

Fig. 2.7. Asymptotic efficiency of the estimator $\hat{m}_{n,k}(a^U)$ for $k = 3, 5$ and 20 and varying α.

According to the definition, for each k, the optimal linear estimator $\hat{m}_{n,k}(a^*)$, with a^* given in (2.76), provides the lowest mean square error in the class of all linear consistent estimators, as $n \to \infty$. In view of (2.74) and (2.77), we have, for the asymptotic mean square error of $\hat{m}_{n,k}(a^*)$:

$$\lim_{n\to\infty} \frac{\mathbf{1}'\Lambda^{-1}\mathbf{1}}{(\kappa_n - m)^2} \mathrm{MSE}(\hat{m}_{n,k}(a^*)) = 1\,, \tag{2.83}$$

for any k. Therefore, for fixed n and k it is natural to define the finite-sample efficiency of an estimator $\hat{m}$ as

$$\frac{(\kappa_n - m)^2}{\mathbf{1}'\Lambda^{-1}\mathbf{1}} / \mathrm{MSE}(\hat{m})\,. \tag{2.84}$$

Since we consider finite samples, it is possible for the efficiency to be slightly greater than 1.

Below, the efficiency of an estimator $\hat{m}$ will be estimated based on taking $R = 10\,000$ estimators of $\hat{m}_j$, where each $\hat{m}_j$ ($j = 1, \ldots, R$) is estimated from an independent sample of size n; that is,

$$\mathrm{MSE}(\hat{m}) \simeq \frac{1}{R}\sum_{j=1}^{R}(\hat{m}_j - m)^2 .$$

Thus, for fixed k, n and R, we use the following definition of efficiency of an estimator $\hat{m}$:

$$\mathrm{eff}(\hat{m}) = \left[\frac{(\kappa_n - m)^2}{\mathbf{1}'\Lambda^{-1}\mathbf{1}}\right] / \left[\frac{1}{R}\sum_{j=1}^{R}(\hat{m}_j - m)^2\right] . \tag{2.85}$$

As $R \to \infty$, the efficiency (2.85) tends to (2.84).
where in our case $m = 0$, $n = 100$, $R = 10\,000$ and k varies.
Fig. 2.8 shows the efficiencies (2.85) computed for $n = 100$, $R = 10\,000$, α=1,2,5,10, and varying k for the following estimators:

- the optimal linear estimator $\hat{m}_{n,k}(a^*)$ (depicted as circles),
- the maximum likelihood estimator (squares),
- the linear estimators $\hat{m}_{n,k}(a^+)$ defined by the vector (2.79) (triangles),
- Csörgő–Mason estimators $\hat{m}_{n,k}(a^{CM})$ (dots),
- the minimum order statistic $\eta_{1,n} = \hat{m}_{n,k}(a^{(0)})$ (bullets).

Fig. 2.8 demonstrates that the mean square error of the optimal linear estimator $\hat{m}_{n,k}(a^*)$ is very close to the asymptotically optimal value of the MSE given by (2.83) for all $\alpha \geq 1$ (sometimes it is even larger than this value). The estimator $\hat{m}_{n,k}(a^*)$ clearly provides the lowest mean square error in the class of estimators considered. The efficiency of the maximum likelihood estimator (MLE) is consistently lower than the efficiency of $\hat{m}_{n,k}(a^*)$, especially when α is small; note that MLE can only be used for $\alpha \geq 2$. Note also that the actual efficiency curves of MLE are rather uneven; they have been considerably smoothed in this figure.

The efficiency of the minimum order statistic decreases monotonically as $k \to \infty$, this is because the estimator is not using $k-1$ out of k order statistics. The efficiency of the linear estimator $\hat{m}_{n,k}(a^+)$ is poor for small k (as the unbiasedness condition (2.72) takes away one degree of freedom for the coefficients) but increases monotonically as k increases. The efficiencies of the minimum order statistic and the m^{Δ} estimators are equal for $k = 2$. This can be verified by considering the asymptotic mean square errors (as $n \to \infty$) of these two estimators at this point. The efficiency of the Csörgő–Mason estimators is poor for small α (note that this estimator is only defined for $\alpha \geq 2$) but gets better when α increases; thus, for $\alpha = 10$ the efficiency of the Csörgő–Mason estimator is basically 1.

The case of small values of α has a particular interest. Unlike the MLE and the Csörgő–Mason estimator, the linear estimators $\hat{m}_{n,k}(a^*)$ and $\hat{m}_{n,k}(a^+)$ are defined in the region $0 < \alpha < 2$ and behave rather well.

Simulation study of the bias of the estimators shows that the bias of the four main estimators (namely, MLE, $\hat{m}_{n,k}(a^*)$, $\hat{m}_{n,k}(a^+)$ and $\hat{m}_{n,k}(a^{CM})$) improves as both k and α increase; for large k and *alpha* this bias is approximately the same; for small α the bias of the Csörgő–Mason estimator is large but the bias of the other three estimators is comparable for all $\alpha \geq 2$ (note again that MLE is properly defined only for $\alpha \geq 2$). See [110] for more simulation results and related discussions.

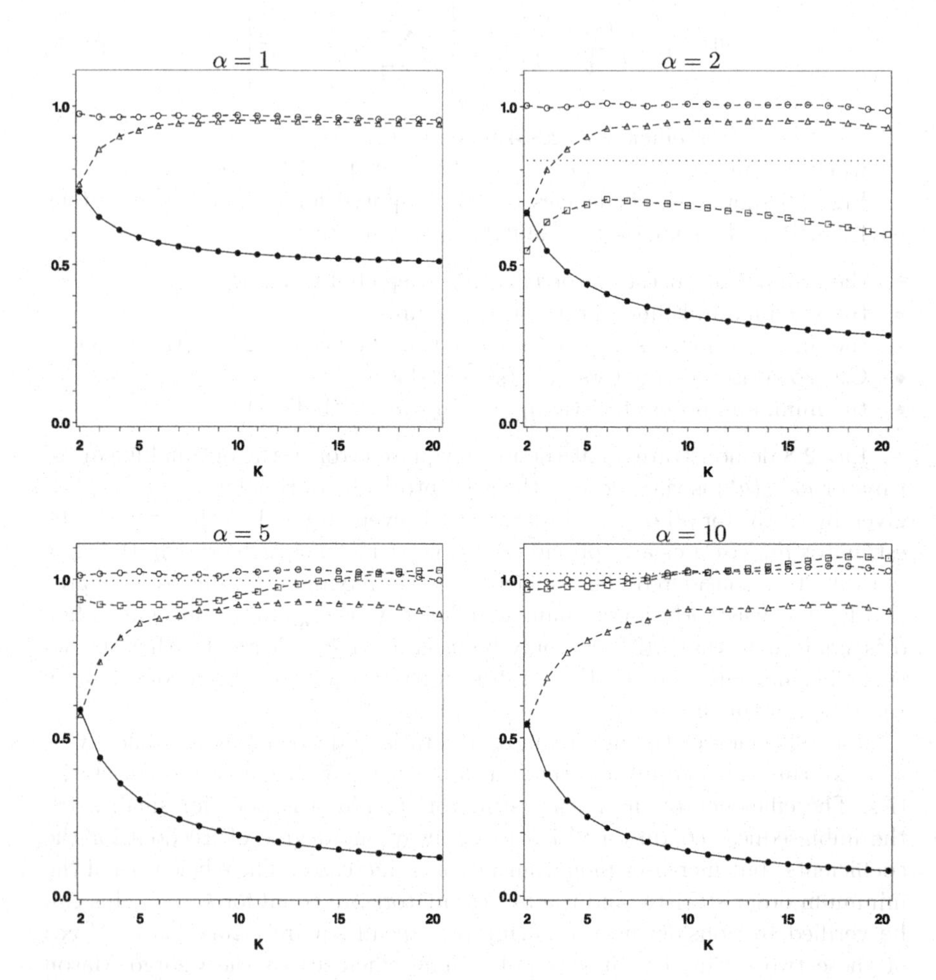

Fig. 2.8. Efficiency as defined in (2.85) of different estimators; $n = 100$, $R = 10\,000$, $\alpha = 1, 2, 5, 10$, against k.

2.4.2 Confidence Intervals and Hypothesis Testing

In global random search one of the most important statistical problems is testing the hypothesis $H_0 : m \le K$ versus the alternative $H_1 : m > K$, where K is some fixed number, $K < \eta_{1,n}$. For instance, K may be the record value of the objective function $f(\cdot)$ attained at another region, see Sect. 2.6.1.

Following the standard route, to construct a test for $H_0 : m \le K$ we construct a one-sided confidence interval for m of a fixed confidence level and reject H_0 if K does not fall into this interval.

The most convenient procedure for constructing confidence intervals for m was proposed in [51]. According to this procedure, the one-sided confidence interval for m is

$$[\eta_{1,n},\, \eta_{1,n} + r_{k,\delta}(\eta_{k,n} - \eta_{1,n})]\,. \tag{2.86}$$

Here

$$r_{k,\delta} = \left((1-\delta^{1/k})^{-1/\alpha} - 1\right)^{-1}, \tag{2.87}$$

the $(1-\delta)$-quantile of the c.d.f. $F_k(u)$ defined in (2.52). Proposition 2.1 of Sect. 2.3.2 then implies that the asymptotic (as $n \to \infty$) confidence level of the interval (2.86) equals $1 - \delta$.

The test corresponding to the confidence interval (2.86), for testing the hypothesis $H_0 : m \le K$, is defined by the rejection region

$$\{(\eta_1, \ldots, \eta_n) \,:\, (\eta_{1,n} - K)/(\eta_{k,n} - \eta_{1,n}) \ge r_{k,\delta}\}\,. \tag{2.88}$$

The first kind error probability of this test asymptotically (as $n \to \infty$) does not exceed δ; this is a consequence of Proposition 2.1 of Sect. 2.3.2.

Different asymptotic expressions for the power function of this procedure can be found in [273], Sect. 7.1.5.

Note that if n is very large (in this case, k may also be chosen large enough), then for constructing the confidence intervals and testing hypotheses one may also use the asymptotic normality (discussed above) of some estimators of m.

2.4.3 Choice of n and k

Let us briefly address the practically important problem of the choice of k and the sufficiency of the sample sample size n for applicability of the methodology considered above to the practice of global optimization.

Theoretically, n should be large enough to guarantee that there are enough, at least k, sample points in the vicinity of the global minimizer. Everything now depends on how we define 'the vicinity of the global minimizer'. This, in turn, depends on the objective function. For example, if the objective function is steep in the vicinity of the global minimizer (as an example, see Fig. 1.1), then the region of attraction of this minimizer is small and there is a high possibility that this region is completely missed. If the region of attraction of the minimizer is not reached, then the statistical inference will be made about some other minimum (perhaps, local).

Another important point is that the vicinity of a minimizer should not be confused with the region of attraction of the minimizer, see Sects. 1.1.2 and 1.1.3. If, for example, the objective function $f(\cdot)$ is the sum of a smooth and slowly varying function $f_1(\cdot)$ and a small irregular function $f_2(\cdot)$ (see Figs. 1.3 and 1.4) then small variations in the values of $f(\cdot)$ due to the presence of $f_2(\cdot)$ can be regarded as small variations in the sample points, which are statistically insignificant provided that k is not too large. On the other hand, if the objective function changes its shape approaching the minimum within the region of attraction of the global minimizer, this would imply that Theorem 2.2 is practically useless. Theoretically, however, as $n \to \infty$ we can select $k \to \infty$ and construct a consistent estimator of the tail index.

The problem of how large the sample size n should be can be approached from the more formal point of view of the rate of convergence in (2.43). There are a number of results concerning the estimation of this convergence rate, see for instance Sect. 2.10 in [86]; note that these estimators depend on different characteristics of behaviour of the c.d.f. $F(t)$ for t close to m.

Theoretically, the value of k should be such that at least k sample points belong to the vicinity of the global minimizer. It may, of course, happen that some of these k points are far from this vicinity. This would narrow the gap between $\eta_{1,n}$ and $\eta_{k,n}$ (in the probabilistic sense, and in comparison to such a gap when all the points belong to the same vicinity of the global minimizer). This would lead, in particular, to the over-estimation of m. Since we know the location of the test points, this may sometimes be corrected as follows: using some prior information about the objective function we can define a region (say, a ball) with the centre at the record point so the points outside the region will not be able to contribute to the set of the k smallest order statistics.

From the theoretical view-point, k should be small relative to the sample size n, which tends to infinity. Typically, the theoretically optimal choice of k is $k \to \infty$ so that $k/n \to 0$ as $n \to \infty$. In practice, however, n is never large enough and therefore small or moderate values of k should be used. Theoretical results of Sect. 2.4 imply that for many procedures a reasonably small value of k, say $k = 5$, is almost as good as the theoretically optimal choice $k \to \infty$, so we do not loose much in the asymptotic efficiency by restricting k to small values.

Another argument in favour of small k is given in Sect. 2.5.2: if the value of the tail index α is not correct (for instance, α has been estimated), then an increase in k (formally, $k \to \infty$) decreases the accuracy of precision in the estimators of m.

If the tail index is unknown, the problem of the choice of k is more serious than when α is known, see Sect. 2.5.1. For example, consistency of estimators of α can only be achieved if $k \to \infty$ as $n \to \infty$. Consideration can be given to a mixed strategy which uses a large number of extreme order statistics to estimate α and a relatively small number of these statistics for estimating m.

2.5 Unknown Value of the Tail Index

Our main objective is making statistical inferences about m based on an independent sample from the c.d.f. $F(\cdot)$ given in (2.13). In Sect. 2.4 we have shown how to make statistical inferences when the value of the tail index α is known. In Sect. 2.5.3 below, we show that the specific form of the c.d.f. (2.13) in many cases enables explicit determination of the value of α.

An alternative approach would be to find an estimator $\hat{\alpha}$ for α and use this estimator in place of the true value of α. However, we will show in Sect. 2.5.2 that this approach leads to a significant drop in precision of statistical inference procedures about m, in comparison to the case of known α.

Additionally, the requirements for the sample size seem to be unrealistic. Indeed, to construct any consistent estimator of α we must have $k = k(n)$ observations with $k(n) \to \infty$ (as $n \to \infty$) belonging to the lower tail of the c.d.f. (2.13), where the approximation (2.44) can be applied. In global optimization problems, however, obtaining more than a few observations in this region is problematic.

In global random search problems, making statistical inference about α is most useful for checking upon one of a few possible exact values of α, say $\alpha = 2/d$ or $\alpha = 1/d$; these expressions for α follow from Theorem 2.3 and related results, see Sect. 2.5.3.

2.5.1 Statistical Inference

In this section, we assume that the conditions of Theorem 2.2 are met but the value of the tail index α is unknown. Unlike the case considered in Sect. 2.4, a satisfactory precision of the statistical inference can only be guaranteed if k is large enough. Therefore, we shall suppose that the parameter k is chosen so that $k = k(n) \to \infty$, $k/n \to 0$, as $n \to \infty$. Also, we shall assume that the condition (2.45) is met.

The standard way of making statistical inference concerning m, when α is unknown, is to construct an estimator $\hat{\alpha}$ of α and to substitute $\hat{\alpha}$ for α in the formulae which determine the statistical procedures for the case of known α.

The topic of making statistical inferences about the value of the tail index is widely discussed in literature including a very recent one, see for example, [13, 32, 46, 61, 65, 133, 134, 149, 183].

Easily readable surveys of standard results concerning different estimators of α and their asymptotic properties can be found in Sect. 6.4.2 of [68] and in Sect. 2.6 of [139]. The two most known estimators are the so-called Hill estimator

$$\hat{\alpha}^{(H)} = \left(\ln \eta_{k,n} - \frac{1}{k} \sum_{j=1}^{k} \ln \eta_{j,n} \right)^{-1}$$

suggested in [118], and the Pickands estimator

$$\hat{\alpha}^{(P)} = \frac{1}{\ln 2} \ln \frac{\eta_{2k,n} - \eta_{k,n}}{\eta_{4k,n} - \eta_{2k,n}}$$

proposed in [186]. Provided that the conditions of Theorem 2.2, along with an additional regularity condition of the type (2.45), are satisfied and $k \to \infty$ as $n \to \infty$, both estimators of α are consistent and asymptotically normal. Their asymptotic properties are similar. The main practical problem is, of course, the choice of k. This problem has been addressed in a number of articles, see e.g. [65]. However, this problem can hardly be adequately resolved in global random search applications as the value of n required to achieve a reasonable precision in statistical inference about $m = \min f$ must be astronomical when the dimension d of A is not very small (recall that one of the main attractive points of the global random search methods is their applicability for solving problems with moderate or large dimension).

Results of Sect. 2.5.2 show that the linear estimators which perform well when the value of α is known become much less precise when the value of α is not known. Their asymptotic properties in the case when $\hat{\alpha}$ is noticeably different from α are poor and, consequently, it is often not worth using these estimators in the case of unknown α.

A slightly different way of making statistical inferences about m is based on making the inferences about m and α simultaneously using the maximum likelihood principle outlined below (see [109, 186, 228] for more details).

Assume that the asymptotic relation (2.42), along with the additional regularity condition (2.45) hold with some $c_0 > 0$ and $\alpha \geq 2$. Then the likelihood function depending on the unknown parameters c_0, α and m is asymptotically, as $n \to \infty$, equal to

$$\begin{aligned} &L(\eta_{1,n}, \ldots, \eta_{k,n};\ m, c_0, \alpha) \\ &= \tfrac{n!}{(n-k)!} (c_0 \alpha)^k \left(1 - c_0 \left(\eta_{k,n} - m\right)^{\alpha}\right)^{n-k} \prod_{j=1}^{k} (\eta_{j,n} - m)^{\alpha - 1} . \end{aligned} \tag{2.89}$$

This asymptotic form of the likelihood function is treated as the exact one. The maximisation of (2.89), with respect to c_0 for fixed $\alpha = \hat{\alpha}$ and $m = \hat{m}$ gives the maximum likelihood estimator for c_0:

$$\hat{c}_0 = \frac{k}{n} \left(\eta_{k,n} - \hat{m}\right)^{\hat{\alpha}} . \tag{2.90}$$

The maximisation of (2.89) with respect to α for fixed $m = \hat{m}$ and the substitution (2.90) for c_0, gives the maximum likelihood estimator for α:

$$\hat{\alpha} = k / \sum_{j=1}^{k-1} \ln(1+\beta_j(\hat{m})), \tag{2.91}$$

where $\beta_j(\hat{m})$ are defined in (2.65). The remaining problem is to define the maximum likelihood estimator $\hat{m}$ for m. It cannot be defined as the global maximizer of the likelihood function $L(\eta_{1,n},\ldots,\eta_{n-k};\, m, \hat{c}_0,\, \hat{\alpha})$, since the global maximum is achieved at $m=\eta_{1,n}$ and equals $+\infty$ (meaning that the proper maximum likelihood estimator of m is $\eta_{1,n}$, which is a poor estimator). According to the proposal of P.Hall [109], $\hat{m}$ is defined as a solution of the likelihood equation, which is

$$1/\sum_{j=1}^{k-1} \ln(1+\beta_j(\hat{m})) \;-\; 1/\sum_{j=1}^{k-1} \beta_j(\hat{m}) \;=\; 1/k \tag{2.92}$$

provided that $\hat{m} \leq \eta_{1,n}$. If there is no solution to the equation (2.92) in the half-interval $(-\infty, \eta_{1,n})$, then $\hat{m}$ is taken as $\eta_{1,n}$; if there is more than one solution of this equation in $(-\infty, \eta_{1,n})$ (that is, the likelihood function is multimodal), then the largest solution is taken. Despite the fact that the estimator does not typically maximize the likelihood function (execpt in the trivial case where $\eta_{1,n}$ is taken as the estimator), it is still called the maximum likelihood estimator. Note that the equation (2.92) is exactly the equation (2.66) with α replaced by $\hat{\alpha}$ of (2.91).

Under the regularity condition (2.45) the maximum likelihood estimator $\hat{m}$ of m is asymptotically normal with mean m and the variance

$$(\alpha-1)^2(1-2/\alpha)(\kappa_n - m)^2 k^{-1+2/\alpha}, \;\; \alpha>2,\; n\to\infty,\; k\to\infty,\; k/n\to 0.$$

This differs from the r.h.s. of (2.67) in the multiplier $(\alpha-1)^2$ only.

Formally, we can avoid estimating α and construct confidence intervals and statistical tests for m using the result proved in [262]. This result says that if the conditions of Theorem 2.2 hold, $k\to\infty$, $k/n\to 0$, $n\to\infty$, then the sequence of random variables

$$\frac{(\ln k)\, \ln[(\eta_{2,n}-m)/(\eta_{1,n}-m)]}{\ln[(\eta_{k,n}-\eta_{3,n})/(\eta_{3,n}-\eta_{2,n})]}$$

converges in distribution to a random variable with the exponential density e^{-t}, $t\geq 0$. Some generalizations of this result can be found in [263].

2.5.2 Using an Incorrect Value of the Tail Index

Confidence intervals

Consider what happens to the level of the one-sided confidence interval (2.86) for the case where

$$r'_{k,\delta} = 1/\left((1-\delta^{1/k})^{1/\vartheta} - 1\right)$$

is being substituted for $r_{k,\delta}$ defined in (2.87); this means that ϑ is being used in place of the true α.

Proposition 2.3. *Let the conditions of Theorem 2.2 hold, $n \to \infty$, k and $\vartheta > 0$ be fixed. Then the asymptotic confidence level of the confidence interval*

$$I' = [\eta_{1,n} - r'_{k,\delta}(\eta_{k,n} - \eta_{1,n})\,,\, \eta_{1,n}] \tag{2.93}$$

is equal to

$$1 - (1 - (1-\delta^{1/k})^{\alpha/\vartheta})^k. \tag{2.94}$$

Proof is given in Sect. 2.7.

Note that if we take $\vartheta = \alpha$, then (2.94) is simplified to $1-\delta$; therefore, Proposition 2.3 generalizes the statement of Sect. 2.4.2 saying that the asymptotic confidence level of the interval (2.86) is equal to $1-\delta$.

Linear estimators of m

Let us now follow [274] and study the consequences of using incorrect values of α while constructing linear estimators of m (using incorrect values of α is inevitable when we do not know the exact value of α and use its estimator instead).

Assume that $\alpha > 1$, $\alpha \neq 2$ and start the investigation with the optimal estimator $\hat{m}_{n,k}(a^*)$. Denote by ϑ $(\vartheta \neq \alpha)$ the value we use to compute $a^* = a^*(\vartheta)$ and by $\Lambda_0 = \Lambda(\vartheta)$ the matrix $\Lambda = \|\lambda_{ij}\|$ defined in (2.68) with ϑ substituted for α.

In view of (2.85) the asymptotic efficiency of the estimator $\hat{m}_{n,k}(a^*(\vartheta))$ is

$$\mathrm{eff}(\hat{m}_{n,k}(a^*(\vartheta))) = \frac{1}{\mathbf{1}'\Lambda^{-1}\mathbf{1}\cdot(a^*(\vartheta))'\Lambda a^*(\vartheta)} = \frac{(\mathbf{1}'\Lambda_0^{-1}\mathbf{1})^2}{\mathbf{1}'\Lambda^{-1}\mathbf{1}\cdot\mathbf{1}'\Lambda_0^{-1}\Lambda\Lambda_0^{-1}\mathbf{1}}.$$

If k is fixed and $|\vartheta - \alpha|$ is small, then the estimator $\hat{m}_{n,k}(a^*(\vartheta))$ is relatively good. For example, if $k = 2$ then

$$\Lambda = \begin{pmatrix} \Gamma(1+2/\alpha) & (1+\frac{1}{\alpha})\Gamma(2+2/\alpha) \\ (1+\frac{1}{\alpha})\Gamma(2+2/\alpha) & \Gamma(2+2/\alpha) \end{pmatrix}, \quad a^*(\vartheta) = \begin{pmatrix} 1+\frac{\vartheta}{2} \\ -\frac{\vartheta}{2} \end{pmatrix},$$

$$\lambda'\Lambda^{-1}\lambda = \frac{2(\alpha+1)}{(\alpha+2)\Gamma(1+2/\alpha)} \quad \text{and} \quad \mathrm{eff}(\hat{m}_{n,k}(a^*(\vartheta))) = \frac{\alpha+2}{\alpha+2+\alpha(1-\frac{\vartheta}{\alpha})^2}.$$

We shall say that the estimator $\hat{m}_{n,k}(a^*(\vartheta))$ is poor if

$$\mathrm{eff}(\hat{m}_{n,k}(a^*(\vartheta))) < \mathrm{eff}(\hat{m}_{n,k}(a^{(0)}))\,; \tag{2.95}$$

that is, the asymptotic efficiency of the estimator $\hat{m}_{n,k}(a^*(\vartheta))$ is worse than the asymptotic efficiency of the simplest estimator $\hat{m}_{n,k}(a^{(0)}) = \eta_{1,n}$. Note that for $k = 2$ we have

$$\mathrm{eff}(\hat{m}_{n,k}(a^{(0)})) = \alpha + 2/(2(\alpha + 1)).$$

The inequality (2.95) cannot be true for $\vartheta < \alpha$. On the other hand, it is easy to see that the estimator $\hat{m}_{n,k}(a^*(\vartheta))$ is poor when $\vartheta > 2\alpha$.

In the case $k > 2$ the situation is not so clear. For instance, for $k = 3$ we have

$$\Lambda = \Gamma(1 + 2/\alpha) \begin{pmatrix} 1 & \frac{\alpha+2}{\alpha+1} & \frac{2(\alpha+2)}{2\alpha+1} \\ \frac{\alpha+2}{\alpha+1} & \frac{\alpha+2}{\alpha} & \frac{2(\alpha+1)(\alpha+2)}{\alpha(2\alpha+1)} \\ \frac{2(\alpha+2)}{2\alpha+1} & \frac{2(\alpha+1)(\alpha+2)}{\alpha(2\alpha+1)} & \frac{(\alpha+1)(\alpha+1)}{\alpha^2} \end{pmatrix},$$

$$a_1^*(\vartheta) = \frac{(\vartheta + 2)(\vartheta + 1)^2}{3\vartheta^2 + 4\vartheta + 2}, \quad a_2^*(\vartheta) = \frac{\vartheta(\vartheta^2 - 1)}{3\vartheta^2 + 4\vartheta + 2},$$

$$a_3^*(\vartheta) = -\frac{\vartheta^2(2\vartheta + 1)}{3\vartheta^2 + 4\vartheta + 2}, \quad \mathrm{eff}(\hat{m}_{n,k}(a^{(0)})) = \frac{(\alpha + 1)(\alpha + 2)}{3\alpha^2 + 4\alpha + 2}.$$

Thus, for given values of α and ϑ, to conclude whether the estimator $\hat{m}_{n,k}(a^*(\vartheta))$ is poor we must compute the value of the two–variate polynomial

$$(a^*(\vartheta))' \left(\frac{1}{\Gamma(1 + 2/\alpha)} \Lambda\right) (a^*(\vartheta)),$$

which depends on α and ϑ, and compare it with 1. The estimator is poor if this value is smaller than 1.

Another interesting case is where k is large. According to [274], for all $\vartheta \neq \alpha$ we have

$$(a^*(\vartheta))' \Lambda\, a^*(\vartheta) \sim (\vartheta - 2)^2 (\alpha - \vartheta)^2 (\vartheta + \alpha\,\vartheta - 2\alpha)^{-2} k^{2/\alpha} \quad \text{as } k \to \infty.$$

In this case the estimator $\hat{m}_{n,k}(a^*(\vartheta))$ is poor (it is asymptotically less efficient than the simplest estimator $\hat{m}_{n,k}(a^{(0)}) = \eta_{1,n}$). The estimator is consistent but the order of convergence (as $k \to \infty$, $n \to \infty$, $k/n \to 0$) of the mean square error $E(m - \hat{m}_{n,k}(a^*(\vartheta)))^2$ to 0 is only $(k/n)^{2/\alpha}$ rather than $(k/n)^{2/\alpha}/k$ for the estimator $\hat{m}_{n,k}(a^{(0)})$.

Thus, if the value of the tail index α is not correct (for instance, α has been estimated), then the increase of k leads to a precision loss in the estimator $\hat{m}_{n,k}(a^*)$. A similar conclusion can be derived for the estimators $\hat{m}_{n,k}(a^+)$ and $\hat{m}_{n,k}(a^{CM})$ since these two estimators are asymptotically equivalent to $\hat{m}_{n,k}(a^*)$ (as $k \to \infty$, $n \to \infty$, $k/n \to 0$).

The situation with the estimator $\hat{m}_{n,k}(a^U(\vartheta))$ is better (that is, this estimator is less sensitive to deviations in α for large k). Indeed, we have as $k \to \infty$:

$$a_1^U(\vartheta) = \frac{b_k}{b_k - b_1} \sim 1 + k^{-1/\vartheta}\Gamma(1+1/\vartheta), \;\; a_k^U(\vartheta) \sim -k^{-1/\vartheta}\Gamma(1+1/\vartheta),$$

$$\lambda_{11} = \Gamma(1+2/\alpha), \quad \lambda_{kk} = \frac{\Gamma(k+2/\alpha)}{\Gamma(k)} \sim k^{2/\alpha},$$

$$\lambda_{k1} = \frac{\Gamma(k+2/\alpha)\Gamma(1+1/\alpha)}{\Gamma(k+1/\alpha)} \sim k^{1/\alpha}\Gamma(1+1/\alpha),$$

$$(a^U)'\Lambda(a^U) = (a_1^U(\vartheta))^2\lambda_{11} + 2a_1^U(\vartheta)a_k^U(\vartheta)\lambda_{k1} + (a_k^U(\vartheta))^2\lambda_{kk}$$

$$\sim \Gamma(1+2/\alpha) - 2\Gamma(1+1/\alpha)k^{1/\alpha-1/\vartheta}\Gamma(1+1/\vartheta) + k^{2/\alpha-2/\vartheta}\Gamma^2(1+1/\vartheta)$$

$$\sim \begin{cases} \Gamma(1+2/\alpha) & \text{for } \vartheta < \alpha, \\ \Gamma(1+2/\alpha) - \Gamma^2(1+1/\alpha) & \text{for } \vartheta = \alpha, \\ k^{2/\alpha-2/\vartheta}\Gamma^2(1-1/\vartheta) & \text{for } \vartheta > \alpha. \end{cases}$$

This implies that for $\vartheta < \alpha$ the asymptotic efficiency of the estimator $\hat{m}_{n,k}(a^U(\vartheta))$ asymptotically (as $k \to \infty$) coincides with the asymptotic efficiency of the simplest estimator $\hat{m}_{n,k}(a^{(0)})$, for $\alpha = \vartheta$ the estimator $\hat{m}_{n,k}(a^U)$ is better than $\hat{m}_{n,k}(a^{(0)})$ but worse than $\hat{m}_{n,k}(a^*)$, and, finally, for $\vartheta > \alpha$ the estimator $\hat{m}_{n,k}(a^U)$ is poor but it is much more asymptotically efficient than $\hat{m}_{n,k}(a^*)$.

2.5.3 Exact Determination of the Value of the Tail Index

Recall that the c.d.f. $F(\cdot)$ arising in global random search problems has the specific form (2.13). As we show in this section, this specific form often enables the determination of the value of the tail index α explicitly. It gives us the possibility of using the simple and efficient techniques of Sect. 2.4, rather than the techniques of Sect. 2.5.1, which require a much larger sample size.

The basic result is the following theorem.

Theorem 2.3. *Assume that the global minimizer x_* of $f(\cdot)$ is unique and Conditions C1 – C4, C8 and C9 of Sect. 2.1.1 along with the condition C10 of Sect. 2.2.1 are met. Assume, in addition, that the representation*

$$f(x) - m = w(\|x-x_*\|)H(x-x_*) + O(\|x-x_*\|^\beta), \;\; \|x-x_*\| \to 0, \quad (2.96)$$

is valid, where $H(\cdot)$ is a positive homogeneous function on $\mathbb{R}^d \backslash \{0\}$ of order $\beta > 0$ (for $H(\cdot)$ the relation $H(\lambda z) = \lambda^\beta H(z)$ holds for all $\lambda > 0$ and $z \in \mathbb{R}^d$) and function $w : \mathbb{R} \to \mathbb{R}$ is positive and continuous. Then the conditions of Theorem 2.2 for the c.d.f. (2.13) are fulfiled and the value of the tail index α is equal to $\alpha = d/\beta$.

Proof of the theorem is given in Sect. 2.7.

The main condition in Theorem 2.3 is (2.96) which characterizes the behaviour of the objective function $f(\cdot)$ in the neighbourhood of its global minimizer. Let us consider two important particular cases of (2.96).

First, let us assume that $f(\cdot)$ is twice continuously differentiable in the vicinity of x_*, $\nabla f(x_*) = 0$ (here $\nabla f(x_*)$ is the gradient of $f(\cdot)$ in x_*) and the Hessian $\nabla^2 f(x_*)$ of $f(\cdot)$ at x_* is non-degenerate. In this case, we can take

$$w(\cdot) = 1, \quad H(z) = -z'[\nabla^2 f(x_*)]z\,,$$

which implies $\beta = 2$ and $\alpha = d/2$.

Assume now that all components of $\nabla f(x_*)$ are finite and non-zero which often happens if the global minimum of $f(\cdot)$ is achieved at the boundary of A. Then we may take $H(z) = z'\nabla f(x_*)$, $w(\cdot) = 1$; this gives $\beta = 1$ and $\alpha = d$.

Consider now two extensions of the basic result.

The following statement demonstrates that if we can assume that the conditions of Theorem 2.2 are met for the c.d.f. (2.13) with some α, then the value of α itself can be determined from assumptions that are weaker than those of Theorem 2.3.

Theorem 2.4. *Assume that the global minimizer x_* of $f(\cdot)$ is unique and Conditions C1–C4, C8 and C9 of Sect. 2.1.1 are met. Assume, in addition, that the conditions of Theorem 2.2 are met for some $\alpha > 0$ and there exist positive numbers ε_0, c_3 and c_4 such that for all $x \in B(x_*, \varepsilon_0)$ the inequality*

$$c_3||x_* - x||^\beta \leq f(x) - m \leq c_4||x_* - x||^\beta$$

is valid. Then $\alpha = d/\beta$.

Proof of the theorem is given in Sect. 2.7.

The next assertion relaxes the uniqueness requirement for the global minimizer.

Theorem 2.5. *Assume that Conditions C1–C4, C7, C8 and C9 of Sect. 2.1.1 along with Condition C10 of Sect. 2.2.1 are met. Let the global minimum $m = \min f$ of $f(\cdot)$ be attained at points $x_*^{(i)}$ $(i = 1, \ldots, l)$ in whose vicinities the tail indexes α_i can be determined. Then the conditions of Theorem 2.2 for the c.d.f.* (2.13) *are fulfilled and the value of the tail index α is $\alpha = \min\{\alpha_1, \ldots, \alpha_l\}$.*

Proof of the theorem is given in Sect. 2.7.

2.6 Some Algorithmic and Methodological Aspects

2.6.1 Using Statistical Inference in Global Random Search

In this section, we consider different ways of using statistical inference procedures in global random search algorithms, discuss the so-called branch and probability bound methods and review the statistical inference procedures in the method of random multistart.

General considerations

Many global random search algorithms consist of several iterations so that at the i-th iteration a particular probability distribution $P = P_i$ is generated to obtain the points where $f(\cdot)$ is to be evaluated – see Algorithm 2.2 of Sect. 2.1.2 and a number of methods in Sect. 3.5. At each iteration of these algorithms and for various subsets Z of A with $P(Z) > 0$, we have independent samples of points which belong to Z and are distributed according to the probability measure P_Z (for a given $Z \in \mathcal{B}$, the measure P_Z is defined as $P_Z(U) = P(U \cap Z)/P(Z)$, $U \subseteq A$), along with the values of the objective function $f(\cdot)$ at these points. For given Z, these values of $f(\cdot)$ form an independent sample from the distribution with the c.d.f.

$$F_Z(t) = P_Z\{x \in Z \, : \, f(x) \leq t\}$$

and the lower bound

$$m_Z = \inf_{z \in Z} f(z) \, .$$

To guarantee that m_Z is indeed the lower bound of $F_Z(\cdot)$ it is sufficient to assume Conditions C1–C3 and C4′ of Sect. 2.1.1 for the set Z and Condition C10 of Sect. 2.2.1 for the measure P.

To decide whether it is worthwhile to place new points in Z we can draw statistical inferences concerning the parameter m_Z and the behaviour of the c.d.f. $F_Z(t)$ in the vicinity of m_Z. Since statistical procedures can be constructed for all sets Z and at various iterations of the algorithms in a similar manner, we can extend all the results of Sect. 2.4 and 2.5 formulated in the case $Z = A$ to the case of a generic $Z \subseteq A$. A wide class of global random search methods based on the statistical inference procedures developed in previous sections, is considered below.

More broadly, the statistical inference procedures of Sect. 2.4 and 2.5 aim to learn about the distance between the current record y_{on} and the unknown target $m = \min f$ and hence can be used for devising various stopping rules in any global random search algorithm presented in the form of Algorithm 2.2 of Sect. 2.1.2. For example, the estimators $\hat{m}$ of m and the confidence intervals for m can be used to define the following stopping rule: if $\hat{m}$ is close enough to the best value of $f(\cdot)$ obtained so far (alternatively, if the confidence interval is small enough), then the algorithm terminates.

The distributions for the new points in algorithms of this kind can differ from the uniform as these distributions are constantly changing. The corresponding algorithms, where the number of iterations is small but the number of points at each iteration is large, constitute a wide class of the so-called genetic random search algorithms, see Sect. 3.5; these algorithms are extremely popular in practice. As the number of points at each iteration is typically large, all the statistical procedures developed above can be used exactly as they are presented. The differences between these algorithms and the branch

and probability bound methods considered below, are:
(a) the subregions are not removed from A; instead, the distributions P_i are adapted; and
(b) the function values that were used in previous iterations cannot be used in subsequent iterations: indeed, the use of them would introduce dependence into the sample $\{f(x_i)\}$; this dependence would be difficult to handle.

Furthermore, the assumption of the independence of points $x_i^{(j)}$ at iteration j in Algorithm 2.2 of Sect. 2.1.2, which is commonly used in practice (see e.g. Sect. 3.5), can be relaxed to allow some dependence in these points and some of the statistical inference procedures developed above can be suitably modified. In Sect. 3.2 we consider in detail the problem of making statistical inference about m for the case of stratified sampling. We will show that a certain reduction in randomness typically leads to more efficient algorithms; note that improving the efficiency of algorithms by reducing the randomness of points is one of the major areas of interest in the theory of Monte-Carlo methods.

Branch and probability bound methods

Branch and bound optimisation methods are widely known. To put it briefly, they consist of several iterations, each including the following stages:

(i) branching the optimisation set into a tree of subsets (more generally, decomposing the original problem into subproblems),
(ii) making decisions about the prospectiveness of the subsets for further search, and
(iii) selecting the subsets that are recognized as prospective for further branching.

To make a decision at stage (ii) prior information about $f(\cdot)$ and values of $f(\cdot)$ at some points in A are used, deterministic lower bounds concerning the minimal values of $f(\cdot)$ on the subsets of A are constructed, and those subsets $Z \subset A$ are rejected (considered as non-prospective for further search) for which the lower bound for $m_Z = \inf_{x \in Z} f(x)$ exceeds an upper bound $\hat{m}$ for $m = \min f$; the minimum among all evaluated values of $f(\cdot)$ in A is a natural upper bound $\hat{m}$ for m. A general recommendation for improving this upper bound is to use a local descent algorithm, starting at the new record point, each time we obtain such a point.

Let us consider a version of the branch and bound technique, which we call 'branch and probability bound'; see [272] and Sect. 4.3 in [273] for a detailed description of this technique and results of numerical experiments. In the branch and probability bound methods, an independent sample from the uniform distribution in the current search region is generated at each iteration and the statistical procedures described in Sect. 2.4.2 for testing the hypothesis $H_0 : m_Z \le \hat{m}$ are applied to make a decision concerning the prospectiveness of sets Z at stage (ii). Rejection of the hypothesis H_0

corresponds to the decision that the global minimum m can not be reached in Z. Naturally, such a rejection may be false. This may result in losing the global minimizer. An attractive feature of the branch and probability bound algorithms is that the asymptotic level for the probability of false rejection can be controlled.

The stages (i) and (iii) above can be implemented in exactly the same fashion as in the classical branch and bound methods. When the structure of A is not too complicated, the following technique has been proven to be convenient and efficient.

Let A_j be a search region at iteration j, $j \geq 1$ (so that $A_1 = A$). At iteration j, in the search region A_j we first isolate a subregion Z_{j1} with centre at the point corresponding to the record value of $f(\cdot)$. The point corresponding to the record value of $f(\cdot)$ over $A_j \backslash Z_{j1}$ is the centre of a subregion Z_{j2}. Similar subregions Z_{ji} $(i = 1, \ldots, I)$ are isolated until either A_j is covered or the hypothesis that the global minimum can occur in the residual set $A_j / \cup_{i=1}^{I} Z_{ji}$ is rejected (the hypothesis can be verified by the procedure described in Sect. 2.4.2). The search region A_{j+1} in the next $(j+1)$–th iteration is naturally either $Z^{(j+1)} = \cup_{i=1}^{I} Z_{ji}$, a hyperrectangle covering $Z^{(j+1)}$, or a union of disjoint hyperrectangles covering $Z^{(j+1)}$. In the multidimensional case the last two ways produce more computationally convenient versions of the branch and probability bound method than the first one.

As the value of the minimum of $f(\cdot)$ over these kind of subsets can often be expected to be attained at the boundary, where all the components of the gradient of the objective function are expected to be non-zero (assuming the objective function is differentiable), the results of Sect. 2.5.3 imply that $\alpha = d$ can be used as the value of the tail index α. For some subregions Z, the value d overestimates the true value of α, but this only affects the power of the test of Sect. 2.4.2 applied for testing the hypothesis $H_0 : m_Z \leq \hat{m}$. On the other hand, the fact that we do not have to estimate α significantly simplifies the problem of making statistical inferences about the minimum of $f(\cdot)$ over the subregions Z_{ji}.

Note also that at subsequent iterations all previously used points can still be used, since they follow the uniform distribution at the reduced regions.

The branch and probability bound methods are rather simple and can easily be realized as computer codes. They are both practically efficient for small or moderate values of d (say, $d < 10$) and theoretically justified in the sense that under general assumptions concerning $f(\cdot)$, they asymptotically converge with a given probability, which can be chosen close to 1. However, as d (and therefore α) increases, the efficiency of the statistical procedures of Sect. 2.4 deteriorates. Therefore, for large d the branch and probability methods are both hard to implement (this is the case for the whole family of branch and bound methods) and their efficiency is poor. As a consequence of this, the use of the branch and probability methods for large dimensions is not recommended.

2.6.2 Statistical Inference in Random Multistart

Random multistart is a global optimization method consisting of several local searches starting at random initial points. In its original form, this method is inefficient as it typically wastes much effort on repeated ascents. However, some of its modifications, such as those using cluster analysis procedures to prevent repeated ascents to the same local extrema, can be quite efficient. These modifications are widely used and have been discussed in a number of papers including [22, 23, 147, 198, 199, 210].

This section mainly follows the paper [278] by R. Zieliński and describes several statistical procedures that can be used to increase the efficiency of the simplest random multistart and some of its modifications. A number of publications have appeared developing the ideas discussed in this section, mostly using the Bayesian inference, see e.g. [16, 17, 20, 21, 114, 261]. However, all the main ideas of the approach were contained in the original paper [278] and there has not been any significant progress in the area since 1981, the time of the publication of [278].

Notation

Let $A \subset \mathbb{R}^d$ satisfy the conditions C1, C2 and C3 of Sect. 2.1.1, $f(\cdot)$ be a continuous function on A with a finite but unknown number l of local minimizers $x_*^{(1)}, \ldots, x_*^{(l)}$, P be a probability measure on A and $\mathcal{A}$ be a local descent algorithm. We shall write $\mathcal{A}(x) = x_*^{(i)}$ for $x \in A$, if when starting at the initial point x the algorithm $\mathcal{A}$ leads to the local minimizer $x_*^{(i)}$.

Set $\theta_i = P(A_i^*)$ for $i = 1, \ldots, l$, where $A_i^* = \{x \in A : \mathcal{A}(x) = x_*^{(i)}\}$ is the region of attraction of $x_*^{(i)}$ (note that A_i^* may depend on the chosen algorithm of local descent). It is clear that $\theta_i > 0$ for $i = 1, \ldots, l$ and $\sum_{i=1}^l \theta_i = 1$.

The method of random multistart is constructed as follows. An independent sample $X_n = \{x_1, \ldots, x_n\}$ from the distribution P is generated and a local optimization algorithm $\mathcal{A}$ is sequentially applied at each $x_j \in X_n$. Let n_i be the number of points x_j belonging to A_i^* (that is, n_i is the number of descents to $x_*^{(i)}$ from the points $x_1, \ldots, x_n$). According to the definition, $n_i \geq 0$ $(i = 1, \ldots, l)$, $\sum_{i=1}^l n_i = n$, and the random vector $(n_1, \ldots, n_l)$ follows the multinomial distribution

$$\Pr\{n_1 = n_1, \ldots, n_l = n_l\} = \binom{n}{n_1, \ldots, n_l} \theta_1^{n_1} \ldots \theta_l^{n_l},$$

where

$$\sum_{i=1}^l n_i = n, \quad \binom{n}{n_1, \ldots, n_l} = \frac{n!}{n_1! \ldots n_l!}, \quad n_i \geq 0 \quad (i = 1, \ldots, l).$$

We consider the problem of drawing statistical inferences concerning the number of local minimizers l, the parameter vector $\theta = (\theta_1, \ldots, \theta_l)$, and the number n_* of trials that guarantees with a given probability that all local minimizers are found.

If l is known, then the problem is reduced to the standard problem of making statistical inferences about the parameters of a multinomial distribution. This problem is well documented in literature, see Chapt. 35 in [129].

The main difficulty is caused by the fact that l is usually unknown. If an upper bound for l is known, then one can apply standard statistical methods; if an upper bound for l is unknown, the Bayesian approach is a natural alternative. Let us first consider the case where the number of local minimizers is bounded.

Bounded number of local minimizers

Let L be an upper bound for l and $n \geq L$. Then $(n_1/n, \ldots, n_l/n)$ is the standard minimum variance unbiased estimate of θ, where n_i/n are the estimators of θ_i's. Of course, for all n and $l > 1$ it may happen, for some i, that $n_i = 0$ but $\theta_i > 0$. So, the above estimator non-degenerately estimates only the θ_i's for which $n_i > 0$.

Let W be the number of n_i's that are strictly positive. Then for given l and $\theta = (\theta_1, \ldots, \theta_l)$ we have

$$\Pr\{W = w \,|\, \theta\} = \sum_{\substack{n_1+\cdots+n_w=n \\ n_i>0}} \; \sum_{1\leq i_1<\cdots<i_w\leq l} \binom{n}{n_1, \ldots, n_w} \theta_{i_1}^{n_1} \ldots \theta_{i_w}^{n_w}.$$

For instance, the probability that all local descents will lead to a single local minimizer is

$$\Pr\{W = 1 \,|\, \theta\} = \sum_{i=1}^{l} \theta_i^n$$

and the probability that all local minima will be found is

$$\Pr\{W = l \,|\, \theta\} = \sum_{\substack{n_1+\cdots+n_l=n \\ n_i>0}} \; \sum_{1\leq i_1<\cdots<i_w\leq l} \binom{n}{n_1, \ldots, n_l} \theta_{i_1}^{n_1} \ldots \theta_{i_l}^{n_l}. \quad (2.97)$$

The probability (2.97) is small if at least one of the θ_i's is small. On the other hand, for any l and θ we can find n_* such that for any given $\gamma \in (0, 1)$ we will have $\Pr\{W = l \,|\theta\} \geq \gamma$ for all $n \geq n_*$. Finding $n_* = n_*(\gamma, \theta)$ is the problem of finding the (minimal) number of points in A such that the probability that all local minimizers will be found is at least γ.

Set $\delta = \min\{\theta_1, \ldots, \theta_l\} \leq 1/l$ and note that

$$\Pr\{W = l|\theta\} \geq \sum_{n_1+\ldots+n_l=n} \binom{n}{n_1, \ldots, n_l} \delta^n = (\delta l)^n \Pr\{W = l \,|\, (\tfrac{1}{l}, \ldots, \tfrac{1}{l})\}.$$

Hence the problem of finding $n_*(\gamma,\theta)$ is reduced to that of finding $n_*(\gamma,\theta_*)$, where $\theta_* = (l^{-1},\dots,l^{-1})$. The latter is easy to approximate as for large n

$$\Pr\{W = l \,|\, \theta_*\} = l^{-n} \sum_{n_1+\cdots+n_l=n} \binom{n}{n_1,\dots,n_l} =$$

$$= \sum_{i=0}^{l} (-1)^i \binom{l}{i} (1-i/l)^n \sim \exp\{-l\exp\{-n/l\}\}, \qquad n\to\infty.$$

By solving the equation $\exp(-l\exp(-n/l)) = \gamma$ with respect to n we obtain the approximation

$$n_*(\gamma,\theta_*) \simeq l\ln l + l\ln(-\ln\gamma)\,. \tag{2.98}$$

This approximation is rather good even for small l and n; see Fig. 2.9, where the exact values of $n_*(\gamma,\theta_*)$ and the approximation (2.98) are given for $\gamma = 0.9$ and $l \le 20$.

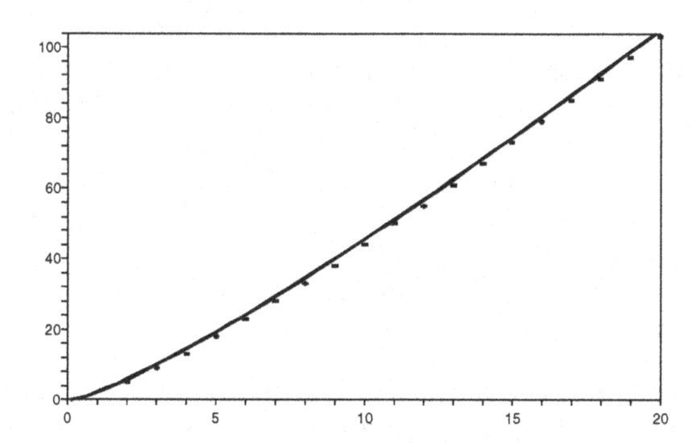

Fig. 2.9. The exact values of $n_*(\gamma,\theta_*)$ (dots) and the approximation $l\ln l + l\ln(-\ln\gamma)$ (solid line) for $\gamma = 0.9$ and $l = 2,\dots,20$.

Bayesian approach

Let α_j ($j = 1,2,\dots$) be the prior probabilities of events that the number l of local minimizers of $f(\cdot)$ is equal to j and let $\lambda_j(d\theta^j)$ be the conditional prior measures for the parameter vector $\theta^j = (\theta_1,\dots,\theta_j)$ under the condition $l\!=\!j$. We shall assume that the measures $\lambda_j(d\theta^j)$ are uniform on the simplices

$$\Theta_j = \Big\{\theta^j = (\theta_1,\dots,\theta_j) : \; \theta_i > 0,\; \sum_{i=1}^{j}\theta_i = 1\Big\}.$$

Thus, the parameter set Θ, on which the vector of unknown parameters $\theta = (\theta_1, \ldots, \theta_l)$ can take its values, has the form $\Theta = \cup_{j=1}^{\infty} \Theta_j$ and the prior measure $\lambda(d\theta)$ on Θ for θ equals

$$\lambda(d\theta) = \sum_{j=1}^{\infty} \alpha_j \lambda_j(d\theta^j). \tag{2.99}$$

It is natural to assume that λ is a probability measure.

Let $d = d(n_1, \ldots, n_W)$ be an estimate of l. The estimate

$$d_* = \arg\min_d \int_{\Theta} \mathrm{E}_\theta \, \mathbf{1}_{[n_1,\ldots,n_W : d \neq l]} \lambda(d\theta)$$

is the optimal Bayesian estimate of l; it can be simplified to

$$d_* = \arg\max_{j \geq W} \alpha_j Q(j, W, n), \tag{2.100}$$

where

$$Q(j, W, n) = \binom{j}{W} \Gamma(j)/\Gamma(n+j).$$

Using a quadratic loss function, the optimal Bayesian estimate for the total P-measure of the domains of attraction of the hidden $l - W$ local minimizers (i.e. of the sum of the θ_i's corresponding to the undiscovered minimizers) is given by

$$\sum_{j=W}^{\infty} \frac{j-W}{n+j} \alpha_j Q(j, W, n) \bigg/ \sum_{j=W}^{\infty} \alpha_j Q(j, W, n).$$

The optimal Bayesian procedure for testing the hypothesis $H_0 : l = W$ under the alternative $H_1 : l > W$ is constructed in a similar way. According to this procedure, H_0 is accepted if

$$c_{01} \sum_{j=W+1}^{\infty} \alpha_j Q(j, W, n) \leq c_{10} \alpha_W \Gamma(W)/\Gamma(n+W),$$

otherwise H_0 is rejected. Here c_{01} is the loss arising after accepting H_0 in the case of H_1's validity and c_{10} is the loss due to accepting the hypothesis H_1 when it is false.

2.6.3 Sampling on Surfaces

Application of any random search algorithm to an optimization problem where the feasible region A is defined by the equality-type constraints requires sampling from probability distributions on the surface defined by these constraints. We show how to reduce the problem of distribution sampling on a surface to the problem of distribution sampling on a subset of $\mathbb{R}^k$ of positive volume (the latter problem is potentially simpler).

Let $X \subset \mathbb{R}^k$ with $0 < \text{vol}(X) < \infty$ and Φ be a continuously differentiable mapping of X into $\mathbb{R}^d$ with $d \geq k$. Using the notation $x = (x_1, \ldots, x_k)$, $z = (z_1, \ldots, z_d)$ and $\Phi = (\varphi_1, \ldots, \varphi_d)$ we can write

$$\begin{cases} z_1 = & \varphi_1(x_1, \ldots, x_k) \\ \vdots & \vdots \\ z_d = & \varphi_d(x_1, \ldots, x_k) \end{cases}$$

simply as $z = \Phi(x)$. For $d > k$, the set

$$A = \Phi(X) = \{z = \Phi(x), x \in X\}$$

is a k-dimensional surface in $\mathbb{R}^d$.

For any $x \in X$ we define

$$d_{ij}(x) = \sum_{l=1}^{d} \frac{\partial \varphi_l(x)}{\partial x_i} \frac{\partial \varphi_l(x)}{\partial x_j} \qquad (i, j = 1, \ldots, k)$$

and

$$D(x) = \sqrt{\det \| d_{ij}(x) \|_{i,j=1}^{k}} \,.$$

The matrix $\| d_{ij}(x) \|_{i,j=1}^{k}$ is non-negative definite for all $x \in X$ so that its determinant is always non-negative.

If $d{=}k$ then $D(x){=}|\partial\Phi/\partial x|$ is the Jacobian of the transformation Φ. Another important particular case is where $d{=}k{+}1$ and $\varphi_j(x){=}x_j$ $(j{=}1, \ldots, k)$; in this case we have

$$D(x) = \left[1 + \sum_{i=1}^{k} \left(\frac{\partial \varphi_{k+1}(x)}{\partial x_i} \right)^2 \right]^{\frac{1}{2}} .$$

Let ds denote the surface measure on the surface $A = \Phi(X)$. As follows from §10, Chapt. 4 in [211], for any Borel-measurable function p defined on A and any $B \subseteq A$ of the form $B = \Phi(U)$, where U is a measurable subset of X, we have

$$\int_B p(s) ds = \int_{\Phi^{-1}(B)} p(\Phi(x)) D(x) dx.$$

Therefore, for any measurable non-negative function $p(\cdot)$ defined on A and satisfying the condition

$$\int_X p(\Phi(x))D(x)dx = 1,$$

the probability measure with density

$$p(\Phi(x))D(x), \quad x \in X$$

induces the probability distribution $p(s)ds$ on the surface $A = \Phi(X)$. In the important particular case where

$$c = \int_X D(x)dx < \infty,$$

the probability distribution with density

$$p_0(x) = \frac{1}{c}D(x), \quad x \in X,$$

induces the uniform distribution ds/c on the surface A.

Thus, the problem of distribution sampling on the surface A is being reduced to the problem of distribution sampling on the set $X \subset \mathbb{R}^k$ with $\mathrm{vol}(X) > 0$. In order to obtain a realization ξ of a random vector in $\mathbb{R}^d$ with distribution $p(s)ds$ on A, it is enough to obtain a realization ζ of a random vector in $X \subset \mathbb{R}^k$ with density $p(\Phi(x))D(x)$ and compute $\xi = \Phi(\zeta)$.

This general methodology was applied in [273], Sect. 6.1, to construct distribution sampling algorithms on various surfaces including ellipsoids, hyperboloids and cones.

2.7 Proofs

Proof of Theorem 2.1.

Fix $\delta > 0$ and find some $\varepsilon > 0$ such that $B(x_*, \varepsilon) \subset W(\delta)$; this is possible as $f(\cdot)$ is continuous in the vicinity of x_*. Define the sequence of independent random variables $\{\zeta_j\}$ on the two-point set $\{0, 1\}$ so that

$$\Pr\{\zeta_j = 1\} = 1 - \Pr\{\zeta_j = 0\} = q_j(\varepsilon)$$

where $q_j(\varepsilon)$ is defined in (2.5).

For each j, the probability of the event $x_j \in B(x_*, \varepsilon)$ is larger than or equal to the probability of the event $\zeta_j = 1$. However, the first part of the Borel's 'zero-one law' (see e.g. [226]) implies that if (2.4) holds, then ζ_j infinitely often takes the value 1; this yields the assertion of the theorem. □

Proof of Proposition 2.1.

Setting $w = (1+1/u)^\alpha - 1$ and using the fact that the joint asymptotic density of

$$\left(\frac{\eta_{1,n} - m}{\kappa_n - m}, \frac{\eta_{k,n} - m}{\kappa_n - m}\right)$$

coincides with the joint density of the random vector $(\nu_1^{1/\alpha}, (\nu_1+\ldots+\nu_k)^{1/\alpha})$, we obtain

$$\Pr\{D_{n,k} \leq u\} \sim \Pr\left\{\frac{\nu_1^{1/\alpha}}{(\nu_1+\cdots+\nu_k)^{1/\alpha} - \nu_1^{1/\alpha}} \leq u\right\} = \Pr\left\{\frac{\nu_1+\cdots+\nu_k}{\nu_1} \geq w\right\}$$
$$= \frac{1}{(k-2)!}\int_0^\infty \left[\int_{wy}^\infty \exp\{-x-y\} \cdot x^{k-2} dx\right] dy = 1 - \left(\frac{w}{w+1}\right)^{k-1}.$$

□

Proof of Proposition 2.2.

The formula (2.53) for the asymptotic moments follows from the fact that $(\eta_{k,n} - m)/(\kappa_n - m)$ converges in distribution (as $n \to \infty$) to the random variable with density (2.48); computing the β-th moment of this distribution with this density immediately gives (2.53).

Proof of (2.54) is similar but more technical. Assume that $1 \leq j < k \leq n$; the case $j = k$ is covered in (2.53) with $\beta = 2$.

Using the fact that the sequence of random vectors (2.49) asymptotically, as $n \to \infty$, has the same density as the vector (2.51), we deduce that the random vector

$$\left(\frac{\eta_{j,n} - m}{\kappa_n - m}, \frac{\eta_{k,n} - m}{\kappa_n - m}\right)$$

asymptotically has the same density as the vector

$$\left(\zeta^{1/\alpha}, (\xi + \zeta)^{1/\alpha}\right),$$

where random variables ξ and ζ are independent and have Gamma-distributions with densities

$$p_\zeta(x) = \frac{1}{\Gamma(j)} x^{j-1} e^{-x} \quad \text{and} \quad p_\xi(x) = \frac{1}{\Gamma(k-j)} x^{k-j-1} e^{-x} \quad (x > 0),$$

respectively. Therefore, as $n \to \infty$, we have

$$\frac{1}{(\kappa_n - m)^2} \mathrm{E}(\eta_{j,n} - m)(\eta_{k,n} - m) \to \mathrm{E}\zeta^{1/\alpha}(\xi + \zeta)^{1/\alpha}$$

$$= \frac{1}{\Gamma(j)\Gamma(k-j)} \int_0^\infty \int_0^\infty z^{1/\alpha}(x+z)^{1/\alpha} z^{j-1} e^{-z} x^{k-j-1} e^{-x} dx dz$$

$$= \frac{1}{\Gamma(j)\Gamma(k-j)} \int_0^\infty \int_0^\infty z^{j-1+2/\alpha}(1+x/z)^{1/\alpha} e^{-z} x^{k-j-1} e^{-x} dx dz$$

$$= \frac{1}{\Gamma(j)\Gamma(k-j)} \int_0^\infty \int_0^\infty (1+t)^{1/\alpha} t^{k-j-1} z^{k-1+2/\alpha} e^{-z(t+1)} dt dz$$

$$= \frac{1}{\Gamma(j)\Gamma(k-j)} \int_0^\infty \frac{t^{k-j-1}}{(t+1)^{k+1/\alpha}} dt \cdot \int_0^\infty u^{k-1+2/\alpha} e^{-u} du = \lambda_{kj} \,.$$

In the process of integration, we have introduced the new variables $t = x/z$ and $u = z(t+1)$; additionally, we have used the formulae

$$\int_0^\infty u^{k-1+2/\alpha} e^{-u} du = \Gamma(k+\frac{2}{\alpha}) \text{ and } \int_0^\infty \frac{t^{k-j-1}}{(t+1)^{k+1/\alpha}} dt = \frac{\Gamma(k-j)\Gamma(j+1/\alpha)}{\Gamma(k+1/\alpha)} \,.$$

□

Proof of Proposition 2.3.

According to Proposition 2.1, the sequence of random variables

$$(\eta_{1,n} - m)/(\eta_{k,n} - \eta_{1,n})$$

converges in distribution to the random variable with the c.d.f.

$$F_k(u) = 1 - \left(1 - \left(1 - \frac{1}{1+u}\right)^{\alpha}\right)^k$$

(note that $r_{k,\delta}$ is the $(1–\delta)$-quantile of this c.d.f.). This implies that as $n \to \infty$, the confidence level of the interval (2.93) can be represented as

$$\Pr\{m \in I'\} = \Pr\left\{\frac{\eta_{1,n} - m}{\eta_{k,n} - \eta_{1,n}} \le r'_{k,\delta}\right\}$$

$$\sim 1 - \left(1 - \left(\frac{1}{1+r'_{k,\delta}}\right)^{\alpha}\right)^k = 1 - \left(1 - \left(1 - \delta^{1/k}\right)^{\alpha/\vartheta}\right)^k .$$

□

3

Global Random Search: Extensions

3.1 Random and Semi-Random Coverings

According to the definition of covering, A is covered by sets $B_1, \ldots, B_n$ if

$$A \subseteq \bigcup_{i=1}^{n} B_i \,. \tag{3.1}$$

We shall only consider the case where the sets B_i are balls

$$B_i = B(x_i, \varepsilon_i, \rho) = \{z \in A : \rho(x_i, z) \leq \varepsilon_i\} \,,$$

where ρ is some metric on A, for all $i \geq 1$, x_i are points in A (centres of the balls) and ε_i are some non-negative numbers (the radii of the balls); more precisely, the sets $B(x_i, \varepsilon_i, \rho)$ are the intersections of the balls in $\mathbb{R}^d$ and the feasible region A.

In optimization problems, the balls B_i are centered at the points where the objective function has previously been evaluated; the radii of the balls are determined through a Lipschitz-type condition about the objective function and the current record value.

The main special case is when ε_i, the radii of the balls, are equal; that is, $\varepsilon_i = \varepsilon > 0$ for all i. In this case, (3.1) becomes

$$A \subseteq \bigcup_{i=1}^{n} B(x_i, \varepsilon, \rho) \,. \tag{3.2}$$

A very important concept related to the coverage of A with balls of equal radius is *dispersion.* We delay its consideration until Sect. 3.1.2 and start this section by considering the case where the points x_i are random (which is always the case in global random search methods).

3.1.1 Covering with Balls and Optimization

Covering with randomly placed balls of fixed radius

Let P be the uniform distribution on A; for any (measurable) subset Z of A with $P(Z) > 0$, we denote the uniform distribution on Z by P_Z. In the case of a general distribution P on A (more precisely, on the measurable space $(A, \mathcal{B})$), the distribution P_Z is defined by $P_Z(U) = P(U \cap Z)/P(Z)$ for any measurable $U \subseteq A$.

The following is the algorithm of Brooks [217, 218].[1]

Algorithm 3.1 (Covering with randomly placed balls of radius ε)

1. *Set* $Z_1 = A$ *and the iteration number* $j = 1$.
2. *Obtain a point* x_j *by sampling from the distribution* P_{Z_j}.
3. *Evaluate the current value of the objective function* $y_j = f(x_j)$ *and the corresponding record value* $y_{oj} = \min_{i=1...j} y_i$.
4. *Check the stopping condition; if the algorithm does not terminate, then set* $Z_{j+1} = Z_j \setminus B(x_j, \varepsilon, \rho)$ *and return to step 2 substituting* $j+1$ *for* j.

The following two stopping conditions at iteration j of Algorithm 3.1 may look natural:

(a) $j = n$, where n is a given number, and
(b) the set A is covered by the balls $B(x_i, \varepsilon, \rho)$, $i = 1, \ldots, j$.

However, both conditions are not totally satisfactory. Indeed, if n in (a) is too large, then A can be covered at an iteration $j < n$ and therefore the algorithm will be unable to reach the iteration n. The stopping condition (b) seems more appropriate than (a) but it is very difficult to implement as it is not clear how to check whether the coverage (3.2) has occurred. Below, we formulate the stopping condition (c), which is a version of (b) but which can be simply implemented.

Let the distributions P_Z be sampled in Algorithm 3.1 using the rejection technique (that is, the distribution P is sampled until a realization falls in Z). Then Algorithm 3.1 differs from PRS (the pure random search algorithm considered in Sect. 2.2.1) in the following detail only: if at the j-th iteration of Algorithm 3.1, a random point uniformly distributed in A falls within a distance ε of one of the previously accepted points x_i, $1 \le i < j$, then this point is rejected, the objective function $f(\cdot)$ is not evaluated at this point and a new random point is generated. The rationale of this simple 'tabu' rule is that we do not want new points to be very close to the points where we have already evaluated the objective function. In this respect, note that many optimization algorithms based on the coverage of A use the 'tabu' rationale and therefore

[1] Arguably, these two papers of Brooks were the first ever papers on the methodology of global random search.

these algorithms are often referred to as 'tabu search' algorithms, see e.g. [95, 185].

Note also that the random points from P in the scheme described above do not have to be independent but even if they are, the points x_j generated by the Algorithm 3.1 are not. We can roughly relate PRS to 'sampling with replacement' from a discrete set and Algorithm 3.1 to 'sampling without replacement' (which creates a sample with dependent elements).

If, when sampling from the distributions P_Z in Algorithm 3.1, we use the rejection technique, then we can use the following substitution for the stopping rule (b):

(c) a fixed number of random points (for example, 10 000) distributed according to P were rejected while trying to obtain the current point x_j.[2]

The stopping condition (c) does not imply full coverage (3.2); it only implies the fact that a large part of A is covered by the balls $B(x_i, \varepsilon, \rho)$ $(i=1, \ldots, j-1)$. This may be enough in some applications. If full coverage of A is required, this can be achieved by increasing ε and using Theorem 3.1 below.

The radius ε of the balls in Algorithm 3.1 determines the accuracy of the required approximation. For instance, if the objective function $f(\cdot)$ belongs to the class of Lipschitz functions $\mathrm{Lip}(A, L, \rho)$, then the values of $f(\cdot)$ at all the points removed from the search region at iteration j (that is, the points in $\cup_{i \leq j} B(x_i, \varepsilon, \rho)$) cannot be smaller than $y_{oj} - \varepsilon L$.

Guaranteeing the full coverage of A

Theorem 3.1. *Let $A = [0,1]^d$, ρ be the Eucledian metric on A, $X_n = \{x_1, \ldots, x_n\}$ be an arbitrary n-point set of points from A, $\{\varepsilon_i\}_{i=1}^n$ be a collection of non-negative numbers and let*

$$C_k = \left\{ \left(\frac{i_1}{k}, \ldots, \frac{i_d}{k} \right), \text{ where } i_j = 0, 1, \ldots, k \ (j = 1, \ldots, d) \right\} \tag{3.3}$$

be a cubic grid in A with $(k+1)^d$ points and the step-length $\frac{1}{k}$ in each coordinate. If all the points of C_k belong to $\bigcup_{i=1}^n B_{\varepsilon_i}(x_i)$, then the cube A is covered by $\bigcup_{i=1}^n B_{\varepsilon_i + r}(x_i)$, where $r = \sqrt{d}/(2k)$; that is,

$$A \subseteq \bigcup_{i=1}^{n} B_{\varepsilon_i + r}(x_i) \, . \tag{3.4}$$

Proof of Theorem 3.1 is given in Sect. 3.6 (see also [275]). Almost exactly the same proof is valid in the case where the grid (3.3) is replaced with the grid

[2] If the reader decides to use the stopping rule (c) then he/she should be very careful of the random number generator used in the related software: for bad generators the new 'random points' never fall into the set Z_j, even when the volume of Z_j is not too small (the author's personal experience).

$$C'_k = \left\{ \left(\frac{2i_1 - 1}{2k}, \ldots, \frac{2i_d - 1}{2k} \right);\ i_j = 1, \ldots, k\ (j = 1, \ldots, d) \right\} \quad (3.5)$$

containing k^d points, which is slightly smaller than the number of points in the grid (3.3). Furthermore, Theorem 3.1 can be easily generalized to sets A other than $[0,1]^d$ and to non-Euclidian metrics.

Coverage of A with randomly placed balls of variable radius

Algorithm 3.1 is non-adaptive in the sense that the information about the objective function obtained in previous iterations of the algorithm is not used. One may significantly increase the efficiency of coverage when additional information about $f(\cdot)$ is available and used in constructing the balls. Let us assume that $f \in \mathrm{Lip}(X, L, \rho)$ and consider the following algorithm suggested by L.Devroye in [63].

Algorithm 3.2 (Covering with randomly placed balls of variable radius)

1. *Set $Z_1 = A$ and the iteration number $j = 1$.*
2. *Obtain a point x_j by sampling from the distribution P_{Z_j}.*
3. *Evaluate the current value of the objective function $y_j = f(x_j)$ and the corresponding record value*

$$y_{oj} = \min_{i=1,\ldots,j} y_i = \min\{y_j, y_{o,j-1}\}.$$

4. *Compute*

$$\varepsilon_i = (y_i - y_{oj} + \delta)/L \ \text{ for all } \ i = 1, \ldots, j \quad (3.6)$$

and set

$$Z_{j+1} = A \setminus \bigcup_{i=1}^{j} B(x_i, \varepsilon_i, \rho).$$

5. *Check a stopping condition; if the algorithm does not terminate, return to step 2 substituting $j+1$ for j.*

From the definition of the Lipschitz condition, if we use a stopping rule similar to (b) above (that is, if we stop when the set A gets covered by the balls $B(x_i, \varepsilon_i, \rho)$, $i = 1, \ldots, j$), then $y_{oj} - m \le \delta$ which means that Algorithm 3.2 finds the minimum with accuracy δ with respect to the values of $f(\cdot)$.

Note that if we set $\delta = 0$, then Algorithm 3.2 becomes the pure adaptive search of Sect. 2.2.3. Additionally, if instead of using the rule (3.6) we set $\varepsilon_i = \varepsilon$ for all i, then Algorithm 3.2 will become identical to Algorithm 3.1.

We can extend the way we obtain the new points x_j in Algorithms 3.1 and 3.2. We can use any distribution P_j supported on the set Z_j rather than just the distribution P_{Z_j}. For example, if we choose the point x_j as the minimizer of the minorant $\underline{f}_j(x)$ and set $\delta = 0$, then Algorithm 3.2 will become the Shubert-Pijavskij algorithm discussed in Sect. 1.2.1.

In order to ensure the convergence of Algorithm 3.2 (with x_j distributed according to a general P_j), it is assumed in [63] that

$$P_j = \alpha_j P_{Z_j} + (1-\alpha_j)G_j \;\text{ with } \alpha_j \geq 0, \text{ and } \sum_{j=1}^{\infty} \alpha_j = \infty, \tag{3.7}$$

where G_j are arbitrary distributions on A (for example, sampling from G_j may correspond to a local descent from the current record point). Note that this convergence result can be easily deduced from Theorem 2.1. Note also that the representation (3.7) is essentially the same as (2.7) and therefore the theoretical rate of convergence of the corresponding algorithm can be extremely slow; indeed, the discussion at the end of Sect. 2.2.2 about the rates of convergence can be applied in this case.

Algorithm 3.2 with x_j distributed according to P_j, which satisfies (3.7), has been further extended in [63] to the case where $f(\cdot)$ is only assumed to be continuous (not necessarily Lipschitz). In this case, defining ε_i according to $\varepsilon_i = \beta_i(y_i - y_{oj})$, $i = 1, \ldots, j$, at iteration n ensures the convergence of the corresponding algorithm if $\beta_i > 0$ for all i and $\beta_i \to 0$ as $i \to \infty$. This convergence result can also be deduced from Theorem 2.1.

Subsequences of infinite sequences versus n-point sequences

Assume that we need to construct a sequence of n points $X_n = \{x_1, \ldots, x_n\}$ possessing certain uniformity properties. There are two ways of achieving this. Firstly, we could choose the first n points of an infinite sequence $X_\infty = \{x_1, x_2, \ldots\}$ as the n-point sequence $X_n = \{x_1, \ldots, x_n\}$. Secondly, we could construct the n-point sequence $X_n = \{x_1, \ldots, x_n\}$ depending on the value of n. The first way is more practical as the stopping rule n (the number of points required to solve a particular problem) can be sequential and worked out during the search procedure. Using the second way, one can often construct sequences with slightly better uniformity characteristics. A celebrated example that illustrates these two ways of constructing n-point sequences is provided by the Halton and Hammersley sequences; these sequences are discussed in Sects. 3.1.2 and 3.1.3.

If we use independent random sampling of points x_j distributed according to some probability measure P, then we can always assume that we have an infinite sequence of random points and we consecutively choose the points from this infinite sequence.

Extensions of Algorithms 3.1 and 3.2 to the case of arbitrary sequences of points and the k-th records

The points x_j in Algorithms 3.1 and 3.2 were assumed to be random but they do not have to be random. Let us modify these two algorithms so that they can be used in the case where we use the rejection technique to generate the points x_j (they should be distributed according to P_{Z_j}) and the points needed to obtain x_j arrive sequentially from some infinite sequence of points (distributed according to the probability measure P in the sense of the definition (3.25)).

Let $f \in \mathrm{Lip}(A, L, \rho)$, $\delta \geq 0$ be a real number determining the required precision, k be a positive integer and $Z_\infty = \{z_1, z_2, \ldots\}$ be a sequence of points in A. In the algorithm below, $y^*_{k,j}$ denotes the k-th smallest value among $f(x_1), \ldots, f(x_j)$; for $k = 1$ (which is the main special case) we obviously have $y^*_{k,j} = y_{oj} = \min\{f(x_1), \ldots, f(x_j)\}$. The usefulness of the idea of using $k > 1$ rather than just simply $k = 1$ can be justified in a similar way to the one in Sect. 2.2.4.; additionally, using $k > 1$ adds extra security when we bound off the subregions of A (when the Lipschitz constant is unknown).

Assume first that the Lipschitz constant L is known.

Algorithm 3.3 (General covering with unequal balls)

1. *Evaluate the objective function $f(\cdot)$ at the first k points $z_1, \ldots, z_k$ of the sequence Z_∞ and set $i = k$, $j = k$, $x_l = z_l$, $\varepsilon_l = \delta/L$ for $l = 1, \ldots, k$. Compute $y^*_{k,j} = \max\{f(x_1), \ldots, f(x_j)\}$.*
2. *Check a stopping condition. If the algorithm does not stop, take the current point $z_{i+1} \in Z_\infty$.*
3. *If $\rho(z_{i+1}, x_l) \leq \varepsilon_l$ for some $l = 1, \ldots, j$, then set $i \to i+1$ and return to Step 2.*
   Alternatively, if $\rho(z_{i+1}, x_l) > \varepsilon_l$ for all $l = 1, \ldots, j$, then go to Step 4.
4. - *Set $j \to j+1$, $x_j = z_{i+1}$ and evaluate $f(x_j)$.*
   - *If $f(x_j) \geq y^*_{k,j-1}$ then set $y^*_{k,j} = y^*_{k,j-1}$, $\varepsilon_j = (f(x_j) - y^*_{k,j} + \delta)/L$, and return to Step 2 with $i \to i+1$.*
   - *If $f(x_j) < y^*_{k,j-1}$ then recompute $y^*_{k,j}$, the k-th smallest value among $\{f(x_1), \ldots, f(x_j)\}$, set $\varepsilon_l = \delta/L + \max\{0, f(x_l) - y^*_{k,j}\}/L$ for all $l = 1, \ldots, j$ and return to Step 2 with $i \to i+1$.*

There are three natural stopping rules in Algorithm 3.3:

(i) the number i of points taken from the sequence Z_∞ has reached a given number;
(ii) the number j of points, where the objective function $f(\cdot)$ has been evaluated, has reached a given number;
(iii) step 4 has not been realized during a given number of successive iterations.

The stopping conditions (i) and (ii) are versions of the condition (a) above, whereas (iii) is a version of (c).

Algorithm 3.3 aims to construct the coverage $\bigcup_i B(x_i, \varepsilon_i, \rho)$ of A with balls of different radii. If the Lipschitz constant L of $f(\cdot)$ is known then Algorithm 3.3 converges in the sense that

$$A \subseteq \bigcup_{i=1}^{j} B(x_i, \varepsilon_i, \rho) \tag{3.8}$$

implies $\max_{x \in A} f(x) - y_{oj} \le \delta$. In practice, for checking (3.8) one may apply Theorem 3.1.

If the Lipschitz constant L is unknown then it can be estimated. The following estimator of L at iteration j is often used:

$$L_j = (1 + r_j) \max_{1 \le s < t \le j} \frac{|f(x_s) - f(x_t)|}{\rho(x_s, x_t)}, \tag{3.9}$$

where $\{r_j\}$ is a non-increasing sequence of non-negative numbers. Provided that $\liminf_{n \to \infty} L_j \ge L$, the convergence of Algorithm 3.3 is guaranteed.

Note that rather than estimating the overall Lipschitz constant L one may prefer to estimate the Lipschitz constants locally [233] using the same estimator (3.9) but restricting the points x_s, x_t to subregions of A. Using local Lipschitz constants may significantly improve the efficiency of the algorithms.

Algorithm 3.3 (for the case $k = 1$) has been extensively tested [275]; the numerical results show that the efficiency of this algorithm is very high, particularly if the sequence Z_∞ is one of the low-dispersion sequences (see Sect. 3.1.2).

3.1.2 Dispersion

In this section, we return to the problem of covering A with balls of equal radius (not necessarily with random centres) and discuss the concept of dispersion, a very important concept in the theory of global optimization. In our discussion we follow different sources; the main one being [174], Chapt. 6.

Dispersion and the problem of optimal covering

The dispersion (or ρ-dispersion) of a n-point sequence $X_n = \{x_i\}_{i=1}^n$ is

$$d(X_n, A, \rho) = \sup_{x \in A} \min_{1 \le i \le n} \rho(x, x_i). \tag{3.10}$$

If for each n the set $X_n = \{x_1, \ldots, x_n\}$ is constructed from the first n points of the sequence $X_\infty = \{x_1, x_2, \ldots\}$, then we write $d_n(X_\infty, A, \rho)$ for $d(X_n, A, \rho)$.

Clearly, the condition $d_n(X_\infty, A, \rho) \to 0$ as $n \to \infty$ is equivalent to the fact that the sequence of points $X_\infty = \{x_1, x_2, \ldots\}$ is everywhere dense in A (provided that the metric ρ is equivalent to the Euclidean metric). Therefore, the dispersions $d(X_n, A, \rho)$ can be considered as characteristics of the denseness of the points of X_n in A. Also, the dispersions are often considered as characteristics of the uniformity of the sequences X_n.

It is easy to see that $d(X_n, A, \rho) \le \varepsilon$ if and only if (3.2) holds; that is, when A is covered by the union of the balls $\bigcup_{i=1}^{n} B(x_i, \varepsilon, \rho)$.

For given ρ and n, the radius of covering of A is defined as

$$d_* = \inf_{X_n} d(X_n, A, \rho).$$

Coverage by balls of equal radius with centres at $X_n^* = \{x_1^*, \dots, x_n^*\}$ is called optimal if

$$d(X_n^*, A, \rho) = d_* \,.$$

The problem of computing the radius of covering and finding the associated optimal covering by Euclidean balls is a famously difficult optimization problem. In realistic cases it is too difficult to find even reasonable approximations to the optimal coverings. One of the classical mathematical problems, see e.g. [50, 204], is finding the asymptotics (as $n \to \infty$) for the radius of covering of the torus $\mathbb{I}^d$ (see Sect. 3.4.1 for the definition of $\mathbb{I}^d$) and finding the associated asymptotically optimum covering schemes. The case $A = \mathbb{I}^d$ should be considered as relatively easy: in this case there is no edge effect to be taken care of. However, even this problem is very difficult and remains unsolved for dimensions $d > 3$. The problem of optimal covering is even more difficult when n is fixed. Even in the case of the torus, the problem has only been solved in the trivial case where $d = 1$ and in the case $d = 2$, where the solution is given by the circumscribed circles of a hexagonal tiling of $\mathbb{R}^2$.

The ρ-dispersion of X_n can also be defined as

$$d(X_n, A, \rho) = \sup_{B(z,\varepsilon,\rho)\cap X_n = \emptyset} \varepsilon, \tag{3.11}$$

where the supremum is taken over all the balls $B(z, \varepsilon, \rho) \subset A$ that do not contain points from X_n. The maximum (multivariate) spacing of X_n with respect to a convex set $B \subset \mathbb{R}^d$ is defined as the largest possible subset $x + rB$ of A which does not intersect X_n. Therefore, the ρ-dispersion can be defined as the radius of the maximum spacing with respect to the unit ball $B = B(0, 1, \rho)$; see also Sect.2.2.2.

Importance of the concept of dispersion in global optimization

The following property explains the importance of the ρ-dispersion in global optimization.

Theorem 3.2. *If (A, ρ) is a compact metric space and $f(\cdot)$ is a continuous function on A, then for any n-point sequence X_n in A with dispersion $d_n = d(X_n, A, \rho)$, we have*

$$y_{on} - m \le \omega(f; d_n) \,. \tag{3.12}$$

Here $\omega(f;\cdot)$ is the modulus of continuity (with respect to ρ) of $f(\cdot)$:

$$\omega_\rho(f;t) = \sup_{\substack{u,v\in A \\ \rho(u,v)\le t}} |f(u) - f(v)| \quad \mathit{for}\; t \ge 0\,; \tag{3.13}$$

if $f \in \mathrm{Lip}(A, L, \rho)$ then $\omega_\rho(f;t) = Lt$.

To prove (3.12), choose $\varepsilon > 0$ and let $y \in A$ be such that $f(y) < m + \varepsilon$. For some k with $1 \le k \le n$, we have $\rho(y, x_k) = \min_{1\le j\le n} \rho(y, x_n)$, where

$x_1, \ldots, x_n$ are the nodes of X_n. It follows that $\rho(y, x_k) \leq d_n$. Furthermore, we have $f(x_k) - f(y) \leq \omega(f; d_n)$, and so

$$m + \varepsilon > f(y) \geq f(x_k) - \omega(f; d_n) \geq m - \omega(f; d_n),$$

which implies (3.12).

Theorem 3.2 implies that if one has a continuous function $f(\cdot)$ on A and intends to find a point close to x_*, the minimizer of $f(\cdot)$, by evaluating values of $f(\cdot)$ at some n points of A to be chosen in the non-adaptive fashion, then they would naturally try to select a sequence X_n with the smallest value of $d(X_n, A, \rho)$. Indeed, if $f \in \text{Lip}(A, L, \rho)$ for some L and n is fixed, then the sequence X_n^* with the minimal value of $d(X_n, A, \rho)$ provides the worst-case optimal global optimization algorithm, which minimizes the value $\sup_f |y_{on} - m|$ in the set of all deterministic algorithms, both sequential and non-sequential. (See Sect. 1.2.1 for more discussion on this topic.)

Dispersions in $A = [0, 1]^d$

Assume that $A = [0, 1]^d$. Then the following two metrics are usually considered in conjunction with dispersion: the standard Euclidean metric ρ_2 and the maximum metric ρ_∞:

$$\rho_2(u, v) = \Big(\sum_{i=1}^{d} (u_i - v_i)^2\Big)^{1/2}, \quad \rho_\infty(u, v) = \max_{1 \leq i \leq d} |u_i - v_i| \qquad (3.14)$$

for $u = (u_1, \ldots, u_d)$ and $v = (v_1, \ldots, v_d)$ in A.

We write $d_n(X_n)$ for the ρ_2-dispersion of a n-point set $X_n = \{x_1, \ldots, x_n\}$ and $d'_n(X_n)$ for the ρ_∞-dispersion of X_n. That is,

$$d_n(X_n) = d(X_n, A, \rho_2), \quad d'_n(X_n) = d(X_n, A, \rho_\infty).$$

If the n-point sequences $X_n = \{x_1, \ldots, x_n\}$ are formed by the first n points of an infinite sequence $X_\infty = \{x_1, x_2, \ldots\}$, then the following notation will be used: $d_n(X_\infty) = d_n(X_n)$ and $d'_n(X_\infty) = d'_n(X_n)$.

Since $\rho_\infty(u, v) \leq \rho_2(u, v) \leq \sqrt{d}\, \rho_\infty(u, v)$ for all $u, v \in \mathbb{R}^d$, it follows that

$$d'_n(X_n) \leq d_n(X_n) \leq \sqrt{d}\, d'_n(X_n). \qquad (3.15)$$

Therefore, the two dispersions have the same order of magnitude as $n \to \infty$.

We have the lower bounds

$$d_n(X_n) \geq \Big(\Gamma(d/2 + 1)\Big)^{1/d} \pi^{-1/2} n^{-1/d}, \quad d'_n(X_n) \geq \frac{1}{2} n^{-1/d}. \qquad (3.16)$$

It is not known (for $d \geq 2$) whether the first inequality in (3.16) is sharp. As shown by Sukharev in [235], the second inequality is sharp (at least, for some n) and the corresponding sequences X_n are constructed as unions of the cubic grids (3.5) with $k = \lfloor n^{1/d} \rfloor$ and sets of arbitrary $n - k^d$ points in A; additionally, no other types of sequences X_n attain the lower bound in (3.16) for d'_n.

Note that one can easily transform point sets and sequences with small dispersions from one domain to another. In doing so we can use the following simple result (see [174], Theorem 6.4): if $X_n = \{x_1, \ldots, x_n\} \subset A$ and $T : (A, \rho) \to (A', \rho')$ is a map from the bounded metric space (A, ρ) onto the metric space (A', ρ') such that there exists a constant $L \geq 0$ with $\rho'(T(x), T(z)) \leq L\rho(x, z)$ for all $x, z \in A$, then $d(X'_n, A', \rho') \leq Ld(X_n, A, \rho)$, where $X'_n = \{T(x_1), \ldots, T(x_n)\} \subset A'$.

Sequences with small asymptotic dispersion

The inequalities (3.16) imply that as $n \to \infty$, the minimal order of decrease of the dispersions $d_n(X_n)$ and $d'_n(X_n)$ is at least $O(n^{-1/d})$. A sequence $X_\infty = \{x_1, x_2, \ldots\}$ of points in A is called a *low-dispersion sequence* if $d_n(X_\infty) = O(n^{-1/d})$ as $n \to \infty$. A few families of low-dispersion sequences X_∞ are known, see e.g. [116, 174] and the end of this section.

The sequence of random points (that is, the sequence $X_\infty = \{x_1, x_2, \ldots\}$ with uniformly distributed i.i.d. random points $x_i \in A$) is not a low-dispersion sequence. Indeed, the result of Deheuvels [62] implies that for the sequence X_∞ of random points in $[0, 1]^d$, the rate of decrease of the dispersion $d_n(X_\infty)$ is $(\ln n)^{1/d} n^{-1/d}$, with probability 1. This is also a consequence of (2.25).

An important achievement in the direction of constructing low-dispersion sequences is summarized in the following statement (which combines Theorems 6.7 and 6.9 of [174]): if $A = [0, 1]^d$ $(d \geq 1)$ then there exists a sequence X_∞ of points in A with

$$\lim_{n \to \infty} n^{1/d} d'_n(X_\infty) = \frac{1}{\ln 4} \simeq 0.7213\,; \tag{3.17}$$

if $A = [0, 1]$, then for any X_∞

$$\limsup_{n \to \infty} n d_n(X_\infty) \geq \frac{1}{\ln 4}\,. \tag{3.18}$$

The inequality (3.18) implies that for $d = 1$ the constant $1/\ln 4$ for $\lim_{n \to \infty} n^{1/d} d'_n(X_\infty)$ is the best possible; it may not however be the best possible for any $d > 2$ (in view of the second inequality in (3.16) this constant belongs to the interval $[1/2, 1/\ln 4]$).

A sequence $X_\infty = \{x_1, x_2, \ldots\}$ of points in $[0, 1]$, where the inequality (3.18) becomes an equality is:

$$x_1 = 1, \;\; x_j = \{\ln(2j - 3)/\ln 2\} \;\; (j = 2, 3, \ldots), \tag{3.19}$$

where $\{\cdot\}$ is the fractional part operation. For this sequence,

$$d_n(X_\infty) = \frac{\ln n - \ln(n-1)}{\ln 4}, \quad \forall\, n \geq 2$$

implying $\lim_{n\to\infty} n d_n(X_\infty) = 1/\ln 4$. Thus, the sequence X_∞ consisting of points (3.19) is the sequence with the smallest possible asymptotic dispersion, in the sense of (3.18).

The d-dimensional sequences X_∞, where the r.h.s. in (3.17) is attained, are constructed on the base of the sequence (3.19) in the following way: for any K, such that $K = k^d$ for some integer $k \geq 1$, the set X_K, containing the first K points of X_∞, is precisely the set of all points of the form $x = (v_1, \ldots, v_d)$ with $v_j \in \{x_1, \ldots, x_k\}$ $(j = 1, \ldots, d)$, where $x_1, \ldots, x_k$ are the first k points of the sequence (3.19). The structure of the sequence X_∞ constructed in this way is rather peculiar. However, as shown in [173], its asymptotic dispersion is much smaller than the dispersion of all known low-discrepancy sequences such as the so-called (t, s)-sequences; for the definition of a low-discrepancy sequence, see Sect. 3.1.3.

Characteristics related to dispersion

There are, of course, other uniformity characteristics related to covering. Let us rewrite the definition of dispersion using the distance between a point $x \in A$ and a set $U \subset A$:

$$\rho(x, U) = \inf_{z \in U} \rho(x, z).$$

Then the ρ-dispersion (3.10) can be rewritten as

$$d(X_n; A; \rho) = \sup_{x \in A} \rho(x, X_n). \tag{3.20}$$

Using this definition, we can consider different modifications of the ρ-dispersion. In particular, we can define

$$d_\alpha(X_n; A; \rho) = \int_{x \in A} (\rho(x, X_n))^{\alpha - 1} \mu(dx), \ \ \alpha > 0\,, \tag{3.21}$$

where averaging over A (with respect to some measure μ) is made rather than maximization, as in (3.20). The family of characteristics (3.21) reflect the uniformity of a sequence X_n better than the discrepancy itself (which is not really a characteristic of uniformity, see the discussion at the end of Sect. 3.1.3).

We can express the ρ-dispersion (3.20) as the limit

$$d(X_n; A; \rho) = \lim_{\alpha \to \infty} \left[d_\alpha(X_n; A; \rho) \right]^{1/\alpha},$$

where we assume that μ in (3.21) is the Lebesque measure on A. The asymptotic behaviour of the criterion (3.21) is well-studied in the case where $\alpha = 2$ and the sequences X_n are formed by the lattice points, see e.g. [50].

Consider a one-dimensional case where $A = [0,1]$, $\rho(x,z) = |x-z|$ and μ is the standard Lebesgue measure (that is, the uniform distribution on A). Assume that the points of X_n are arranged in non-decreasing order and include 0 and 1: $0 = x_1 \le x_2 \le \ldots \le x_n = 1$. Then (3.21) can be rewritten as

$$d_\alpha(X_n; A; \rho) = \int_0^1 (\rho(x, X_n))^{\alpha-1} dx = \sum_{i=1}^{n-1} \int_{x_i}^{(x_i+x_{i+1})/2} (x - x_i)^{\alpha-1} dx$$

$$+ \sum_{i=1}^{n-1} \int_{(x_i+x_{i+1})/2}^{x_{i+1}} (x_{i+1} - x)^{\alpha-1} dx = \frac{2^{1-\alpha}}{\alpha} S_\alpha(X_n)$$

where

$$S_\alpha(X_n) = \sum_{i=1}^{n-1} (x_{i+1} - x_i)^\alpha .$$

For the uniform grid $X_{n+1} = \{0, \frac{1}{n}, \ldots, \frac{n-1}{n}, 1\}$ we have $S_\alpha(X_{n+1}) = n^{1-\alpha}$ for all $n > 1$. In the case where the points of X_n are i.i.d.r.v. with uniform distribution on $[0,1]$ (with added points 0 and 1), we have (see [67])

$$\lim_{n\to\infty} n^{\alpha-1} S_\alpha(X_n) = \Gamma(\alpha+1), \text{ with probability one,}$$

where $\Gamma(\cdot)$ is the Gamma-function. This implies, in particular, that random points on an interval have the smallest possible order of decrease of the characteristic $d_\alpha(X_n; A; \rho)$ defined in (3.21), for any $\alpha > 0$.

The Rényi and Tsallis entropies of order α ($\alpha \ge 0$, $\alpha \ne 1$) of the partition of $[0,1]$ generated by $X_n = \{x_1, \ldots, x_n\}$ are expressed through $S_\alpha(X_n)$, the sum of α-th powers of $x_{i+1} - x_i$, as follows:

$$H_\alpha^{\text{Rén}}(X_n) = \frac{1}{1-\alpha} \ln S_\alpha(X_n), \quad H_\alpha^{\text{Ts}}(X_n) = \frac{1 - S_\alpha(X_n)}{1-\alpha}. \tag{3.22}$$

As $\alpha \to 1$, the limiting value for both entropies is the Shannon entropy $H_1(X_n) = -\sum_{i=1}^{n-1} (x_{i+1} - x_i) \ln(x_{i+1} - x_i)$. The Rényi and Tsallis entropies are very important tools in many applied areas including information theory, physics, statistical mechanics, dynamical systems, probabilistic number theory and dynamical search, see e.g. [53, 101, 128, 193].

Halton and Hammersley sequences and their dispersion

Many low-discrepancy sequences are known. Among them, the Halton sequences (along with the associated Hammersley sequences) are the most known. These sequences are not the best but deserve consideration as they are very simple and classical.

Let an integer $R \geq 2$ be fixed. Then any non-negative integer K may be uniquely represented as

$$K = \sum_{j=0}^{M} a_j R^j, \quad \text{where } 0 \leq a_j \leq R-1, \ 0 \leq M < \infty . \tag{3.23}$$

Define the mapping $\phi_R : \{0, 1, \ldots\} \to [0, 1]$ by

$$\phi_R(K) = \sum_{j=0}^{M} a_j R^{-j-1},$$

where K, R, M and a_j $(j = 0, 1, \ldots, M)$ are as defined in (3.23).

The d-dimensional Halton sequences are the sequences $X_\infty^{(\mathrm{Halt})} = \{x_1, x_2, \ldots\}$ consisting of the points

$$x_{j+1} = (\phi_{R_1}(j), \phi_{R_2}(j), \ldots, \phi_{R_d}(j)), \quad j = 0, 1, \ldots, \tag{3.24}$$

where R_i, $i = 1, \ldots, d$, are pairwise relatively prime (usually taken to be the first d primes 2,3,5,...) and $\min_i R_i \geq 2$. In the case $d = 1$ the sequences (3.24) are called van der Corput sequences.

For given n, the d-dimensional Hammersley sequences are the sequences $X_n^{(\mathrm{Hamm})} = \{x_1, \ldots, x_n\}$ consisting of the points

$$x_{j+1} = \left(\frac{j}{n}, \phi_{R_1}(j), \ldots, \phi_{R_{d-1}}(j)\right), \quad j = 0, 1, \ldots, n-1,$$

where R_i, $i = 1, \ldots, d-1$, are pairwise relatively prime (usually taken to be the first $d-1$ primes) and $\min_i R_i \geq 2$. Unlike the Halton sequences which are infinite, the Hammersley sequences are finite and explicitly depend on n, the number of points.

Let $A = [0, 1]^d$, $d \geq 2$, and let $R_1, \ldots, R_d$ be integers ≥ 2 that are pairwise relatively prime. Then we have the following bounds for the dispersions of the Halton and Hammersley sequences:

$$d'_n(X_n^{(\mathrm{Hamm})}) < d'_n(X_\infty^{(\mathrm{Halt})}) < n^{-1/d} \max_{1 \leq i \leq d} R_i \quad \text{for all } n \geq 1;$$

$$d_n(X_n^{(\mathrm{Hamm})}) < d_n(X_\infty^{(\mathrm{Halt})}) < n^{-1/d} \left[R_1^4 + \sum_{i=2}^{d} R_i^2 \right]^{1/2} \quad \text{for } n \geq \prod_{i=1}^{d} R_i .$$

Both the Halton and Hammersley sequences possess the minimal order of magnitude (which is $O(n^{-1/d})$ as $n \to \infty$) for the dispersions. Of course, the respective constants in $O(n^{-1/d})$, $n \to \infty$, are not optimal for these sequences and are much larger than the constant $1/\ln 4$ (for the dispersion d'_n) in the r.h.s. of (3.18).

3.1.3 Uniform Sequences and Discrepancies

In this section, we discuss the concept of discrepancy. There is extensive literature devoted to the study of discrepancies, see e.g. [66, 116, 174]. For us, the main importance of discrepancy is related to the fact that there is a link between the discrepancies and dispersions. This section tries to explore and explain this link.

Uniform sequences

Let P be a probability distribution on A. An infinite sequence $X_\infty = \{x_1, x_2, \ldots\} \subset A$ is called uniform on A (with respect to P) if

$$\lim_{n\to\infty} \frac{1}{n} \sum_{j=1}^{n} f(x_j) = \int_A f(x) P(dx) \tag{3.25}$$

for any Riemann-integrable function $f(\cdot)$ given on A. If the elements x_n of the sequence X_∞ are random then the convergence with probability one is assumed in (3.25).

One can check whether a particular sequence is uniform by testing the validity of (3.25) for the classes of functions that are more narrow than the class of all Riemann-integrable functions on A. In particular, a sequence X_∞ is uniform if (3.25) holds for the class of continuous functions on A or even for the family of indicator functions

$$\mathbf{1}_U(x) = \begin{cases} 1 & \text{if } x \in U, \\ 0 & \text{otherwise} \end{cases} \tag{3.26}$$

for all Jordan-measurable subsets U of $A \subset \mathbf{R}^d$ (Jordan-measurability of U means that $U \in \mathcal{B}$ and the volume of the boundary of U is zero). In this case (3.25) becomes

$$\lim_{n\to\infty} \frac{1}{n} \sum_{j=1}^{n} \mathbf{1}_U(x_j) = P(U) \tag{3.27}$$

for all Jordan-measurable $U \subset A$.

In the case where $A = [0,1]^d$ and P is the uniform distribution on A we may only consider the family of hyper-rectangulars of the form

$$[0,b) = [0,b_1) \times \ldots [0,b_d), \quad 0 < b_i \leq 1 \ \ (i = 1, \ldots, d)\,. \tag{3.28}$$

The condition of uniformity (3.25) can then be written as

$$\lim_{n\to\infty} \frac{\{\text{number of } x_j, 1 \leq j \leq n, \text{ such that } x_j \in [0,b)\ \}}{n} = b_1 \times \ldots \times b_d \tag{3.29}$$

for all $b = (b_1, \ldots, b_d)$, $0 < b_i \leq 1 \ \ (i = 1, \ldots, d)$. That is, if $A = [0,1]^d$ and P is the uniform distribution on A then the sequence X_∞ is uniform on A if and only if the asymptotic relation (3.29) is valid.

Finite sequences

In practice, the number of elements x_n to be used is always finite and one is usually interested in the uniformity of a finite sequence $X_n = \{x_1, \ldots, x_n\}$, where n can be either fixed or not. If n is not fixed it is more reasonable to talk about a family $\{X_n\}_{n=1}^{\infty}$ of n-point sequences rather than of one sequence.

Unlike the case of infinite sequences, where the concept of uniformity is strictly defined, in the finite case the meaning of this concept is less clear. There is no general definition for the uniformity of n-point sequences (for fixed n) but there are many uniformity characteristics. These characteristics include the various discrepancies considered below.

One can use sequences with good uniformity characteristics for different purposes. The most widely used purpose is the estimation of integrals:

$$\int_A f(x)P(dx) \simeq \frac{1}{n}\sum_{j=1}^{n} f(x_n), \tag{3.30}$$

where $f(\cdot)$ is a function from some functional class $\mathcal{F}$. In view of the similarity with the Monte Carlo method, deterministic sequences used for the estimation of integrals as in (3.30) are often called quasi-random sequences and the elements which they consist of are called quasi-random points.

If the functional class $\mathcal{F}$ consists of indicator functions, the approximation formula (3.30) has the form

$$\frac{1}{n}\sum_{j=1}^{n} \mathbf{1}_U(x_n) \simeq P(U), \quad U \in \mathcal{M}, \tag{3.31}$$

where $\mathcal{M} \subset \mathcal{B}$ is a family of subsets of A.

Discrepancies

Discrepancies are characteristics of uniformly measuring the precision of the estimators (3.31). The forms of the discrepancies depend on the choice of $\mathcal{M}$ and on the approach used to compute the inaccuracy of the estimators. The worst-case and the Bayesian approaches are the standard approaches most often used.

Using the worst-case approach, we obtain the following definition: a discrepancy of an n-point sequence $X_n = \{x_1, \ldots, x_n\}$ with respect to a family of sets $\mathcal{M} \subset \mathcal{B}$ is

$$D_n(\mathcal{M}, X_n, P) = \sup_{U \in \mathcal{M}} \left| \frac{1}{n}\sum_{j=1}^{n} \mathbf{1}_U(x_n) - P(U) \right| . \tag{3.32}$$

By suitable specializations of the family $\mathcal{M}$ in (3.32), we obtain different discrepancies. The two most important are the so-called 'star discrepancy' and 'extreme discrepancy'; the latter is often referred to simply as 'discrepancy'.

Let A be $[0,1]^d$. The *star discrepancy* $D_n^*(X_n) = D_n^*(x_1,\ldots,x_n)$ of the point set X_n is $D_n^*(X_n) = D_n(\mathcal{I}^*, X_n, \mu)$, where μ is the Lebesgue measure (uniform distribution) on A and $\mathcal{I}^*$ is the family of all subintervals of $[0,1]^d$ of the form (3.28). We obtain the definition of extreme discrepancy $D_n(X_n) = D_n(x_1,\ldots,x_n)$ if we use the set of hyper-rectangles of the form

$$[a,b) = [a_1,b_1) \times \ldots [a_d,b_d), \;\; 0 < a_i \le b_i \le 1 \;\; (i=1,\ldots,d) \tag{3.33}$$

as the family of sets $\mathcal{M}$ in (3.32). That is, $D_n(X_n) = D_n(\mathcal{M}, X_n, \mu)$, where μ is the Lebesgue measure on $A = [0,1]^d$ and $\mathcal{M}$ is the set of hyper-rectangles (3.33). The star discrepancy $D_n^*(X_n)$ and the extreme discrepancy $D_n(X_n)$ are related by the inequalities

$$D_n^*(X_n) \le D_n(X_n) \le 2^d D_n^*(X_n), \;\; \forall n, \tag{3.34}$$

so that the order of their decrease is the same as $n \to \infty$. Other discrepancies are often also of interest, see [66, 116, 174].

Using the criterion of uniformity based on (3.29), we can reformulate the definition of uniformity as follows: a sequence X_∞ is uniform on A if and only if $D_n^*(X_\infty) \to 0$ as $n \to \infty$ (or $D_n(X_\infty) \to 0$ as $n \to \infty$).

The star discrepancy characterizes not only the error of the approximation (3.31) for the family of sets (3.28), it also bounds the error of the approximation (3.30) for the set of continuous functions. More precisely, we have the following classical result (see e.g. [174]): if $A = [0,1]^d$, $\rho = \rho_\infty$ and $f(\cdot)$ is a continuous function on A, then for any $X_n = \{x_1,\ldots,x_n\} \subset A$ we have

$$\left| \frac{1}{n} \sum_{j=1}^{n} f(x_n) - \int_A f(x)dx \right| \le 4\omega_{\rho_\infty}(f; D_n^*(X_n)^{1/d}), \tag{3.35}$$

where $\omega_\rho(f;\cdot)$ is the modulus of continuity of $f(\cdot)$ defined in (3.13). If $d=1$, then the multiplier 4 in the right-hand side of (3.35) can be dropped.

Low-discrepancy sequences

It is widely believed (but not yet proven for $d \ge 3$) that, in the d-dimensional case with $A = [0,1]^d$, the star discrepancy of any n-point sequence X_n and any infinite sequence X_∞ satisfies

$$D_n^*(X_n) \ge Bn^{-1}(\ln n)^{d-1}, \;\; D_n^*(X_\infty) \ge Bn^{-1}(\ln n)^d, \tag{3.36}$$

where the constant B depends only on d.

Correspondingly, a sequence X_∞ is called a low-discrepancy sequence if $D_n^*(X_\infty) = O((\ln n)^d/n)$ as $n \to \infty$. Similarly, a family of n-point sequences $\{X_n\}_n$ is called low-discrepancy (family of sequences) if $D_n^*(X_n) = O((\ln n)^{d-1}/n)$ as $n \to \infty$.

Random sequences X_∞ consisting of i.i.d. uniform random points are not low-discrepancy sequences. Indeed, the law of the iterated logarithm proved by J.Kiefer [137] states that

$$\limsup_{n\to\infty} \frac{\sqrt{2n}D_n^*(X_\infty)}{\sqrt{\ln\ln n}} = 1, \quad \text{with probability 1.}$$

That is, for the random sequences X_∞ we have $D_n^*(X_\infty) = O(\sqrt{\ln\ln n}/\sqrt{n})$ as $n \to \infty$, with probability 1.

Many low-discrepancy sequences and families of sequences are known, see [116, 174]. An example of a low-discrepancy sequence is the Halton sequence $X_\infty^{(\mathrm{Halt})}$, whose star-discrepancy satisfies

$$D_n^*(X_\infty^{(\mathrm{Halt})}) \le V(R_1,\dots,R_d)n^{-1}(\ln n)^d + O(n^{-1}(\ln n)^{d-1}) \quad \text{for all } n \ge 2,$$

where $V(R_1,\dots,R_d) = \prod_{i=1}^d (R_i-1)/(2\ln R_i)$. The minimum value of this coefficient is obtained when $R_1,\dots,R_d$ are the first d primes.

Similarly, for the Hammersley n-point sequence $X_n^{(\mathrm{Hamm})}$ we have

$$D_n^*(X_n^{(\mathrm{Hamm})}) \le V(R_2,\dots,R_{d-1})n^{-1}(\ln n)^{d-1} + O(n^{-1}(\ln n)^{d-2}).$$

The coefficients $V(\cdot,\dots,\cdot)$ in $O(n^{-1}(\ln n)^v)$ (where $v = d$ or $d-1$ and $n \to \infty$) increase super-exponentially as $d \to \infty$ for the Holton and Hammersley sequences. This makes these sequences practically useless for large dimension d. For large d, other low-discrepancy sequences (with much smaller coefficients in the leading terms) are recommended, see [116, 174]. However, neither of these sequences can perform well if the dimension d is very large.

Relations between discrepancies and dispersion

Assume that $A = [0,1]^d$ and ρ is either ρ_2 or ρ_∞, see (3.14). There is the following relationship between the dispersion d_n' and the extreme discrepancy D_n:

$$d_n'(X_n) \le \frac{1}{2}D_n(X_n)^{1/d} \quad \text{for any } X_n = \{x_1,\dots,x_n\} \subset A. \tag{3.37}$$

To prove (3.37), let B be the largest subcube of $A = [0,1]^d$ such that $B \cap X_n = \emptyset$. Then $\mathrm{vol}(B) \le D_n(X_n)$ and $2d_n'(X_n) \le [\mathrm{vol}(B)]^{1/d}$. This implies (3.37).

Using the inequalities (3.15), (3.34) and (3.37), we may write a few more upper bounds for the dispersions through discrepancies. These upper bounds imply that if the discrepancy is small, then the dispersion of this sequence cannot be large. In particular, if a sequence X_∞ is uniform (that is, $D_n(X_\infty) \to 0$ as $n \to \infty$), then $d_n(X_\infty) \to 0$ as $n \to \infty$.

Questions naturally arise as to whether there are useful upper bounds for the discrepancies through the dispersions and whether the relation $d_n(X_\infty) \to 0$ implies $D_n(X_\infty) \to 0$ (as $n \to \infty$). The answers to both questions are negative. The reason is that the discrepancies are global characteristics of uniformity whereas the dispersions are local. For example, the star discrepancy $D_n^*(X_n)$ is the L_∞-distance between the multivariate c.d.f. of the uniform distribution on A and the empirical c.d.f. related to the sample X_n. In mathematical statistics, $D_n^*(X_n)$ is known as the Kolmogorov-Smirnov statistic.

Consider the one-dimensional case with $A = [0,1]$. Let $X_n = \{x_1, \ldots, x_n\}$ be a set of n points such that $0 = x_1 \leq x_2 \leq \ldots \leq x_n = 1$. The values $p_i = x_{i+1} - x_i$ are called spacings ($i = 1, \ldots, n-1$). The dispersion of X_n is $d_n(X_n) = \max_{1 \leq i < n} p_i$. Provided that we keep the maximum spacing between the points of X_n fixed, we can move other points of X_n freely; this will not affect the dispersion but will affect the discrepancy.

Furthermore, consider X_n and some other n-point sequence X'_n constructed so that the sets of the spacings corresponding to the sequences X_n and X'_n are the same. Then all the characteristics (3.21), including the dispersion, are the same for both sequences, but the discrepancies of X_n and X'_n are generally different. The largest value of the star discrepancy, for a fixed value of any of the characteristics (3.21), is attained when the spacings p_i ($i = 1, \ldots, n-1$) are arranged in the order of either increase or decrease.

A specific example showing different sides of dispersion and discrepancy is the sequence $X_\infty = \{x_1, x_2, \ldots\}$ consisting of points (3.19). This is the sequence with the smallest asymptotic dispersion. However, this sequence is not uniform as $D_n(X_\infty)$ does not tend to 0 as $n \to \infty$. This discussion extends to the d-dimensional case, for the sequence X_∞ (considered in Sect. 3.1.2), where the r.h.s. in (3.17) is attained.

We conclude this discussion by stating that unlike discrepancies, which are characteristics of the uniformity of points, dispersions are more characteristics of the denseness than of the uniformity.

3.2 Comparison of Stratified and Independent Sampling Schemes

This section is based on the results of [69, 138] (see also Sect. 4.2.8 in [271], Sect. 4.4 in [273] and [274]) and establishes the superiority of the stratified sampling procedure over the simplest independent sample in global optimization problems.

3.2.1 Stratified Sampling

Definition of sampling schemes

Let P be a probability measure on A. Consider a partition $\mathcal{R}_k$ of A into k disjoint connected subsets of positive measure:

$$\mathcal{R}_k : A = \bigcup_{i=1}^{k} A_i, \quad A_i \subseteq A, \; q_i = P(A_i) > 0 \; \text{ for } i = 1, \ldots, k, \quad A_i \cap A_j = \emptyset \; (i \neq j).$$

Since P is a probability measure, we have $\sum_{i=1}^{k} q_i = 1$. Let us define the probability measures P_i on A_i by

$$P_i(B) = P(B \cap A_i)/q_i \text{ for } B \subseteq A$$

(of course, we assume $B \in \mathcal{B}$). Given a partition $\mathcal{R}_k$ and a collection of integers $L = \{l_1, \ldots, l_k\}$ such that $\sum_{i=1}^{k} l_i = n$, the *stratified sample* $X_{k,L}$ can be defined as

$$X_{k,L} = (x_{1,1}, \ldots, x_{1,l_1}, \ldots, x_{k,1}, \ldots, x_{k,l_k}), \tag{3.38}$$

where for each $i=1, \ldots, k$, the points $x_{i,1}, \ldots, x_{i,l_i}$ are random, independent and distributed according to the distribution P_i on A_i. (In practice, an additional randomization of the order of appearance of $x_{i,j}$ can be useful as well.) We shall call the stratified sample (3.38) proper if the number of points in A_i is proportional to $q_i = P(A_i)$, that is

$$l_i = nq_i \quad \text{for all } i = 1, \ldots, k. \tag{3.39}$$

The joint probability distribution of the random vector (3.38) is

$$Q_{k,L}(d\,X_{k,L}) = \prod_{i_1=1}^{l_1} P_1(dx_{1,i_1}) \times \ldots \times \prod_{i_k=1}^{l_k} P_k(dx_{k,i_k}).$$

The pair $\mathcal{P}_{k,L} = (f_*[X_{k,L}], Q_{k,L})$ with $k > 1$ corresponds to the stratified sampling on A, and $\mathcal{P}_1 = \mathcal{P}_{1,n} = (f_*[X_{1,n}], Q_{1,n})$ corresponds to the independent sampling from the distribution P. Here for a given function $f(\cdot)$ and a sample X we denote the record value of $f(\cdot)$ as $f_*[X] = \min_{x_i \in X} f(x_i)$; clearly, for any X we have $f_*[X] \geq m = \min_{x \in A} f(x)$.

Dominance criteria

Let $\mathcal{F}$ be a class of functions on A and $\Phi_f(\mathcal{P})$ be a comparison criterion for the procedures $\mathcal{P}$ for a fixed $f \in \mathcal{F}$. According to the general concept of domination, we say that $\mathcal{P}$ dominates $\mathcal{P}'$ in $\mathcal{F}$ if $\Phi_f(\mathcal{P}) \leq \Phi_f(\mathcal{P}')$ for every $f \in \mathcal{F}$ and there exists at least one function $f^* \in \mathcal{F}$ such that $\Phi_{f^*}(\mathcal{P}) < \Phi_{f^*}(\mathcal{P}')$.

Below, we consider two related dominance criteria:

(i) the p-th moment of the difference $f_*[X] - m$, where $p > 0$ is fixed and the sample size n tends to infinity (asymptotic dominance), and
(ii) the c.d.f. of the record value $f_*[X]$ (stochastic dominance).

The dominance of the stratified sampling with respect to the criterion (i) with $p = 2$ will imply, in particular, that the linear estimators of Sect. 2.4.1 constructed on the base of the stratified sample have smaller mean square error than the same estimators constructed from the independent sample, see Sect. 3.2.2 for the related discussion.

An important feature of the criterion (ii) is that it is a finite sample (rather than asymptotic) criterion. The dominance of the stratified sampling over the independent sampling with respect to the criterion (ii) means that for any sample size n the random variable $f_*[X] - m$ is concentrated closer to zero in the case of the proper stratified sampling. This means, roughly speaking, that for any objective function and any n, the gap between $m = \min f$ and the record value $f_*[X]$ is smaller (in the sense of the stochastic domination) when we use the proper stratified sample; the same is true for the distance between the minimizer x_* and its approximation.

3.2.2 Asymptotic Criteria

In the present section we only consider the proper stratified sampling procedures $\mathcal{P}_{k,l} = \mathcal{P}_{k,L}$, where $P(A_i) = 1/k$ and $l_i = l$ for all $i = 1, \ldots, k$. We also assume that $n = kl$, l=const, and $k \to \infty$; that is, the number of subsets in the partition $\mathcal{R}_k$ tends to infinity but the number of points in each subset stays constant.

As a criterion for the comparison of the procedures, consider the p-th moment of the random variable $f_*[X_{k,l}] - m$, where $m = \min_{x \in A} f(x)$ and $X_{k,l} = X_{k,L}$ for $L = (l, \ldots, l)$:

$$\Phi_f(k,l) = \mathbf{E}(f_*[X_{k,l}] - m)^p, \quad p > 0. \tag{3.40}$$

Theorem 3.3. *Assume that the measure P and the functional class $\mathcal{F}_*$ of continuous functions $f(x) = \varphi(x - x_*)$ with a unique global minimizer $x_*(f)$ are such that the condition (2.42), for $F = F_f$, is satisfied with some $\alpha > 0$ and the point x_* has a certain distribution $R(dx)$ on A, which is equivalent to the standard Lebesgue measure on A. Let also n=kl, l=const, $k \to \infty$. Then for every $p > 0$ with R-probability one*

$$\frac{\mathbf{E}(f_*[X_{k,l}] - m)^p}{\mathbf{E}(f_*[X_{1,n}] - m)^p} = r(l, p/\alpha) + o(1), \quad n \to \infty, \tag{3.41}$$

where

$$r(l, p/\alpha) = l^{p/\alpha} \Gamma(l+1)/\Gamma(p/\alpha + l + 1).$$

Proof of the theorem is given in Sect. 3.6. A key result in the proof is the asymptotic formula (3.90), where an extreme value distribution for the stratified sampling is derived. The c.d.f. of this extreme value distribution is

$$\Psi_{\alpha,l}(z) = \begin{cases} 0 & \text{for } z \le 0 \\ z^\alpha / l & \text{for } 0 < z \le l^{1/\alpha}. \end{cases} \tag{3.42}$$

This c.d.f. depends on two parameters: α, the tail index, and l, the number of points in each strata. The formulae (3.42) and (2.40) imply that as $l \to \infty$

the asymptotic distribution for the properly normalized record value $f_*[X_{k,l}]$ in the case of a stratified sample is the same as in the case of an independent sample.

For fixed l, p and α, the value $r(l, p/\alpha)$ defines the asymptotic inefficiency of the independent sampling procedure with respect to the stratified sampling according to the dominance criterion (i) above. One can easily verify that $r(l,\beta) < 1$ for all $l, \beta>0$, the function $r(l,\beta)$ is strictly increasing as a function of l, strictly decreasing as a function of β and $\lim_{\beta\to 0} r(l,\beta) = \lim_{l\to\infty} r(l,\beta) = 1$, see Fig. 3.1.

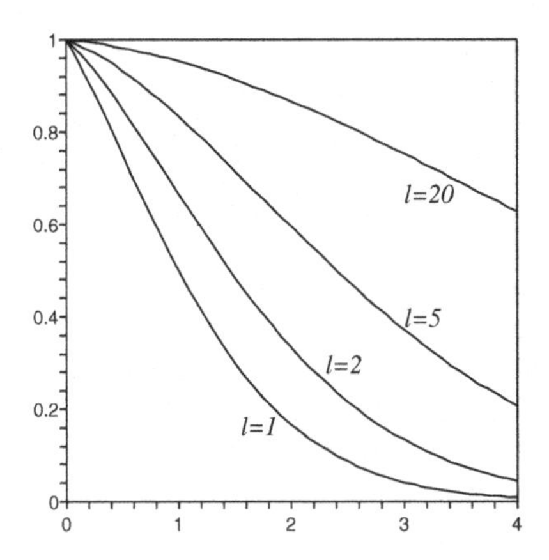

Fig. 3.1. Function $r(l,\beta)=l^\beta\Gamma(l+1)/\Gamma(\beta+l+1)$ as a function of β for $l = 1, 2, 5, 20$.

The properties of $r(l,\beta)$ imply, for example, that the stratified sampling procedure with $l = 1$ (that is, with the maximum possible stratification) is the most efficient. Additionally, the formulae (2.74) and (3.41) with $p = 2$ imply that for any linear estimator $\hat{m}_{n,k_0}(a)$ with $k_0 < l$, the ratio of the mean square errors $E(\hat{m}_{n,k}(a) - m)^2$ for the stratified sample $X_{k,l}$ and the independent sample $X_{1,n}$ is $r(l, 2/\alpha) + o(1)$, $n \to \infty$, which is smaller than 1 for large n; see below for some details.

Efficiency of statistical inference concerning m when the underlying sample is stratified

Several statistical procedures considered in Sect. 2.4 for the case of an independent sample can be applied without any change to the case where the underlying sample is stratified and the number of order statistics used is $k_0 < l$. This concerns the construction of the linear estimators $\hat{m}_{n,k_0}(a)$, the confidence intervals (2.86), and the procedure of testing the hypothesis $H_0: \ m \geq K$ defined by the rejection region (2.88).

Theorem 3.3 implies that under the conditions of this theorem and known α we have

$$E(\hat{m}_{n,k_0}(a) - m) \sim (\kappa_n - m)\, r(l, 1/\alpha)\, a^T b\,,$$
$$E(\hat{m}_{n,k_0}(a) - m)^2 \sim (\kappa_n - m)^2\, r(l, 2/\alpha)\, a^T \Lambda a$$

for any $k_0 < l$, where $\hat{m}_{n,k_0}(a)$ is the linear estimator constructed using the stratified sample and the function $r(l, \beta)$ is defined in Theorem 3.3.

This implies that the linear estimators constructed through the stratified sample are more accurate than the corresponding estimators in the case of the independent sample. Since for any p and α the minimum $\min_l r(l, p/\alpha)$ is achieved at $l = 1$, the value $l = 1$ provides the most efficient estimators. When $p = 2$, the minimum is equal to $r(1, 2/\alpha) = 1/\Gamma(2+2/\alpha)$. This value is the maximum possible gain in mean square error for linear estimation problems and it is achieved when we use the maximum possible stratification of the search region.

In the case of a proper stratified sample with $l \geq 2$ one can construct confidence intervals for m and hypothesis tests concerning m using $k_0 = 2$ minimal order statistics in exactly the same way as in the case of an independent sample; this follows from Lemma 4.4.1 in [273]. The ratio of the average lengths of the confidence intervals (2.86) constructed for the stratified and independent samples approaches $r(l, 1/\alpha) < 1$ as the sample size n tends to infinity. Similar coefficient appears in the ratio of the power functions of the tests with the rejection region (2.88) for the hypothesis $H_0 : m \geq K$ constructed for the stratified and independent samples. The problem of construction of the confidence intervals for m and hypothesis tests concerning m using $k_0 > 2$ minimal order statistics from a stratified sample is more difficult, see Sect. 4.4.2 in [273].

3.2.3 Stochastic Dominance with Respect to Record Values

Let us consider the stochastic dominance when the criterion $\Phi_f(\mathcal{P})$ is the c.d.f. of the record value $f_*[X] = \min_{x_i \in X} f(x_i)$:

$$F_{f,\mathcal{P}}(t) = P(f_*[X] \leq t), \;\; t \in (\inf f, \sup f). \tag{3.43}$$

In this case, the dominance of a procedure $\mathcal{P}$ over $\mathcal{P}'$ in $\mathcal{F}$ means that $F_{f,\mathcal{P}}(t) \geq F_{f,\mathcal{P}'}(t)$ for all real t and $f \in \mathcal{F}$, and there exists $f^* \in \mathcal{F}$ such that $F_{f^*,\mathcal{P}}(t) > F_{f^*,\mathcal{P}'}(t)$ for all $t \in (\inf f^*, \sup f^*)$.

Theorem 3.4. *Let $\mathcal{R}_k$ be a fixed partition of A into $k \leq n$ subsets, $\mathcal{F} = C^p(A)$ for some $0 \leq p \leq \infty$ and $\mathcal{P}_{k,L} = (f_*[X_{k,L}], Q_{k,L})$ correspond to the stratified sampling such that $L = \{l_1, \ldots, l_k\}$, $l_i \geq 0$, $\sum_{i=1}^k l_i = n$. Then*
(i) if the stratified sample $X_{k,L}$ is proper, that is (3.39) holds, then the stratified sampling random search procedure $\mathcal{P}_{k,L}$ stochastically dominates the independent random sampling procedure $\mathcal{P}_1$ in $\mathcal{F}$, with respect to the criterion (3.43);

(ii) if (3.39) *does not hold for at least one i, then $\mathcal{P}_{k,L}$ does not stochastically dominate $\mathcal{P}_1$; moreover, there exists $f^* \in \mathcal{F}$ such that for some t $F_{k,L}(f^*,t) < F_1(f^*,t)$, where $F_{k,L}(f,t) = F_{f,\mathcal{P}_{k,L}}(t)$ and $F_1(f,t) = F_{f,\mathcal{P}_1}(t)$ are the c.d.f.'s* (3.43) *for the stratified and independent sampling procedures, respectively.*

The proof of the theorem is given in Sect. 3.6.

Similar to (i) in Theorem 3.2, it can be shown that if $k' < k$, $\mathcal{R}_k$ is a subpartition of the partition $\mathcal{R}_{k'}$, $X_{k,L}$ is a proper stratified sample and $\mathcal{P}_{k,L}$ and $\mathcal{P}_{k',L'}$ are the pairs corresponding to the stratified samples $X_{k,L}$ and $X_{k',L'}$, then $\mathcal{P}_{k,L}$ stochastically dominates $\mathcal{P}_{k',L'}$ in $\mathcal{F} = C^p(A)$ for every $0 \le p \le \infty$. This implies that the stratified sample $X_{k,L}$ with the maximum stratification, that is, when $P(A_i) = 1/k$ and $L = (1, \ldots, 1)$, generates the best possible procedure $\mathcal{P}_{k,L}$, with respect to the stochastic dominance based on the c.d.f. (3.43). This result is in full agreement with the results of the previous section.

3.3 Markovian Algorithms

Markovian algorithms of global optimization are based on sampling from Markov chains. As the principles of theoretical analysis of these algorithms are relatively simple, many papers have appeared studying various aspects of Markovian algorithms including their convergence and rate of convergence. As a result, the theory of Markovian algorithms of global optimization is a relatively advanced part of the theory of global random search. However, Markovian algorithms are often practically inefficient. The main reason for this inefficiency is the poor use of information about the objective function collected at previous iterations. Indeed, at iteration n of any Markovian algorithm only the value $y_n = f(x_n)$ (and, perhaps, the record value y_{on}) of the objective function $f(\cdot)$ is used to construct a rule for selecting the new observation point x_{n+1}; this is not an efficient use of the information about $f(\cdot)$ contained in the data $\{x_1, \ldots, x_n; y_1, \ldots, y_n\}$.

We start by making some general remarks about constructing Markovian algorithms in Sect. 3.3.1. In Sect. 3.3.2 we then briefly discuss the celebrated *simulated annealing* method. Our main special case of Markovian algorithms is considered in Sect. 3.4; we shall call the corresponding class of algorithms *Markov monotonous search.*

3.3.1 Construction of Markovian Algorithms

Markovian structure

Any global random search algorithm is defined by a rule which constructs probability distributions $P_{n+1}(\cdot) = P_{n+1}(\cdot\,|x_1, y_1, \ldots, x_n, y_n)$ for generating the points x_{n+1} $(n = 0, 1, \dots)$, given the set of previous points $x_i \in A$ and the corresponding objective function values $y_i = f(x_i)$, $i = 1, \dots, n$. The Markovian property of the algorithm means that for all $n = 1, 2, \dots$ the distributions P_{n+1} depend only on x_n, y_n, which are the last point and corresponding observation; that is,

$$P_{n+1}(\cdot \mid x_1, f(x_1), \dots, x_n, f(x_n)) = P_{n+1}(\cdot \mid x_n, f(x_n)) . \tag{3.44}$$

Alternatively, one may define a Markovian algorithm in terms of the record value $y_{on} = \min_{i \le n} y_i$ and the corresponding record point x_{on} with $f(x_{on}) = y_{on}$; in this case,

$$P_{n+1}(\cdot \mid x_1, f(x_1), \dots, x_n, f(x_n)) = P_{n+1}(\cdot \mid x_{on}, y_{on}) . \tag{3.45}$$

Effectively, this means that the last point, x_n, is always the record point.

Simulated annealing algorithms considered in Sect. 3.3.2 use the rule (3.44) whereas the Markov monotonous search of Sect. 3.4 is based on the rule (3.45).

One can also use both pairs, (x_n, y_n) and (x_{on}, y_{on}), for constructing the distribution $P_{n+1}(\cdot)$:

$$P_{n+1}(\cdot \mid x_1, f(x_1), \dots, x_n, f(x_n)) = P_{n+1}(\cdot \mid x_n, y_n, x_{on}, y_{on}) . \tag{3.46}$$

The additional information about the objective function that is used in (3.46) may increase the practical efficiency of an algorithm keeping the structure of the algorithm relatively simple. This is what is done in several versions of the simulated annealing algorithms. Strictly speaking, the algorithms that non-trivially use y_{on} or x_{on} (in addition to x_n and y_n) are not Markovian but their structure is very similar to the structure of Markovian algorithms.

General scheme

A general scheme incorporating and conveniently describing many of Markovian global random search algorithms is as follows.

Algorithm 3.4 (A general scheme of Markovian algorithms).

1. *By sampling from a given distribution P_1 on A, obtain a point x_1. Evaluate $y_1 = f(x_1)$, set iteration number $n = 1$.*
2. *Obtain a point z_n in $\mathbb{R}^d$ by sampling from a distribution $Q_n(x_n, \cdot)$, which may depend on n and x_n (perhaps, on x_{on} as well).*
3. *If $z_n \notin A$, return to Step 2. Otherwise evaluate $f(z_n)$ and set*

$$x_{n+1} = \begin{cases} z_n & \text{with probability } p_n , \\ x_n & \text{with probability } 1 - p_n . \end{cases} \tag{3.47}$$

Here p_n is the acceptance probability; this probability may depend on n, x_n, z_n, y_n, $f(z_n)$ and, perhaps, on x_{on} and y_{on} as well.

4. Set

$$y_{n+1} = \begin{cases} f(z_n) & \text{if } x_{n+1} = z_n, \\ y_n & \text{if } x_{n+1} = x_n. \end{cases}$$

5. Check a stopping criterion. If the algorithm does not stop, substitute $n+1$ for n and return to Step 2.

Particular choices of the initial probability distribution P_1, transition probability $Q_n(x, \cdot)$, and acceptance probabilities $p_n(x, z, y, f(z))$ lead to specific Markovian global random search algorithms. The most well-known among them is the celebrated 'simulated annealing' which is considered in the next section. Note however that in the last few years the popularity of the simulated annealing technique has significantly decreased as many practitioners have finally realized that the practical efficiency of this technique is relatively low.

3.3.2 Simulated Annealing

Similarity to a physical phenomenon and relation to the Metropolis method

The name of the algorithm originated from its similarity to the physical procedure called annealing used to remove defects from metals and crystals by heating them locally near the defect to dissolve the impurity and then slowly re-cooling them so that they could find a basic state with a lower energy configuration. According to the physical interpretation, a point $x \in A$ corresponds to a configuration of the atoms of a substance and $f(x)$ determines the energy of the configuration. Because of a large number of atoms and possible arrangements, there could be many configurations where the energy reaches local minimum (that is, local minimizers of f).

The simulated annealing method can be considered as a version of the classical *Metropolis method* introduced in the seminal paper [159]. The Metropolis method simulates the behaviour of an ensemble of atoms in equilibrium at a given temperature. This method gave rise to the popular technique called MCMC (Markov Chain Monte Carlo), which proved to be very useful in Bayesian statistics and other areas (see e.g. [56, 93, 202] for a description of various heuristics and practical applications of MCMC).

The main difference between the methods of MCMC and simulated annealing lies in their different objectives: in global optimization the main aim is the convergence to the vicinity of the global minimizer of $f(\cdot)$ whereas in MCMC one tries to construct a Markov chain with a given stationary distribution and good mixing properties. Hence, it is customary to use an adaptation in optimization algorithms making the corresponding Markov chains time-heterogeneous. On the other hand, in the majority of the MCMC methods the Markov chains are time-homogeneous; this makes the analysis and implementation easier.

As usual, we assume that the feasible region A is a subset of $\mathbb{R}^d$. However, all the discussions below can easily be adapted and applied in the case where A is a discrete set (which is a more common case in literature on simulated annealing).[3]

Simulated annealing, description of the algorithm

A general simulated annealing algorithm is Algorithm 3.4 with acceptance probabilities

$$p_n = \min\{1, \exp(-\beta_n \Delta_n)\} = \begin{cases} 1 & \text{if } \Delta_n \leq 0, \\ \exp(-\beta_n \Delta_n) & \text{if } \Delta_n > 0, \end{cases} \tag{3.48}$$

where $\Delta_n = f(z_n) - f(x_n)$ and $\beta_n \geq 0$ $(n = 1, 2, \ldots)$.

The choice (3.48) for the acceptance probability p_n means that any 'promising' new point z_n (for which $f(z_n) \leq f(x_n)$) is accepted unconditionally; a 'non-promising' point (for which $f(z_n) > f(x_n)$) is accepted with probability $p_n = \exp\{-\beta_n \Delta_n\}$. As the probability of acceptance of a point which is worse than the preceding one is always greater than zero, the search trajectory may leave a neighbourhood of a local and even a global minimizer. Note however that the probability of acceptance decreases if the difference $\Delta_n = f(z_n) - f(x_n)$ increases. This probability also decreases if β_n increases. In the limiting case, where $\beta_n = \infty$ for all n, the simulated annealing algorithm becomes the Markov monotonous search of Sect. 3.4.

The standard version of the simulated annealing can be described as follows. An initial point $x_1 \in A$ is chosen arbitrarily. Let x_n $(n \geq 1)$ be the current point, $y_n = f(x_n)$ be the corresponding objective function value, β_n be a positive parameter, and let ξ_n be a realization of a random vector having some probability distribution Φ_n (if $A \subset \mathbb{R}^d$ then it is natural to choose Φ_n as an isotropic distribution on $\mathbb{R}^d$). Then one should check the inclusion $z_n = x_n + \xi_n \in A$ (otherwise return to obtaining a new realization ξ_n), evaluate $f(z_n)$ and use (3.47) and (3.48) for defining the new point x_{n+1}.

In short, the standard version of the simulated annealing algorithm is Algorithm 3.4 where a special form of the probabilities $Q_n(x_n, \cdot)$ is used and the acceptance probabilities p_n are chosen according to (3.48). Very often, the probabilities $Q_n(x_n, \cdot)$ are chosen so that they do not depend on the iteration number n. If $A \subseteq \mathbb{R}^d$, the transition probabilities $Q_n(x_n, \cdot)$ are typically selected so that the corresponding transition densities $q_n(x, z)$ are symmetric (this means that the distributions Φ_n are isotropic); in this case, $q_n(x, z) = q_n(z, x)$ $\forall x, z$. Moreover, very often these densities $q_n(x, z)$ are bounded away from 0:

[3] if A is discrete, then the uniform measure on A replaces the Lebesgue measure and for any function g defined on A the symbol $\int g(x)dx$ stands for $\sum_{x_i \in A} g(x_i)$.

$$q_n(x, z) \geq c > 0 \;\; \forall n, \; \forall x, z \in A. \tag{3.49}$$

If A is discrete, the condition (3.49) becomes

$$Q_n(x, z) \geq c > 0 \;\; \forall n, \; \forall x, z \in A. \tag{3.50}$$

Of course, one may use many other transition probabilities $Q_n(x_n, \cdot)$ and other forms of the acceptance probabilities. Many heuristical arguments have been offered to improve the efficiency of the simulated annealing algorithms, see e.g. [1, 146, 206, 258, 259]. There is, however, a problem with the algorithms involving many heuristic and problem-related details: these algorithms become less transparent theoretically and the comparison of their efficiency with the efficiency of other global optimization techniques becomes very difficult.

Simulated annealing, convergence

The main theoretical aim in constructing optimization algorithms is to guarantee their convergence. In the simulated annealing algorithms, one may be interested in either

(i) the convergence of the record values y_{on} to m as $n \to \infty$, see (1.1), or
(ii) the convergence of the whole sequence of y_n to m as $n \to \infty$.

Of course, the convergence (ii) is stronger than the convergence (i). On the other hand, if $\beta_n \to \infty$ as $n \to \infty$, then the simulated annealing algorithm asymptotically does not move away from the points with small function values, and therefore the convergence (i) immediately implies (ii), see [14, 23]. Note that the Markov chains created by the corresponding simulated annealing algorithms are not time-homogeneous.

In terms of the sequence of the points $\{x_n\}$, the convergence (i) means that the sequence of record points x_{on} converges to the set A_* of global minimizers of $f(\cdot)$, see (1.2). If the global minimizer x_* of $f(\cdot)$ is unique, then the condition (ii) means that the sequence of probability distributions of the points x_n weakly converges to the delta-measure concentrated at x_*.

Guaranteeing the convergence (i) is easy. For example, this convergence automatically holds if the transition probabilities satisfy either (3.49) (when $A \subset \mathbb{R}^d$) or (3.50) (when A is discrete). More general conditions for the convergence (i) follow from Theorem 2.1 and other results of Sect. 2.1.3.

If the conditions of the types (3.49) and (3.50) cannot be guaranteed, then to ensure the convergence (ii), we must assume that $T_n = 1/\beta_n$ tends to zero very slowly. For a general simulated annealing algorithm (in both cases, when $A \subset \mathbb{R}^d$ and when A is discrete), one of the conditions needed to achieve the convergence is $T_n \geq c/\ln(2 + n)$, where c is a sufficiently large constant depending on $f(\cdot)$ and A, see e.g. [89, 107, 108, 145, 162].

The theoretical rate of convergence of the simulated annealing is very slow; this convergence is based on the convergence of the pure random search which is contained within the simulated annealing algorithms in the sense of the representation (2.7). This implies that the discussion at the end of Sect. 2.2.2 concerning the rate of convergence of general global random search algorithms, can be applied in this situation in full.

From the view-point of the theoretical results on the rate of convergence, the Markov monotonous search algorithms considered in Sect. 3.4 are much more promising than the simulated annealing algorithms.

Time-homogeneous simulated annealing, Gibbs distribution

The expression (3.48) for the acceptance probability p_n is derived using the analogy to the annealing process. In statistical mechanics, the probability that the system will transit from a state with energy E_0 to a state with energy E_1, where $\Delta E = E_1 - E_0 > 0$, is $\exp(-\Delta E/KT)$, where $K = 1.38 \cdot 10^{-16} erg/T$ is the Boltzmann constant and T is the absolute temperature. Therefore, $\beta = 1/KT$ and the lower the temperature, the smaller the probability of transition to a higher energy state.

If $\beta_n = \beta = 1/KT$, that is

$$p_n(x, z, y, \eta) = p(x, z) = \min\{1, \exp[\beta(f(x) - f(z))]\}, \tag{3.51}$$

and the transition probabilities do not depend on the iteration number n (that is, $Q_n(\cdot, \cdot) = Q(\cdot, \cdot)$), then the sequence $\{x_n\}$ constitutes a time-homogeneous Markov chain, converging in distribution (under very general conditions on Q and f) to a stationary distribution with density

$$\pi_\beta(x) = \exp\{-\beta f(x)\} \Big/ \int_A \exp\{-\beta f(z)\} dz . \tag{3.52}$$

This distribution is often called the Gibbs distribution (sometimes, it is also called the Boltzmann distribution).

As $T \to 0$ (or $\beta \to \infty$), the Gibbs density $\pi_\beta(\cdot)$ defined by (3.52) tends to concentrate on the set A_* of global minimizers of $f(\cdot)$ subject to some mild conditions; this holds, for example, if the feasible region A and the objective function $f(\cdot)$ satisfy the conditions C1–C4 and C9 of Sect. 2.1.1; see also [90]. If additionally the global minimizer x_* of $f(\cdot)$ is unique, then the Gibbs distribution converges to the δ-measure concentrated at x_* for $T \to 0$. We illustrate this in Fig. 3.2, for the function (1.3) with $k = 3$, where we plot the Gibbs densities (3.52) corresponding to this function with $\beta = 1$ and $\beta = 3$.

Numerically, if T is small (i.e. β is large), then the points x_n obtained by a time-homogeneous simulated annealing method have the tendency to concentrate around the global minimizer(s) of $f(\cdot)$. Unfortunately, the time required to approach the stationary Gibbs distribution increases exponentially with β and may reach very large values for large β (this is confirmed by numerical results). This phenomenon can be explained by the fact that for large β (i.e., small T) a time-homogeneous simulated annealing method tends to be like a local random search algorithm that rejects the majority of unprofitable steps, and therefore its globality features are poor.

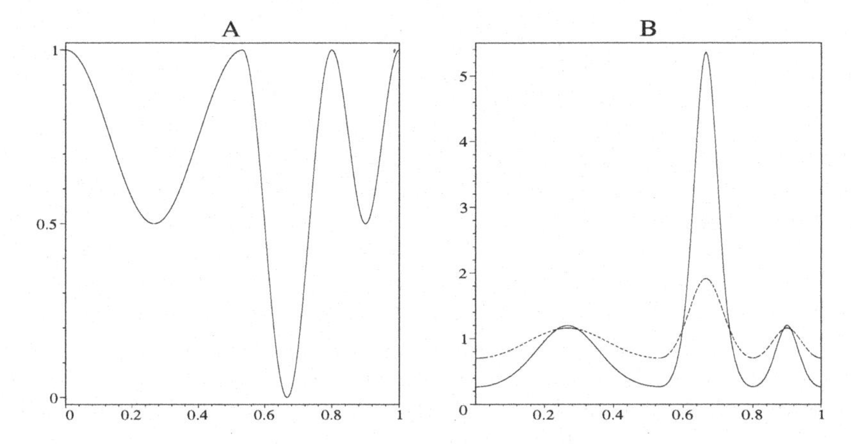

Fig. 3.2. (A) Graph of the function (1.3) with $k = 3$; (B) the Gibbs densities (3.52) corresponding to this function with $\beta = 1$ (dotted line) and $\beta = 3$ (solid line).

MCMC

The time-homogeneous simulated annealing method is a particular case of the above mentioned Metropolis (or Metropolis–Hastings) algorithm that uses the acceptance probabilities

$$p(x,z) = \min\left\{1, \frac{w(z)q(z,x)}{w(x)q(x,z)}\right\} \tag{3.53}$$

rather than (3.51). Here $w(\cdot)$ is an arbitrary summable positive function on A and $q(z,x)$ is the transition density; that is, the density corresponding to the transition probability $Q(z,\cdot)$. In (3.53), we set $p(x,z) = 1$ if $w(x)q(x,z) = 0$.

We obtain the expression (3.51) if we use (3.53) with $w(x) = \exp(-\beta f(x))$ and the symmetric transition density $q(x,z) = q(z,x)\ \forall x,z$. The points $\{x_n\}$ generated by the Metropolis-Hastings algorithm converge in distribution to the stationary distribution with density

$$\varphi_w(x) = w(x) \Big/ \int w(z)dz \tag{3.54}$$

which is proportional to the function $w(\cdot)$ and generalizes (3.52). An important property of this algorithm and perhaps the main reason for the popularity of MCMC in general is the fact that one does not need to know the normalizing constant $\int w(z)dz$ in (3.54) to construct Markov chains with the stationary density (3.54).

Formula (3.53) is not the only way of choosing the acceptance probabilities of the above described Markov chains that make (3.54) the stationary density of the chain. For example, if the transition density $q(\cdot,\cdot)$ is symmetric, then one can use

$$p(x,z) = \begin{cases} g\left(\frac{w(z)}{w(x)}\right) & \text{if } w(z) \geq w(x), \\ \frac{w(z)}{w(x)} g\left(\frac{w(z)}{w(x)}\right) & \text{if } w(z) < w(x), \end{cases}$$

where $g : [0, \infty) \to [0, 1]$ is an arbitrary measurable function, $g \neq 0$. However, as shown by Peskun in [184], the choice (3.53) is optimum in the sense that the variance of the Monte-Carlo estimators of the integral $\int_A h(z)w(z)dz$ is minimal for any integrable function h, where the minimum is taken over all possible choices of the acceptance probability $p(\cdot, \cdot)$.

A serious limitation of the classical MCMC algorithms constraining their efficiency is the fact that these algorithms generate Markov chains that are homogeneous in time. Adapting the rules of generating new points x_n is therefore not allowed in these algorithms. Of course, the use of adaptation may significantly increase practical efficiency of the MCMC algorithms; see, for example, [205, 247].

Publishing books and papers on the methodology, implementation and practical applications of MCMC algorithms has been very fashionable during the past 10-15 years, see e.g. [56, 93, 99, 100, 202, 227]. The areas of application of MCMC, especially in Bayesian statistics, are huge. Also, the convergence and the convergence rate of the MCMC algorithms are thoroughly investigated theoretically, see [15, 158, 203, 238]. However, there is a diversity of opinions about the applicability of MCMC algorithms in practice. Many authors claim that they successfully apply MCMC algorithms in very large dimensions (as $d \simeq 1000$); a much smaller group of authors (usually, more interested in theoretical issues than in inventing heuristics and straightforward applications) have serious concerns about the quality of MCMC methods even in small dimensions; the author of this section belongs to the latter group.

3.4 Markov Monotonous Search

Markov monotonous search can be considered as Algorithm 3.4 where the acceptance probabilities p_n are

$$p_n = \begin{cases} 1 & \text{if } f(z_n) \leq f(x_n), \\ 0 & \text{if } f(z_n) > f(x_n). \end{cases}$$

As these probabilities can be written in the form (3.48) with $\beta_n = \infty$, $\forall n$, Markov monotonous search can be considered as the limiting case of the simulated annealing algorithm. This search, however, has specific features that are very different from the features of simulated annealing. These features deserve special investigation. Such an investigation has been recently carried out by A.Tikhomirov and V.Nekrutkin, see e.g. [167, 168, 242, 243, 244, 245]. This section surveys some of their results.

3.4.1 Statement of the Problem

Optimization space

Let A be a feasible region and ρ be a metric on A. We shall call the pair (A, ρ) the optimization space. In this section, we consider the following two feasible regions A: $A = \mathbb{R}^d$ and $A = \mathbb{I}^d = [0,1)^d$.

For $A = \mathbb{R}^d$, we shall use the metrics

$$\rho(x,y) = \rho_p(x,y) = \begin{cases} \left(\sum_{i=1}^d |x_i - y_i|^p\right)^{1/p} & \text{if } 1 \le p < \infty\,, \\ \max_{1\le i\le d} |x_i - y_i| & \text{if } p = \infty\,, \end{cases}$$

where $x = (x_1, \ldots, x_d)$ and $y = (y_1, \ldots, y_d)$.

For $A = \mathbb{I}^d$, the corresponding metrics are

$$\rho(x,y) = \rho_p(x,y) = \begin{cases} \left(\sum_{i=1}^d (\varrho(x_i,y_i))^p\right)^{1/p} & \text{if } 1 \le p < \infty, \\ \max_{1\le i\le d} \varrho(x_i,y_i) & \text{if } p = \infty \end{cases} \tag{3.55}$$

with $\varrho(x_i, y_i) = \min\{|x_i - y_i|, 1 - |x_i - y_i|\}$. For any of the metrics (3.55) the optimization space $(\mathbb{I}^d, \rho)$ is the d-dimensional torus.

The reason for choosing these optimization spaces is to avoid the 'edge effects', which would make statements of all the results more difficult. In many respects, $\mathbb{R}^d$ and the torus are very similar; in these spaces it is easy to define symmetric search algorithms (see below). On the other hand, as we will be interested in the order of convergence, choosing a torus rather than a cube does not seem to be a principal issue: indeed, for $x_* \in (0,1)^d$ sufficiently small neighbourhoods of the torus and cube coincide, and the results for the rate of convergence can automatically be transferred from the torus to the unit cube.

If a particular result holds for any optimization space among those listed above, then we shall simply denote the space as (A, ρ). The notations $(\mathbb{R}^d, \rho)$ and $(\mathbb{I}^d, \rho)$ have similar sense. We shall use the notation $B_\varepsilon(x) = \{y \in A : \rho(x,y) \le \varepsilon\}$ for the closed ball. Note that the volume of this ball, $\mathrm{vol}(B_\varepsilon(x))$, does not depend on x (vol is the d-dimensional Lebesgue measure). Set $\varphi(\varepsilon) = \mathrm{vol}(B_\varepsilon(x))$ and note that for all sufficiently small ε

$$\varphi(\varepsilon) = c\varepsilon^d, \tag{3.56}$$

where $c = c(d, \rho)$ is a constant depending on the dimension d and the metric ρ. For $A = \mathbb{R}^d$, the equality (3.56) holds for all $\varepsilon > 0$. For the torus we have $\varphi(\varepsilon) = 1$ when $\varepsilon > \mathrm{diam}(A)$.

Objective function

We shall always assume that the objective function $f : A \mapsto \mathbb{R}$ is measurable, bounded from below and satisfies the following conditions:

F1: function $f(\cdot)$ attains the global minimum m at a single point x_*;

F2: function $f(\cdot)$ is continuous at the point x_*;

F3: for all $\varepsilon > 0$, we have $\inf\{f(x) \colon x \in B_\varepsilon^c(x_*)\} > f(x_*)$, where $U^c = A \setminus U$ for any $U \subset A$.

Note that the condition F3 and the convergence $f(x_n) \to f(x_*)$ imply $\rho(x_n, x_*) \to 0$.

One more condition on $f(\cdot)$ will be introduced later on.

Markov monotonous search

Let $\{x_n\}_{n \geq 1}$ be any sequence (either finite or infinite) of random points in A. If this sequence forms a Markov chain (that is, if for all n the distribution of x_{n+1} conditional on $x_1, \dots, x_n$ coincides with the distribution of x_{n+1} conditional on x_n only), then we say that $\{x_n\}_{n \geq 1}$ is a Markov (random) search. If, in addition, for all $n \geq 1$ we have $f(x_{n+1}) \leq f(x_n)$ with probability 1, then we shall say that $\{x_n\}_{n \geq 1}$ is a *Markov monotonous search.* Convergence of this type of algorithms was first considered in [7]. This section concentrates on studying the rate of convergence of Markov monotonous search algorithms.

Below, we shall consider Markov monotonous searches with transition densities belonging to a particular class of densities. Let us describe this class. Set

$$W_x = \{y \in A : f(y) \leq f(x)\},$$

and consider a Markov chain $\{x_n\}_{n \geq 1}$ (generally, non-homogeneous) with initial point $x_1 = x$ and transition probabilities

$$R_n(x, \cdot) = \delta_x(\,\cdot\,) P_n(x, W_x^c) + P_n(x, \,\cdot \cap W_x), \tag{3.57}$$

where $\delta_x(\,\cdot\,)$ is the probability measure concentrated at the point x and $P_n(x, \cdot\,)$ are Markov transition probabilities; that is, $P_n(x, \cdot\,)$ is a probability measure for all $n \geq 1$ and $x \in A$, and $P_n(\cdot, U)$ is $\mathcal{B}$-measurable for all $n \geq 1$ and $U \in \mathcal{B}$ (where $\mathcal{B}$ is the Borel σ-algebra of subsets of A). Obviously, $R_n(x, W_x) = 1$; this implies that the inequalities $f(x_{n+1}) \leq f(x_n)$ hold with probability 1 for all $n \geq 1$.

For simplicity of references, let us now formulate the algorithm for generating N points of a Markov monotonous search.

Algorithm 3.5.

1. *Set $x_1 = x$, set iteration number $n = 1$.*
2. *Generate η as a realization of the probability measure $P_n(x_n, \cdot\,)$.*
3. *If $f(\eta) \leq f(x_n)$, then set $x_{n+1} = \eta$, otherwise set $x_{n+1} = x_n$.*

4. If $n < N$, then set $n = n+1$ and return to Step 2; otherwise terminate the algorithm.

We shall consider the case where the transition probabilities $P_n(x, \mathrm{d}y)$ have *symmetric* densities $p_n(x,y)$ of the form

$$p_n(x,y) = g_n\big(\rho(x,y)\big), \tag{3.58}$$

where g_n are non-increasing functions of a positive argument. In this case, $p_n(x, x+y) = p_n(0,y)$ for $A = \mathbb{R}^d$ and $p_n(x, x \oplus y) = p_n(0,y)$ for $A = \mathbb{I}^d$, where $\oplus$ denotes the operation of coordinate-wise summation modulo 1. The functions g_n are not arbitrary, of course. In particular, for $A = \mathbb{R}^d$ they must satisfy the condition

$$\int_{(0,\infty)} r^{d-1} g_n(r)\mathrm{d}r < \infty\,.$$

Markov monotonous search with transition densities of the form (3.58) will be called *Markov symmetric monotonous search.*

Below, we shall write P_x and E_x for the probabilities and expectations related to the search of Algorithm 3.5 starting at a point $x \in A$.

The aim of search

There are generally two different tasks in optimization: approximation of the minimal value $m = \min f$ of the objective function $f(\cdot)$ with given accuracy δ, and approximation of the minimizer x_* with given precision $\varepsilon > 0$.

In the first case (approximation with respect to the function value) we are interested in hitting the set

$$W(\delta) = \big\{x \in A : f(x) \le m + \delta\big\}\,;$$

note that the monotonous search never leaves the set $W(\delta)$ after reaching it.

In the second case (approximation with respect to the argument) our aim is to reach the set $B_\varepsilon(x_*)$. It can happen, however, that after reaching the set $B_\varepsilon(x_*)$ at iteration n, a search algorithm leaves it at a subsequent iteration. In order to avoid complications related to this phenomenon, we introduce the sets

$$M_r = \big\{x \in B_r(x_*) : f(x) < f(y) \text{ for all } y \in B_r^c(x_*)\big\}.$$

It is easy to see that the sets M_r have the following properties:
a) if $r_2 < r_1$, then $M_{r_2} \subset M_{r_1}$, and
b) if $x \in M_r$ and $y \notin M_r$, then $f(x) < f(y)$.

Thus, any monotonous search does not leave the set M_r after reaching it. To study the approximation with respect to the argument, we shall study the moment the algorithm reaches the set M_ε for the first time; as above, ε is the required precision with respect to the argument. Respectively, the closeness of

x to x_* is expressed through the quantity $\delta(x) = \inf\{\delta \geq 0 : x \in M_\delta\}$ rather than the distance $\rho(x, x_*)$.

The following proposition establishes a relation between the sets $W(\cdot)$ and M_r.

Proposition. *Set* $\Delta = \Delta(\varepsilon) = \sup\{\rho(x, x_*) : x \in W(\varepsilon)\}$. *Then*

1) $\bigcup_{0<r<\Delta} M_r \subset W(\varepsilon) \subset M_\Delta \subset B_\Delta(x_*)$, *and*

2) if f *is continuous in* $B_\Delta(x_*)$, *then for* $\Delta < \mathrm{diam}(A)$ *we have* $W(\varepsilon) = M_\Delta$.

One more condition on the objective function $f(\cdot)$ we are going to use is the following:

F4: $\bigcup_{r>0} M_r = A$.

This condition guarantees that the initial point x belongs to M_r for some r. In view of the monotonicity of the search algorithm, the points x_n never leave the set M_r and hence cannot go infinitely far away from x_*. Let us illustrate this condition.

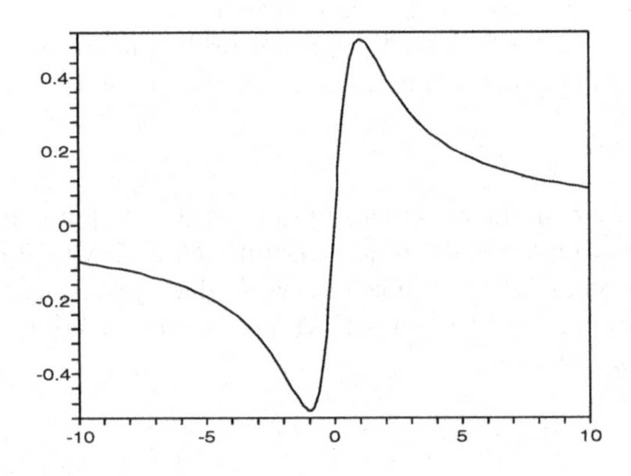

Fig. 3.3. Plot of the function $f(x) = x/(x^2 + 1)$.

It is easy to see that for the optimization space $(\mathbb{R}, \rho)$ the function $f(x) = x/(x^2 + 1)$ with $m = -0.5$ and $x_* = -1$ (see Fig. 3.3) satisfies conditions F1 – F3 but does not satisfy F4. Indeed, in this case for all $r > 0$ we have $M_r = [-(r + 1), -1/(r + 1)]$ and $\bigcup_{r>0} M_r = (-\infty, 0)$. Hence, for $f(x) = x/(x^2+1)$ the point '$+\infty$' has the role of an 'infinitely distant local extremum' with an infinitely large attraction region. Condition F4 guarantees that this kind of situation cannot happen. Note that for $A = \mathbb{I}^d$ we have $\mathrm{diam}(A) < \infty$ and the condition F4 automatically holds since $M_{\mathrm{diam}(A)} = A$.

Information about the objective function

The main information used about the objective function $f(\cdot)$ will be contained in the so-called *asymmetry coefficient*

$$F^f(r) = F(r) = \mathrm{vol}(M_r)/\mathrm{vol}(B_r(x_*))\,.$$

This coefficient 'compares' the behaviour of $f(\cdot)$ with the F-*ideal* uniextremal function which has an asymmetry coefficient $F^f = 1$. In particular, the asymmetry coefficient codes the information about the local minima of $f(\cdot)$.

The conditions imposed on $f(\cdot)$ guarantee that $F^f(r) > 0$ for all $r > 0$. The functions $f(\cdot)$ such that $\liminf F^f(r) > 0$ as $r \to 0$, will be called *non-degenerate*. In particular, if for $(\mathbb{R}^d, \rho_2)$ or $(\mathbb{I}^d, \rho_2)$ the function $f(\cdot)$ is twice continuously differentiable in some neighbourhood of x_* and the Hessian $(\partial^2 f(x_*))$ is a non-degenerate matrix, then

$$\lim_{r\to 0} F^f(r) = \sqrt{\prod_{i=1}^{d} \lambda_{\min}/\lambda_i} > 0,$$

where λ_i are eigenvalues of the matrix $(\partial^2 f(x_*))$ and $\lambda_{\min} = \min \lambda_i > 0$.

In some cases we shall use the function

$$V^f(r) = V(r) = \mathrm{vol}\,(M_r) = F^f(r)\mathrm{vol}\,(B_r(x_*)) \tag{3.59}$$

instead of $F^f(r)$ (we will call $V^f(r)$ the asymmetry function).

Characteristics of the random search algorithms

Let $N = \infty$ in Algorithm 3.5 and denote $\tau_\varepsilon = \min\{n \geq 1 : x_n \in M_\varepsilon\}$. Since we always assume that in order to generate the transition probabilities P_n we do not need to evaluate the objective function $f(\cdot)$, we only need one function evaluation at each iteration $x_n \mapsto x_{n+1}$ of Algorithm 3.5. Hence the distribution of the random variable τ_ε provides us with very useful information about the quality of a particular random search algorithm. Indeed, in τ_ε iterations of Algorithm 3.5 the objective function $f(\cdot)$ is evaluated τ_ε times and, in view of the monotonicity of the search, $\mathsf{P}_x(\tau_\varepsilon \leq n) = \mathsf{P}_x(x_n \in M_\varepsilon)$.

We shall study $\mathsf{E}_x\tau_\varepsilon$ as functions of the required precision ε, as $\varepsilon \to 0$. The quantity $\mathsf{E}_x\tau_\varepsilon$ can be interpreted as the average number of iterations of a search algorithm required to reach the set M_ε. Most of the results that are true for $\mathsf{E}_x\tau_\varepsilon$, can also be extended to

$$n(x,\varepsilon,\gamma) = \min\{n : \mathsf{P}_x(x_n \in M_\varepsilon) \geq \gamma\} = \min\{n : \mathsf{P}_x(\tau_\varepsilon \leq n) \geq \gamma\},$$

the number of iterations such that the set M_ε has reached with a probability of at least γ.

Note that for many local optimization algorithms (such as steepest descent) the number of iterations has the order $O(|\ln \varepsilon|)$, $\varepsilon \to 0$. In global optimization problems the order for the number of iterations is typically worse; it is $O(1/\varepsilon^\alpha)$ for some $\alpha > 0$. Below, we shall indicate versions of the Markov monotonous symmetric search such that $\mathsf{E}_x\tau_\varepsilon$ (as well as $n(x,\varepsilon,\gamma)$) has the order $O(|\ln \varepsilon|^\alpha)$ with some $\alpha \geq 1$ (in the main cases, where the objective function is non-degenerate in a neighbourhood of the global minimizer, we will have $\alpha = 2$). This is achieved by means of a clever choice of the transition probabilities P_n in (3.57).

Let us start by showing that for a symmetric search it is not possible to obtain a rate better than $O(|\ln \varepsilon|)$ as $\varepsilon \to 0$.

3.4.2 Lower Bounds

Theorem 3.5. *Consider any Markov monotonous symmetric search in* (A, ρ) *defined by* (3.57) *and* (3.58). *Set*

$$\tau'_\varepsilon = \min\{n \ge 1 \colon x_n \in B_\varepsilon(x_*)\},\; n'(x,\varepsilon,\gamma) = \min\{n \ge 1 \colon \mathsf{P}_x(x_n \in B_\varepsilon(x_*)) \ge \gamma\}.$$

Assume that $f(\cdot)$ *satisfies the condition F1 and* $\rho = \rho_\infty$.
1. If $A = \mathbb{R}^d$ *then for* $\varepsilon < \rho(x, x_*)$ *we have*

$$\mathsf{E}_x \tau_\varepsilon \ge \mathsf{E}_x \tau'_\varepsilon \ge \ln\left(\rho(x,x_*)/\varepsilon\right) + 2, \tag{3.60}$$

$$n(x,\varepsilon,\gamma) \ge n'(x,\varepsilon,\gamma) \ge \gamma\big(\ln\left(\rho(x,x_*)/\varepsilon\right) + 2\big). \tag{3.61}$$

2. If $A = \mathbb{I}^d$ *then for* $\varepsilon < \rho(x, x_*) \le 1/4$ *we have the inequalities* (3.60), (3.61) *and for* $\varepsilon < 1/4 < \rho(x, x_*)$ *we have*

$$\mathsf{E}_x \tau_\varepsilon \ge \mathsf{E}_x \tau'_\varepsilon \ge \ln\left(\sqrt{\rho(x,x_*)}/\varepsilon\right) + 1 - \ln 2,$$

$$n(x,\varepsilon,\gamma) \ge n'(x,\varepsilon,\gamma) \ge \gamma\left(\ln\left(\sqrt{\rho(x,x_*)}/\varepsilon\right) + 1 - \ln 2\right).$$

Similar statements can be formulated for $\rho = \rho_2$.

For the proof of Theorem 3.5 and extensions, see [168, 241].

3.4.3 Homogeneous Search

In this section, we consider a homogeneous Markov symmetric search where the transition densities $p_n(x, y) = p(x, y)$ do not depend on the iteration number n and have the form

$$p(x,y) = g\big(\rho(x,y)\big). \tag{3.62}$$

Here g is a non-increasing non-negative left-continuous function defined on $(0, \infty)$ for $A = \mathbb{R}^d$ and on $(0, \mathrm{diam}(A)]$ for $A = \mathbb{I}^d$. Additionally, g must be normalized so that

$$\int_{(0,\mathrm{diam}(A))} g(r)\mathrm{d}\varphi(r) = 1\,; \tag{3.63}$$

only in this case $p(x, y)$ defined in (3.62) becomes a transition density.

We omit all the proofs and refer to [239, 243, 244, 245]; in these papers the reader can find some more material on the topic.

Integral bounds

Set $\delta(x) = \inf\{r \geq 0 : x \in M_r\}$, and note that the conditions F1 – F4 imply $\delta(x) < \infty$ and $x \in M_{\delta(x)}$. We assume that $\varepsilon < \delta(x)$; if $\varepsilon \geq \delta(x)$ then $x \in M_\varepsilon$ no search is required.

For $0 < a < \delta$, $r > 0$, set $h(r) = g\big(\min(2r, \mathrm{diam}(A))\big)$ and

$$I(\delta, a; f, g) = 1/\big(V(\delta)h(\delta)\big) + \int_{(a,\delta]} 1/h(r)\mathrm{d}\big(-1/V(r)\big)\,, \tag{3.64}$$

where $V(\cdot)$ is the asymmetry function defined in (3.59) and the integral in the r.h.s. in (3.64) is the Lebesgue-Stieltjes integral. Note that if g is strictly positive at the point $\min(2\delta, \mathrm{diam}(A))$, then the functions V and h are monotonous and bounded away from zero on $[a, \delta]$. Under these conditions I is finite.

Theorem 3.6. *Let $f(\cdot)$ satisfy conditions F1 – F4. Then for any symmetric monotonous homogeneous Markov search defined through* (3.57), (3.62) *and starting at $x \in A$ we have*

$$\mathsf{E}_x\tau_\varepsilon \leq I(\delta(x), \varepsilon; f, g) + 1. \tag{3.65}$$

The quantity $I(\delta(x), \varepsilon; f, g)$ depends on the objective function $f(\cdot)$ through the asymmetry function $V(\cdot)$, the initial point x, the required precision ε and the function $g(\cdot)$. Set

$$F_{\varepsilon,x} = \inf\{F^f(r) : \varepsilon \leq r < \delta(x)\} \tag{3.66}$$

and note that the conditions F1 – F4 yield $F_{\varepsilon,x} > 0$.

Theorem 3.7. *Under the assumptions of Theorem 3.6, for any $\varepsilon < \delta(x)$ we have*

$$I\big(\delta(x), \varepsilon; f, g\big) \leq F_{\varepsilon,x}^{-1} I\big(\delta(x), \varepsilon; f^\star, g\big),$$

where $F_{\varepsilon,x}$ is defined in (3.66) *and $f^\star$ is the F-ideal objective function (for the function $f^\star$, the asymmetry coefficient is equal to 1).*

I-optimal search

The next problem is to construct the density g^* minimizing $I(\delta(x), \varepsilon; f, g)$, for fixed x, ε and $f(\cdot)$. Such a density g^* will be called I-optimal.

In the case $F^f = \text{const}$, the I-optimal density g^* can be written exactly.

Theorem 3.8. *Suppose $F^f \equiv F_0 > 0$. Let $\delta_1 = \delta_1(x) = \min\{\delta(x), \mathrm{diam}(\mathbb{I})^d/2\} \leq 1/4$ for $(A, \rho) = (\mathbb{I}^d, \rho)$ and $\delta_1 = \delta(x)$ for $(A, \rho) = (\mathbb{R}^d, \rho)$. If $\varepsilon < \delta_1/\sqrt[d]{4}$, then the I-optimal g^* has the form*

$$g^*(r) = \frac{1}{c\lambda} \begin{cases} (a\varepsilon)^{-d} & \text{if } 0 < r \le a\varepsilon, \\ r^{-d} & \text{if } a\varepsilon < r \le b, \\ b^{-d} & \text{if } b < r \le 2\delta_1, \\ 0 & \text{otherwise,} \end{cases}$$

where $a = 2\sqrt[d]{2}$, $b = 4\delta_1/a$, $c = c(d, \rho)$ *is defined in* (3.56) *and*

$$\lambda = \lambda(d, \varepsilon, \delta_1) = d \ln(\delta_1/\varepsilon) + 2 - \ln 4.$$

Thus we have

$$I(\delta(x), \varepsilon; f, g^*) = 2^d \lambda^2 / F_0.$$

Let $\delta(x)$ be known (it depends on x and x_*), the optimization space be $(\mathbb{I}^d, \rho)$ and assume the additional condition $F^f(r) \ge F_0 > 0$ for all $r > 0$. Then as shown in [239, 245], for $\varepsilon < 1/(4\sqrt[d]{4})$ and $\delta_1(x) < 1/4$, there exists $g = g^*$ (depending on $\delta(x)$ and ε) such that

$$\mathsf{E}_x \tau_\varepsilon \le I(\delta(x), \varepsilon; f, g^*) + 1 \le \frac{2^d d^2}{F_0} \ln^2 \varepsilon + 1,$$

which is of order $O(\ln^2 \varepsilon)$ as $\varepsilon \to 0$. A similar result holds for $A = \mathbb{R}^d$.

Transition densities independent of the initial point x

In this section, we shall still use the inequality (3.65) but only consider the functions $g(\cdot)$ that do not depend on the location of x with respect to x_*. This will allow us to formulate the whole class of Markov homogeneous symmetric monotonous search algorithms with $\mathsf{E}_x \tau_\varepsilon = O(\ln^2 \varepsilon)$ as $\varepsilon \to 0$.

Consider a Markov homogeneous symmetric monotonous search algorithm starting at $x \in A$ and defined through (3.57) and (3.62). Let $\psi(r)$ be some left-continuous monotonously non-increasing strictly positive function defined on $(0, \infty)$ for $A = \mathbb{R}^d$ and on $(0, \mathrm{diam}(A)]$ for $A = \mathbb{I}^d$ and satisfying the condition $\psi(r) r^d \to 1$ as $r \to 0$. For the case $A = \mathbb{R}^d$ we shall additionally assume that $\int_b^\infty \psi(r) r^{d-1} \mathrm{d}r < \infty$ for any $b > 0$.

Choose some $a > 0$, which is a parameter of the search algorithm, and define for $\varepsilon < (\mathrm{diam}(A))/a$,

$$g_\varepsilon(r) = \frac{1}{\Lambda(\varepsilon)} \begin{cases} \psi(a\varepsilon) & \text{if } 0 < r \le a\varepsilon, \\ \psi(r) & \text{if } r > a\varepsilon, \end{cases} \tag{3.67}$$

where the multiplier $1/\Lambda(\varepsilon)$ provides the normalization condition (3.63) for all $\varepsilon > 0$. Then for any non-degenerate function $f(\cdot)$ satisfying conditions F1 – F4 we have

$$\mathsf{E}_x \tau_\varepsilon \le F_{\varepsilon,x}^{-1} O(\ln^2 \varepsilon) = O(\ln^2 \varepsilon), \quad \text{as } \varepsilon \to 0, \tag{3.68}$$

with $F_{\varepsilon,x}$ defined in (3.66).

Consider the case $(A,\rho) = (\mathbb{I}^d, \rho_\infty)$. In (3.67), we take $a = 2\sqrt[d]{2}$ and

$$\psi(r) = \begin{cases} r^{-d} & \text{if } r \le 1/a, \\ a^d & \text{if } 1/a < r \le 1/2 \end{cases}$$

(we will then have $\Lambda(\varepsilon) = d\ln(1/4\varepsilon) + 2 - \ln 4$). Then for $\varepsilon < 1/(4\sqrt[d]{4})$ the inequality (3.68) can be written more precisely as

$$\mathsf{E}_x\tau_\varepsilon \le \frac{2^d\big(d\ln(1/\varepsilon) + 2 - (d+1)\ln 4\big)}{F_{\varepsilon,x}} \left(2 + d\left(\ln \frac{\delta_2(x)}{\sqrt[d]{2}\,\varepsilon}\right)^+\right) + 1,$$

where $t^+ = \max\{t, 0\}$ and

$$\delta_2(x) = \min\left\{\delta(x), 1/\left(4\sqrt[d]{2}\right)\right\}.$$

Consider now the optimization space $(\mathbb{R}^d, \rho)$. Let $f(\cdot)$ satisfy conditions F1 – F4 and $g = g_\varepsilon$ be given by (3.67). Introduce $\nu(r) = \psi(r)r^k$, $b = \max\{a, 2\}$ and note that $\nu(r) \to 1$ as $r \to 0$. As shown in [245], in this case for any $\varepsilon < \delta = \delta(x)$ we have

$$I(\delta,\varepsilon; f, g) \le \frac{2^d}{F_{\varepsilon,x}} \left(\nu(2\varepsilon) + d\int_{a\varepsilon}^{\infty} \frac{\nu(r)}{r}\,\mathrm{d}r\right) \left(\frac{(b/2)^d - 1}{\nu(b\varepsilon)} + \frac{1}{\nu(2\delta)} + d\int_{b\,\varepsilon}^{2\delta} \frac{\mathrm{d}r}{\nu(r)r}\right).$$

Search algorithms that do not depend on the required precision

The transition densities above depend on the required precision ε. Consider the case where the transition density does not depend on ε. Consider a Markov homogeneous symmetric monotonous search algorithm starting at some point $x \in A$ and defined through (3.57) and (3.62).

Additionally, let g have the form

$$g(r) = \nu(r)r^{-d},$$

where

$$\nu(r) = \frac{c}{(e + d|\ln r|)\ln^t(e + d|\ln r|)}, \qquad t > 1,$$

and c is the normalizing constant so that

$$\int_0^{\mathrm{diam}(A)} \nu(r)r^{-d}\,\mathrm{d}\varphi(r) = 1,$$

where $\varphi(r) = \mathrm{vol}\big(B_r(x)\big)$. Then, as shown in [240, 245],

$$\mathsf{E}_x\tau_\varepsilon \le F_{\varepsilon,x}^{-1} O\left(\ln^2\varepsilon \,\ln^t(|\ln\varepsilon|)\right),$$

where $F_{\varepsilon,x}$ is defined in (3.66). This implies that a Markov homogeneous symmetric monotonous random search algorithm with transition densities that do not depend on the required precision is asymptotically almost as good as a general Markov search can be. Therefore, its asymptotic rate of convergence is just marginally worse than the rate of convergence of a standard descent algorithm (e.g., steepest descent) for an ordinary local optimization problem.

3.5 Methods of Generations

Simulated annealing and other Markovian global random search algorithms discussed in Sects. 3.3 and 3.4 make use of some information about the objective function gained during the process of search. This information though is limited to the last observation only. As many Markovian algorithms Markovian algorithm have proven to be more practically efficient than the non-adaptive search algorithms, the possibility to use information even at one point leads to a visible improvement in efficiency.

The next step is to construct selection rules for the observation points incorporating both a relatively simple structure of the algorithms and a possibility to use more information about the objective function. The algorithms discussed below do exactly this: they are the algorithms transforming one group of points (current generation) to another group of points (next generation) by certain probabilistic rules. We shall call these algorithms 'methods of generations'. Note that this class of algorithms includes the popular 'genetic algorithms' as a subclass.

If all the generations (that is, groups of points) have the same population size (that is, the same number of observation points), then the algorithms are Markovian algorithms in the product space A^n (here n is the size of each population) rather than in the original space A for the ordinary Markovian algorithms. The technique of Markov chains can therefore be applied to study the asymptotic behaviour of these algorithms.

This section is organized as follows. In Sect. 3.5.1 we consider the general form of the algorithms In Sect. 3.5.2. various heuristics including the celebrated genetic algorithms are discussed. Sect. 3.5.3 deals with the most interesting case when each population has the same size n and the algorithms are time-homogeneous Markov chains on the space A^n.

3.5.1 Construction of Algorithms and Their Convergence

General form

We shall use the general scheme of global random search algorithms written in the form of Algorithm 2.2. We assume that the construction of the probability distributions P_{j+1} does not involve the points $x_{l_i}^{(i)}$ ($l_i = 1, \ldots, n_i$; $i = 1, \ldots, j{-}1$) and the results of the objective function evaluation at these points. That is to say, the probability distributions P_{j+1} are constructed using only the points of the j-th iteration $x_l^{(j)}$ ($l = 1, \ldots, n_j$) and the results of the objective function evaluation at these points. The corresponding algorithms will be called 'methods of generations'. For convenience of references, let us formulate the general form of these algorithms.

Algorithm 3.6 (Methods of generations: general form)

1. *Choose a probability distribution P_1 on the n_1–fold product set $A \times \ldots \times A$, where $n_1 \geq 1$ is a given integer. Set iteration number $j = 1$.*
2. *Obtain n_j points $x_1^{(j)}, \ldots, x_{n_j}^{(j)}$ in A by sampling from the distribution P_j. Evaluate the objective function $f(\cdot)$ at these points.*
3. *Check a stopping criterion.*
4. *Using the points $x_l^{(j)}$ $(l = 1, \ldots, n_j)$ and the objective function values at these points, construct a probability distribution P_{j+1} on the n_{j+1}–fold product set $A \times \ldots \times A$, where n_{j+1} is some integer that may depend on the search information.*
5. *Substitute $j+1$ for j and return to Step 2.*

We will call the set of points of the j-th iteration

$$x_1^{(j)}, \ldots, x_{n_j}^{(j)} \tag{3.69}$$

the 'parent generation' and the related set of points of the $(j+1)$-th iteration

$$x_1^{(j+1)}, \ldots, x_{n_{j+1}}^{(j+1)} \tag{3.70}$$

the 'generation of descendants', or 'children'.

To define a specific method of generations, one has to define

(a) the stopping rule,
(b) the rules for computing the numbers n_j (population sizes), and
(c) the rules for obtaining the population of descendants (3.70) from the population of parents (3.69).

Stopping rules

- The simplest stopping rule is based on counting the total number of points generated.
- If the objective function satisfies the Liptchitz-type condition, then for defining the stopping rule one can use the recommendations of Sect. 3.1.
- If a local descent is routinely used in obtaining the descendants, then the statistical techniques of Sect. 2.6.2 may be employed.
- If the population sizes n_j are large enough and all the descendants (3.70) are generated using the same probabilistic rule (which is typical, for example, in the heuristics of the methods of generations), then to devise the stopping rule one can use the statistical procedures of Sects. 2.4 and 2.5 to infer about the minimal value of $f(\cdot)$ in A and in subsets of A.

Choice of the population sizes n_j

Provided that $n_j > 1$ for at least one j, Algorithm 3.6 becomes more general than any Markovian algorithm of global random search including Algorithm 3.4 and Algorithm 3.5. Therefore, the methods of generations have more flexibility than the Markovian algorithms in adapting the search for increasing the efficiency of this search.

The choice of population sizes n_j is an important tool in creating efficient methods of generations for various classes of optimization problems. Large values of n_j are used for increasing the globality of the search. Small n_j make the search more local. In the limiting case, when for some j_0 we have $n_j = 1$ ($j \geq j_0$), the search (after iteration j_0) becomes Markovian, which is almost local.

There are no general restrictions on the ways of choosing n_j's. We distinguish the following four ways of selecting the population sizes n_j's:

(i) n_j's are random and depend on the statistical information gathered during the search;
(ii) the sequence of n_j is non-increasing: $n_1 \geq n_2 \geq \ldots \geq n_j \geq \ldots$;
(iii) $n_j = n$ for all j;
(iv) the sequence of n_j is non-decreasing: $n_1 \leq n_2 \leq \ldots \leq n_j \leq \ldots$

Using the rule (i) may lead to very efficient algorithms. If one wants to use the rule (i), we advice to use the statistical procedures and recommendations of Sects. 2.4 and 2.5.

From the practical view-point, the choice (ii) seems very natural. Indeed, in the first few iterations of Algorithm 3.6 we need to make the search more global and therefore it is normal to choose the first few n_j large. As search progresses, it is natural to assume that this search reduces the uncertainty about the minimizer/s and and it is therefore natural to narrow the search area. In this way, it is may be a good idea to keep reducing the population sizes and pay more and more and more attention to the local search (to accelerate local convergence to the minimizers).

The choice (iii) is the most convenient from the theoretical point of view. This case is studied in Sect. 3.5.3. Note that the value of n does not have to be large. Even in the case $n = 2$, Algorithm 3.6 already significantly generalizes the Markovian algorithms of Sects. 3.3 and 3.4; at the same time, its structure is still relatively simple.

Using the choice (iv) does not seem natural. However, it looks that there is no other way of achieving the convergence of all the points (3.69) (as $j \to \infty$) to the neighbourhood of the global minimizer/s in the case when the objective function is evaluated with random error, see [273], Sect. 5.2. Indeed, as long as we approach the global minimizer, we need to diminish the effect of random errors; this can only be done if $n_j \to \infty$ as j increases.

Obtaining the population of descendants from the population of parents

The major distinction between different versions of Algorithm 3.6 is related to the rules which define the way of obtaining the population of descendants (3.70) from the population of parents (3.69).

A very convenient way to obtain the population of descendants (3.70) is to obtain each descendant separately using the same probabilistic rule (this would imply that the descendants are independent random points conditionally the parent populations are given). In many algorithms, each descendant $x_l^{(j+1)}$ has only one parent $x_i^{(j)}$, where $i \in \{1,\dots,n_j\}$ is typically a number computed with a help of some probabilistic rule (see below for related heuristics). Then to move from the parent $x_i^{(j)}$ to the current descendant $x_l^{(j+1)}$, one has to perform an operation which is called 'mutation' in genetic algorithms. In probabilistic language, it corresponds to sampling some probability distribution $Q_j(x_i^{(j)},\cdot)$ which is called transition probability. Transition probabilities define the way of choosing the point of the next generation in the neighborhood of a chosen point from the parent generation.

For the sake of simplicity, the transition probabilities $Q_j(x,\cdot)$ are often chosen so that for sampling $Q_j(x,\cdot)$ one samples uniform distribution on A with a small probability $p_j \ge 0$ and an arbitrary distribution depending on x with probability $1-p_j$. For example, this distribution can be the uniform distribution on either a ball or a cube with centre at x and volume depending on j; it can even correspond to performing several iterations of a local descent algorithm starting at x. In either case, the condition $\sum_{j=1}^{\infty} p_j = \infty$ guarantees the convergence of the algorithm, see Sect. 2.1.3.

If the evaluations of the objective function are error-free, then the transition probabilities $Q_j(x,\cdot)$ can be chosen so that

$$Q_j(x,U) = \int \mathbf{1}_{[\zeta\in U, f(x)\ge f(\zeta)]} T_j(x,d\zeta) + \mathbf{1}_U(x) \int \mathbf{1}_{[f(\zeta)>f(x)]} T_j(x,d\zeta) \quad (3.71)$$

where $T_j(x,\cdot)$ are some transition probabilities. To choose a realization z of a random point with the distribution $Q_j(x,\cdot)$ as defined through (3.71), we have to obtain a random point $\zeta \in A$ with distribution $T_j(x,\cdot)$ and define

$$z = \begin{cases} \zeta & \text{if } f(\zeta) \le f(x), \\ x & \text{otherwise.} \end{cases}$$

The corresponding versions of Algorithm 3.6 become direct generalizations (for the case of arbitrary n_j) of the Markov monotonous search algorithms studied in Sect. 3.4. The transition probabilities $T_j(x,\cdot)$ in (3.71) can naturally be chosen through their transition densities defined as in (3.72).

One cannot use (3.71) for defining the transition probabilities $Q_j(x,\cdot)$ when the objective function evaluations are subject to noise. In the presence of noise, the transition probabilities can naturally be defined through the corresponding transition densities $q_j(x,z)$[4] which have the form

$$q_j(x,z) = c_j(x)\varphi\left((z-x)/\beta_j\right) \quad (3.72)$$

where $\beta_j > 0$, $\varphi(\cdot)$ is a continuous, symmetric at 0, bounded density in $\mathbb{R}^d$ and $c_j(x) = 1/\int_A \varphi\left((z-x)/\beta_j\right)dz$ is the normalization constant taking care of the boundary effect.

[4] the transition density corresponding to the transition probability $Q_j(x,\cdot)$ is a function $q_j(x,z)$ such that $Q_j(x,U) = \int_{z\in U} q_j(x,z)dz$ for all $U \in \mathcal{B}$

To obtain a realization z of the random point in A with the transition probability $Q_j(x, \cdot)$ defined through its transition density (3.72), one must obtain a realization ζ of the random point in $\mathbb{R}^d$ distributed with the density $\varphi(\cdot)$, check the inclusion $x + \beta_k \zeta \in A$ (otherwise, to obtain a new realization ζ) and set $z = x + \beta_k \zeta$.

Conditions guaranteeing convergence

Assume that the conditions C1 – C9 of Sect. 2.1.1 concerning the feasible region and the objective function are met. Then the transition probabilities $Q_j(x, \cdot)$ can easily be chosen to satisfy the condition (2.6) which guarantees the convergence of Algorithm 3.6 in the sense of Theorem 2.1. For example, if these probabilities have transition densities (3.72),

$$\inf_{x,z \in A} \phi(x - z) > 0 \quad \text{and} \quad \sum_{j=1}^{\infty} \beta_j n_{j+1} = \infty, \tag{3.73}$$

then the condition (2.6) is met and Algorithm 3.6 converges.

Similarly, if the transition probabilities are (3.71) where the transition densities of $T_j(x, \cdot)$ have the form (3.72) and satisfy (3.73), then the condition (2.6) is met and Algorithm 3.6 converges. Assume, additionally, that the global minimizer x_* is unique and $\beta_j \to 0$ as $j \to \infty$. Then for many versions of Algorithm 3.6 (as Algorithms 3.7 – 3.9 considered below) the sequence of distributions of points $x_i^{(j)}$ weakly converges (as $j \to \infty$) to the delta-measure concentrated at the global minimizer x_*. To achieve this convergence, the choice of the points $x_i^{(j)}$ from the parent generation should be made in such a way that the points with smaller values of the objective function should have larger chances of being selected as parents than the points with larger values of $f(\cdot)$.

This can be easily done if the observations of the objective function are error-free. Achieving the convergence of the sequence of distributions of points $x_i^{(j)}$ to the delta-measure concentrated at x_* is more difficult (but still possible) when the observations of $f(\cdot)$ are subject to noise. Consider, for example, Algorithm 3.10 below and assume that the transition probabilities $Q_j(x, \cdot)$ have transition densities of the form (3.72) and the conditions (3.73) are met (these conditions guarantee the convergence of the algorithm in the sense of Theorem 2.1). The next condition to be met is $\sum_{j=1}^{\infty} \beta_j < \infty$; this condition guarantees that the points $x_i^{(j)}$ do not move away from x_* after they reach the vicinity of this point (see [273], Sect. 5.2, for exact formulation of the convergence results). It seems that the condition $\sum_{j=1}^{\infty} \beta_j < \infty$ contradicts to (3.73). It does not since $n_j \to \infty$ (as $j \to \infty$); see choice (iv) of the sequence of population sizes $\{n_j\}$. The condition $n_j \to \infty$ as $j \to \infty$ diminishes the influence of random errors for large j; it is a necessary condition of convergence in the case when there are errors in observations. Note that the study of convergence of the methods of generations performed in [273] for the case when there are errors in observations is based on the asymptotic representation (3.78).

3.5.2 Special Cases; Genetic Algorithms

There are many classes of global random search algorithms that can be considered as particular versions of the method of generations represented in the form of Algorithm 3.6. These classes clearly include pure random search and pure adaptive search of any order (see Sect. 2.2), Markovian algorithms (see Sect. 3.3, 3.4) and even the random multistart algorithm considered in Sect. 2.6.2. Indeed, this algorithm can be represented as Algorithm 3.6 with two iterations: the first population consists of i.i.d. random points in A and the second population consists of the local minimizers reached from the points of the first population.

Special cases and associated heuristics

Let us now describe less obvious heuristics that can be put in the context of the methods of generations and are often used in practice of global random search. Note that in the following two algorithms each descendant has only one, specially selected, parent from the previous population.

Algorithm 3.7

1. *Sample some number n_1 times a distribution P_1, obtain points $x_1^{(1)}, \ldots, x_{n_1}^{(1)}$; set iteration number $j = 1$.*
2. *From all previously obtained points*

$$x_{l_i}^{(i)} \quad (l_i = 1, \ldots, n_i;\ i = 1, \ldots, j) \tag{3.74}$$

   choose k points $x_{1*}^{(j)}, \ldots, x_{k*}^{(j)}$ having the smallest values of $f(\cdot)$.
3. *Check a stopping criterion.*
4. *Determine some non-negative integers $n_{j,l}$ $(l = 1, \ldots, k)$ so that $\sum_{l=1}^{k} n_{j,l} = n_{j+1}$. To obtain the points (3.69) sample $n_{j,l}$ times some distributions (transition probabilities) $Q_j(x_{l*}^{(j)}, \cdot)$ for $l = 1, \ldots, k$.*
5. *Substitute $j+1$ for j and return to Step 2.*

Algorithm 3.7 have some resemblance to the pure adaptive search of order k. The most known version of Algorithm 3.7 is when $k = 1$, that is in the subsequent iterations search is restricted only to the neighbourhood of the record point; in this case, it looks like a Markov monotonous search algorithm of Sect. 3.4. Although various modifications of Algorithm 3.7 can be successfully used in practical applications, it is difficult to study the non-asymptotic globality properties of this algorithm. The following algorithm generalizes Algorithm 3.7 by introducing a probabilistic model into the method of obtaining points of the next generation. This method will coincide with our basic probabilistic model in the case when the objective function is evaluated without errors.

Algorithm 3.8

1. *Sample n_1 times a distribution P_1, obtain points $x_1^{(1)}, \ldots, x_{n_1}^{(1)}$; set iteration number $j = 1$.*
2. *Check a stopping criterion.*
3. *Construct an auxiliary non-negative function $f_j(\cdot)$ using the results of evaluation of $f(\cdot)$ at all previously obtained points.*
4. *Sample the distribution*

$$P_{j+1}(\cdot) = \sum_{l=1}^{n_j} p_l^{(j)} Q_j(x_l^{(j)}, \cdot) \tag{3.75}$$

where

$$p_l^{(j)} = f_j(x_l^{(j)}) \Big/ \sum_{i=1}^{n_j} f_j(x_i^{(j)})$$

and thus obtain the points (3.69) of the next iteration.
5. *Substitute $j+1$ for j and return to Step 2.*

The distribution (3.75) is sampled by the superposition method: the discrete distribution

$$\varepsilon_j = \left\{ \begin{matrix} x_1^{(j)}, \ldots, x_{n_j}^{(j)} \\ p_1^{(j)}, \ldots, p_{n_j}^{(j)} \end{matrix} \right\} \tag{3.76}$$

is sampled first; this is followed by sampling the distribution $Q_j(x_t^{(j)}, \cdot)$, where $x_t^{(j)}$ is the realization obtained from sampling the distribution (3.76). As this procedure is repeated n_{j+1} times, the point $x_l^{(j)}$ $(l = 1, \ldots, n_j)$ is chosen on average $n_{j+1}p_l^{(j)}$ times while sampling the distribution (3.76).

Since the functions $f_j(\cdot)$ are arbitrary, they may be chosen in such a way that these average values $n_{j+1}p_l^{(j)}$ are equal to the numbers $n_{j,l}$ of Algorithm 3.7. Allowing for this fact and for the possibility of using quasi-random points from the distribution (3.76), we can conclude that Algorithm 3.7 is a special case of Algorithm 3.8.

In theoretical studies of Algorithm 3.8 (more precisely, of Algorithm 3.9, which is a generalization of Algorithm 3.8 to the case when the functions $f_j(\cdot)$ are evaluated with random errors) it will be assumed that the discrete distribution (3.76) is sampled in a standard way, i.e. independent realizations of a random variable are generated with this distribution. In practical calculations, it is more advantageous to generate quasi-random points from this distribution by means of the following procedure that is well known in the regression design theory (see e.g. [73]) as the construction procedure of exact designs from approximate designs. For $l = 1, \ldots, n_j$, set $r_l^{(j)} = \lfloor n_{j+1}p_l^{(j)} \rfloor$, the greatest integers which are smaller than or equal to $n_{j+1}p_l^{(j)}$,

$$n_{(j)} = \sum_{l=1}^{n_{j+1}} r_l^{(j)}, \quad n^{(j)} = n_{j+1} - n_{(j)}, \quad \alpha_l^{(j)} = n_{j+1} p_l^{(j)} - r_l^{(j)} \ (l = 1, \ldots, n_j).$$

Then we have

$$\varepsilon_j = (n_{(j)}/n_{j+1})\varepsilon_j^{(1)} + (n^{(j)}/n_{j+1})\varepsilon_j^{(2)},$$

where

$$\varepsilon_j^{(1)} = \left\{ \begin{array}{ccc} x_1^{(j)} & ,\ldots, & x_{n_j}^{(j)} \\ r_1^{(j)}/n_{(j)} & ,\ldots, & r_{n_j}^{(j)}/n_{(j)} \end{array} \right\}, \quad \varepsilon_j^{(2)} = \left\{ \begin{array}{ccc} x_1^{(j)} & ,\ldots, & x_{n_j}^{(j)} \\ \alpha_1^{(j)}/n^{(j)} & ,\ldots, & \alpha_{n_j}^{(j)}/n^{(j)} \end{array} \right\}.$$

Instead of sampling (3.76) n_{j+1} times, we choose $r_l^{(j)}$ times the points $x_l^{(j)}$ for $l = 1, \ldots, n_j$ and sample $n^{(j)}$ times the distribution $\varepsilon_j^{(2)}$.

The above procedure reduces the indeterminacy in the selection of points $x_l^{(j)}$ in whose vicinity the next generation points are chosen according to $Q_j(x_l^{(j)}, \cdot)$. If we use this procedure, these points $x_l^{(j)}$ include several record points from the preceding iteration. Note that the procedure can be applied to any set of weights $p_l^{(j)}$ (and is, therefore, applicable in the case when evaluations of $f(\cdot)$ are subject to random noise).

The efficiency of different versions of Algorithm 3.8 depends on the choice of the functions $f_j(\cdot)$. These functions should be non-negative and reflect the properties of $-f(\cdot)$ (e.g., be on the average larger, where $f(\cdot)$ is great and smaller, where $f(\cdot)$ is large). A natural choice of the functions $f_j(\cdot)$ is

$$f_j(x) = \exp\{-\alpha_j f(x)\}, \quad \alpha_j > 0 \tag{3.77}$$

or simply $f_j(x) = \max\{0, -\alpha_j f(x)\}$. The choice of the values α_j should depend on the prior information about $f(\cdot)$, various estimates of $f(\cdot)$ constructed during the search and the compromise between the required 'peakness' and smoothness of $f_j(\cdot)$.

If the objective function $f(\cdot)$ is evaluated with random error, then statistical estimates $\hat{f}_j(\cdot)$ of $f_j(\cdot)$ has to be used in place of $f_j(\cdot)$ in Algorithm 3.8; see Sect. 5.2 in [273] for detailed consideration and convergence study of the corresponding methods. The convergence study is based on the asymptotic representation (3.78).

The measures $P_{j+1}(\cdot)$, $j = 1, 2, \ldots$, defined through (3.75), are the distributions of random points $x_i^{(j+1)}$ conditioned on the results of previous evaluations of $f(\cdot)$. Denote the corresponding unconditional (average) distributions as $P(j+1, n_j; \cdot)$. Then, as shown in [273], Corollary 5.2.2, for any $j = 1, 2, \ldots$ the distributions $P(j+1, n_j; \cdot)$ converge in variation as $n_j \to \infty$ to the limiting distributions $P_{j+1}(\cdot)$ and

$$P_{j+1}(dx) = \left[\int P_j(dz) f_j(z) \right]^{-1} \int P_j(dz) f_j(z) Q_j(z, dx). \tag{3.78}$$

This is a general result valid in the case when $f(\cdot)$ is evaluated with random error (the errors have to be concentrated on a bounded interval).

Genetic algorithms

The following algorithm is a modification of Algorithm 3.8 for the case when each descendant has two parents rather than one (as in Algorithms 3.7 and 3.8). Different heuristics for the case of two parents have received a great popularity because of sexual analogues, see e.g. [115, 151, 161, 185, 195]. Our claim, though, is that these algorithms are often practically less efficient than the simpler versions of the methods of generations where each descendant has only one parent. Note however that the genetic algorithms (similarly to the case of the simulated annealing algorithms) are usually applied for solving discrete optimization problems.

Algorithm 3.9.

1. *Sample n_1 times a distribution P_1, obtain points $x_1^{(1)}, \ldots, x_{n_1}^{(1)}$; set iteration number $j = 1$.*
2. *Check a stopping criterion.*
3. *Construct an auxiliary function $f_j : A \times A \to (0, \infty)$ using the results of evaluating $f(\cdot)$ at all previously obtained points.*
4. *Sample the distribution*

$$P_{j+1}(\cdot) = \sum_{i=1}^{n_j} \sum_{l=1}^{n_j} p_{il}^{(j)} Q_j \left(x_i^{(j)}, x_l^{(j)}, \cdot \right) \tag{3.79}$$

where

$$p_{il}^{(j)} = f_j \left(x_i^{(j)}, x_l^{(j)} \right) \bigg/ \sum_{u=1}^{n_j} \sum_{v=1}^{n_j} f_j \left(x_u^{(j)}, x_v^{(j)} \right)$$

and thus obtain the points (3.69) of the next iteration.

5. *Substitute $j+1$ for j and return to Step 2.*

The measures $Q_j(y, z, \cdot)$ in (3.79) are the transition probabilities from two points $y, z \in A$ of previous generation to a point of the next generation; for fixed $y, z \in A$, $Q(y, z, \cdot)$ is a probability measure on A. Different heuristics of how two parents, that is y and z, are to be selected correspond to different construction rules of functions $f_j(y, z)$; heuristics describing what kind of children these two parents produce correspond to the choice of the transition probability $Q(y, z, \cdot)$. One of the natural choices of $f_j(y, z)$ is

$$f_j(y, z) = \exp\{-\alpha_j [f(y) + f(z)]\} \tag{3.80}$$

which is an extension of (3.77).

Unlike the case when each descendant has only one parent, to define the transition probabilities $Q_j(y, z, \cdot)$ it is not enough to define the algorithm of sampling in a neighbourhood of a given point ('mutation'); we also need to define the ways of how two parents interact (so-called 'recombination'). Many different ways of defining recombination have been suggested. However, unlike the mutation (which is a perfectly understandable procedure in terms of the geometry of original space), the recombination does not have natural geometrical meaning, at least if in the case $A \subset \mathbb{R}^d$. Indeed, choosing a point (descendant) in a neighbourhood of a given point (parent) is a natural operation. On the other hand, there is no naturally defined neighbourhood of two points in $\mathbb{R}^d$, especially if they are located far away from each other.

3.5.3 Homogeneous Transition Probabilities

In the present section we consider the case where $n_k = n$ is constant and the transition probabilities $Q_k(z, dx)$ are time-homogeneous, that is $Q_k(z, dx) = Q(z, dx)$. In this case the search algorithms above behave like ordinary Markov chains and the study of their asymptotic properties reduces to convergence study to stationary distributions and to study of properties of these stationary distributions.

Let us introduce some notations that will be used below.

Let A be a compact metric space; $\mathcal{B}$ be the σ-algebra of Borel-subsets of A; $\mathcal{M}$ be the space of finite signed measures, i.e. regular (countable) additive functions on $\mathcal{B}$ of bounded variation; $\mathcal{M}_+$ be the set of finite measures on $\mathcal{B}$ ($\mathcal{M}_+$ is a cone in the space $\mathcal{M}$); $\mathcal{M}^+$ be the set of probability measures on $\mathcal{B}(\mathcal{M}^+ \subset \mathcal{M}_+)$; $C_+(A)$ be the set of continuous non-negative functions on A ($C_+(A)$ is a cone in $C(A)$, the space of continuous functions on A); $C^+(A)$ be the set of continuous positive functions on A ($C^+(A)$ is the interior of the cone $C_+(A)$); a function K: $A \times \mathcal{B} \to \mathcal{R}$ be such that $K(\cdot, U) \in C_+(A)$ for each $U \in \mathcal{B}$ and $K(x, \cdot)) \in \mathcal{M}_+$ for each $x \in A$. The analytical form of K may be unknown, but it is required that for any $x \in A$ a method be known for evaluating realizations of a non–negative random variable $y(x)$ such that

$$Ey(x) = g(x) = K(x, A), \qquad \operatorname{var} y(x) \le \sigma^2 < \infty,$$

and of sampling the probability measure $Q(x, dz) = K(x, dz)/g(x)$ for all $x \in \{x \in A : g(x) \neq 0\}$.

Denote by $\mathcal{K}$ the linear integral operator from $\mathcal{M}$ to $\mathcal{M}$ by

$$\mathcal{K}\nu(\cdot) = \int \nu(dx) K(x, \cdot). \tag{3.81}$$

The conjugate operator $Ł = \mathcal{K}^* : C(A) \to C(A)$ is defined as follows

$$Łh(\cdot) = \int h(x) K(\cdot, dx). \tag{3.82}$$

As it is known from the general theory of linear operators, any bounded linear operator mapping from a Banach space into $C(A)$ is representable as (3.82) and $\|\text{Ł}\| = \|\mathcal{K}\| = \sup g(x)$. Moreover, the operators $\mathcal{K}$ and Ł are completely continuous in virtue of the compactness of A and continuity of $K(\cdot, U)$ for all $U \in \mathcal{B}$.

The theory of linear operators in a space with a cone implies that a completely continuous and strictly positive operator Ł has eigen-value λ that is maximal in modulus, positive, simple and at least one eigen-element belonging to the cone corresponds to it; the conjugate operator Ł* has the same properties.

In the present case, the operator Ł is determined by (3.82). It is strictly positive if for any non-zero function $h \in C_+(A)$ there exists $m = m(h)$ that Ł$^m h(\cdot) \in C^+(A)$ where Ł^m is the operator with kernel

$$\int \dots \int K(\cdot, dx_1)K(x_1, dx_2)\dots K(x_{m-2}, dx_{m-1})K(x_{m-1}, \cdot).$$

Thus, if the operator Ł $= \mathcal{K}^*$ is strictly positive (which is assumed to be the case), the maximal in modulus eigen-value λ of $\mathcal{K}$ is simple and positive; a unique eigen-measure P in $\mathcal{M}^+$ defined by

$$\lambda P(dx) = \int P(dz)K(z, dx) \tag{3.83}$$

corresponds to it and λ is expressed in terms of this measure as

$$\lambda = \int g(x)P(dx). \tag{3.84}$$

It is evident from (3.83) and (3.84) that if $\lambda \neq 0$, then the necessary and sufficient condition that P is a unique in $\mathcal{M}^+$ eigen-measure of $\mathcal{K}$ is as follows: P is a unique in $\mathcal{M}^+$ solution of the integral equation

$$P(dx) = \left[\int g(z)P(dz)\right]^{-1} \int P(dz)K(z, dx). \tag{3.85}$$

Theorem 3.9. *Let the conditions C1 – C9 of Section 2.1.1 be satisfied; assume that $Q(z, dx) \geq c_2\mu(dx)$ for μ-almost all $z \in A$ where $c_2 > 0$. Then*

1. *for any $n = 1, 2, \dots$ the random elements $a_k = (x_1^{(k)}, \dots, x_n^{(k)})$, $k = 1, 2, \dots$, constitute a homogeneous Markov chain with stationary distribution $R_n(dx_1, \dots, dx_n)$, the random elements with this distribution being symmetrically dependent;*
2. *for any $\varepsilon > 0$ there exists $n_* \geq 1$ such that for $n \geq n_*$ the marginal distribution*

$$R_{(n)}(dx) = R_n(dx, A, \dots, A)$$

differs in variation from $P(dx)$ at most by ε.

Let us now demonstrate that under some conditions the eigen-measures $P(\cdot)$ of linear operators (3.81) are close to $\varepsilon^*(dx)$ that is δ-measure concentrated at the global minimizer of $g(\cdot)$. In another words we demonstrate that the problem of determining the global minimizer of $g(\cdot)$ can be regarded as a limit case of determining the eigen-measures P of integral operators (3.81) with kernels $K_\beta(x,dz) = g(x)Q_\beta(x,dz)$ where the Markovian transition probabilities $Q_\beta(x,dz)$ weakly converge to $\varepsilon_x(dz)$ for $\beta \to 0$.

In order to relieve the presentation of unnecessary *details*, assume that $A = \mathbb{R}^d$, $\mu = \mu_n$ and that $Q_\beta(x,dz)$ are chosen by (3.72) with $\beta_k = \beta$, i.e.

$$Q_\beta(x,dz) = \beta^{-n}\varphi((z-x)/\beta)\mu_n(dz). \tag{3.86}$$

Lemma 3.10. *Let the transition probability $Q = Q_\beta$ have the form* (3.86), *where φ is a continuously differentiable distribution density on $\mathbb{R}^d$,*

$$\int \|x\|\varphi(x)\mu_n(dx) < \infty$$

$g(\cdot)$ be positive, satisfy the Lipschitz condition with a constant L, attain the global maximum at the unique point x_*, and $g(x) \to 0$ for $\|x\| \to \infty$. Then for any $\varepsilon > 0$ and $\delta > 0$, there exists $\beta > 0$ such that $P(B(\delta)) \geq 1 - \varepsilon$ where P is the probabilistic solution of (3.85).

Heuristically, the statement of Lemma 3.10 can be illustrated by the following reasoning. In the case studied, $P(dx)$ has a density $p(x)$ that may be obtained as the limit (for $k \to \infty$) of recurrent approximations

$$p_{k+1}(x) = s_{k+1} \int p_k(z)g(z)\varphi_\beta(x-z)\mu_n(dz) \tag{3.87}$$

where

$$\varphi_\beta(x) = \beta^{-n}\varphi(x/\beta), \qquad s_{k+1} = 1 \bigg/ \int p_k(z)g(z)\mu_n(dz).$$

(3.87) implies that p_{k+1} is a kernel estimator of the density $s_{k+1}p_k f$, where the parameter β is called *window width.* One can anticipate that for a small β the asymptotic behaviour of densities (3.87) should not differ very much from that of distribution densities $const\, f^k(x)$, which converge to $\varepsilon^*(dx)$.

numerical calculations have revealed the fact that the problems resembling realistic ones (for not *too bad* functions f) the eigen-measures $P = P_\beta$ explicitly tend, for small β, to concentrate mostly within a small vicinity of the global minimizer x_* (or the minimizers). Moreover, the tendency mentioned manifests itself already for not very *small* β (say, of the order of 0.2 to 0.3, under the unity covariance matrix of the distribution with the density φ).

now, the possibility of using the homogenous methods of generations of this section for searching the global maximum of $g(\cdot)$ is based on that all search points in these algorithms have asymptotically the distribution $P(dx)$ that can be brought *near* to a distribution concentrated at the set $A^* = \{\arg\max f\}$ of global minimizers.

Mention that the algorithms of independent random sampling of points in A can also be classified as methods of generations if one assumes that $Q(x, dz) = P_1(dz)$. For these algorithms $P(dx) = P_1(dx)$ and the points generated by them, therefore, do not tend to concentrate in the vicinity of A^* and, from the viewpoint of asymptotic behaviour, they are defeated by those methods of generations whose stationary distributions $P(dx)$ are concentrated near to A^*. (This way, the situation here is similar to the situation concerning the simulated annealing method).

3.6 Proofs

Proof of Theorem 3.1.

To prove Theorem 3.1 we need the following Lemma.

Lemma 3.1. *Let $B_r(O)$ be a $d-$dimensional Eucledian ball in $\mathbb{R}^d$ of radius r with centre at a point O, C be a cube in $\mathbb{R}^d$ with edge length h and z be a point in $\mathbb{R}^d$. If $z \in C \bigcap B_r(O)$ then $C \subset B_{r+h\sqrt{d}}(O)$.*

Proof of Lemma 3.1. Let $z \in C \bigcap B_r(O)$ and x be any point in C then the inequality $\rho(z, x) \le h\sqrt{d}$ is valid (this inequality becomes the equality if and only if x and z are opposite vertices in C). By the triangle inequality we have $\rho(O, x) \le \rho(O, z) + \rho(z, x) \le r + h\sqrt{d}$ implying $C \subset B_{r+h\sqrt{d}}(O)$. □

Proof of Theorem 3.1. Let us take any subcube $A' \subset A$, constructed by the neighbouring points of the cubic grid C_k; these points are the vertices of A'. The length of the edges of the cube A' is $1/k$. Denote by z' the centre of a cube A'.

Let us relate the subcube A_z with the main diagonal zz' to each vertex z of the cube A'. Considering all subcubes A' of A we therefore consider all points z of the grid C_k and $\bigcup_{A_z} = A$. Note that the edge length of the subcubes A_z is equal to $h = 1/(2k)$.

Since $z \in B_{\varepsilon_j}(x_j)$ for some $x_j \in X_n$ and corresponding ε_j, by Lemma 3.1 the cube A_z belongs to $B_{\varepsilon_j+h\sqrt{d}}(x_j) \subset \bigcup_{i=1}^n B_{\varepsilon+h\sqrt{d}}(x_i)$. Since any subcube A_z is covered by $\bigcup_{i=1}^n B_{\varepsilon_i+h\sqrt{d}}(x_i)$ and the union of all subcubes A_z is exactly A, the whole cube A is also covered completely; that is, (3.4) holds. □

Proof of Theorem 3.3.

If the condition (2.42) holds for $F = F_f$, then for the independent sample we derive in a standard way

$$\mathbf{E}(f_*[X_{1,n}] - m)^p \sim (\kappa_n - m)^p \,\Gamma(p/\alpha + 1)\,, \quad n \to \infty\,. \tag{3.88}$$

Consider now the stratified sample. Recall that $f(\cdot)$ is a continuous function and its global minimum is attained at a single point $x_* = x_*(f)$. Denote by $A_{i_\diamond}$ that set among the collection of sets $\{A_1, \dots, A_k\}$, which contains the point x_*. Conditions of the theorem imply that the probability that the point x_* is on the boundary of $A_{i_\diamond}$ is zero. This implies that for certain $t_0 > m$ the set $\{x : f(x) < t_0\}$ is a subset of $A_{i_\diamond}$. For all $t \in [m, t_0]$ we have the following expression for the c.d.f. $F_\diamond(t) = P_{i_\diamond}(f(x) \le t)$:

$$F_\diamond(t) = kP(f(x) \le t) = kF(t). \tag{3.89}$$

Let $\kappa_{\diamond l}$ be the $1/l$-quantile of the c.d.f. $F_\diamond(t)$. For k large enough, $\kappa_{\diamond l}$ belongs to the set where the representation (3.89) is valid, that is $\{x : f(x) < \kappa_{\diamond l}\} \subseteq A_{i_\diamond}$.

The representation (3.89) yields $\kappa_n = \kappa_{\diamond l}$. Indeed,

$$F_\diamond(\kappa_{\diamond l}) = 1/l \Leftrightarrow kF(\kappa_{\diamond l}) = 1/l \Leftrightarrow F(\kappa_{\diamond l}) = \frac{1}{kl} = \frac{1}{n}.$$

As $F(\cdot)$ satisfies (2.42), for all z, $0 < z < l^{1/\alpha}$, we have

$$F_\diamond(m+(\kappa_n - m)z) \sim k\left(1 - \exp\left\{-\frac{z^\alpha}{n}\right\}\right) \sim \frac{z^\alpha}{l} \quad \text{as } k \to \infty. \tag{3.90}$$

Hence, for all $y \in (0, 1)$ we obtain

$$F_\diamond^{-1}(y) - m \sim (\kappa_n - m)\,(ly)^{1/\alpha}, \quad k \to +\infty,$$

and for every $p > 0$ we obtain as $k \to \infty$:

$$\mathbf{E}(f_*[X_{k,l}] - m)^p = l \int_{-\infty}^{+\infty} (x-m)^p\,(1 - F_\diamond(x))^{l-1}\, dF_\diamond(x) =$$

$$l \int_0^1 (F_\diamond^{-1}(y) - m)^p (1-y)^{l-1} dy \sim l \int_0^1 \left((\kappa_n - m)(ly)^{1/\alpha}\right)^p (1-y)^{l-1} dy$$

$$= (\kappa_n - m)^p \frac{\Gamma(p/\alpha + 1) l^{p/\alpha} \Gamma(l+1)}{\Gamma(p/\alpha + l + 1)}.$$

This and (3.88) yield (3.41) completing the proof of the theorem. □

Proof of Theorem 3.3.

Let $f(\cdot)$ be an arbitrary function in $\mathcal{F}$. Then the c.d.f. $F_1(f, t)$ for the independent sampling procedure $\mathcal{P}_1 = (f_*[X_{1,n}], Q_{1,n})$ is

$$F_1(f,t) = \Pr\{f_*[X_{1,n}] \le t\} = 1 - \Pr\{f(x_{1,j}) > t, j=1,\dots,n\}$$
$$= 1 - (\Pr\{f(x_{1,j}) > t\})^n = 1 - P^n(A_t)$$

where $A_t = f^{-1}((t,\infty))$ is the inverse image of the set (t,∞). Since $\{A_i\}_{i=1}^k$ is a complete system of events, we have

$$P(A_t) = \sum_{i=1}^{k} P(A_t \cap A_i) = \sum_{i=1}^{k} \beta_i$$

where $\beta_i = P(A_t \cap A_i)$, $i = 1,\ldots,k$, and $\sum_{i=1}^k \beta_i = P(A_t) \le 1$. We thus have

$$F_1(f,t) = 1 - \left(\sum_{i=1}^{k} \beta_i\right)^n .$$

Similarly, for the stratified sampling procedure $\mathcal{P}_{k,L} = (f_*[X_{k,L}], Q_{k,L})$, the c.d.f. $F_{k,L}(f,t)$ can be written as

$$F_{k,L}(f,t) = 1 - \Pr\{f(x_{1,1})>t,\ldots,f(x_{1,l_1})>t,\ldots,f(x_{k,l_k})>t\} =$$

$$1 - \prod_{i=1}^{k} [\Pr\{f(x_{i,1})>t\}]^{l_i} = 1 - \prod_{i=1}^{k} [P(\{f(x_{i,1})>t\} \cap A_i)/q_i]^{l_i} = 1 - \prod_{i=1}^{k} \left(\frac{\beta_i}{q_i}\right)^{l_i} .$$

For every $i = 1,\ldots,k$, we set

$$\gamma_i = l_i/n \text{ and } \alpha_i = \beta_i/q_i = P(A_t \cap A_i)/P(A_i) .$$

We have $0 < \gamma_i < 1$, $0 \le \alpha_i \le 1$ for $i = 1,\ldots,k$, $\sum_{i=1}^k \gamma_i = 1$; note also that the vector $\alpha = (\alpha_1,\ldots,\alpha_k)$ may get any value in the interior of the cube $[0,1]^k$ depending on $f(\cdot)$ and t.

Thus, the following two inequalities are equivalent:

$$F_{k,L}(f,t) \ge F_1(f,t) \iff \prod_{i=1}^{k} \alpha_i^{\gamma_i} \le \sum_{i=1}^{k} q_i \alpha_i ;$$

we rewrite this in a more convenient form

$$F_{k,L}(f,t) \ge F_1(f,t) \iff \ln\left(\sum_{i=1}^{k} q_i\alpha_i\right) \ge \sum_{i=1}^{k} \gamma_i \ln \alpha_i . \qquad (3.91)$$

Similar equivalence takes place when $\ge$ in (3.91) is replaced with the strict inequality $>$.

Let us now prove (i). If (3.39) holds then $\gamma_i = q_i$ for all $i = 1,\ldots,k$ and the validity of the second inequality in (3.91), for every $\alpha \in [0,1]^k$ and thus for every $f \in \mathcal{F}$, follows from the concavity of the logarithm. Consider a function $f^* \in \mathcal{F}$ such that $0 \le f^*(\cdot) \le 1$, $f^*(x) = 1$ for all $x \in A_1$ and $\min_{x\in A_2} f^*(x) = 0$. Then $\alpha_1 = P(A_t \cap A_1)/P(A_1) = 1$ and $\alpha_2 = P(A_t \cap A_2)/P(A_2) < 1$ for all $t \in (0,1) = (\inf f, \sup f)$. Therefore the values α_i are not all equal each other and the strict concavity of the logarithm implies the strict inequality in (3.91).

To prove (ii), we assume that (3.39) does not hold, which implies that there exists $i_0 \leq k$ such that $\gamma_{i_0} < q_{i_0}$. Consider a function $f^* \in \mathcal{F}$ such that $0 \leq f^*(\cdot) \leq 1$, $f^*(x) = 1$ for all $x \in A \setminus A_{i_0}$ and $\min_{x \in A_{i_0}} f^*(x) = 0$. Then $\alpha_j = 1$ for all $j \neq i_0$ and α_{i_0} gets all possible values in $(0, 1)$ depending on t. It is straightforward to show that for this function $F_{k,L}(f^*, t) < F_1(f^*, t)$ for all t sufficiently close to 1, which proves (ii). $\square$

4

Methods Based on Statistical Models of Multimodal Functions

4.1 Statistical Models

4.1.1 Introduction

In many global optimization problems the information about properties of an objective function is very scarce. Such problems can be described as 'black box' optimization problems. They are frequently attacked by heuristic methods. Theoretically justified methods also can be developed in the general framework of rational decision making under uncertainty. The models of functions under uncertainty developed in probability theory are stochastic functions: random processes in the case of functions of one variable, and random fields in the case of functions of many variables. Assuming that the objective function is a sample function of a random process/field it would be attractive to construct the method of the best average performance with respect to the chosen statistical model. Stochastic models of objective functions are also helpful for the theoretical research on average complexity of global optimization problems; see e.g. [36].

The first paper on global optimization based on arguments of average optimality with respect to a statistical model of the objective function was the paper by Kushner [140]. The one-dimensional case has been considered, and the model used was a Wiener process. The constructed algorithm seemed very promising, and its extension to the case of observations in the presence of noise was straightforward [141]. Despite the attractiveness of the approach it was not free from some disadvantages. Some controversy with respect to the acceptability of the Wiener process for a statistical model is discussed below.

The choice of a statistical model for constructing global optimization algorithms should be justified by the usual arguments of adequacy and simplicity. This is not always an easy task. The theory of stochastic functions is well developed from the point of view of probabilistic analysis. However, the properties known from the theory of stochastic functions are not always helpful in constructing global optimization algorithms. On the other hand, important properties for global optimization are largely outside of the interests of probability theoreticians. For example, a numerically tractable stochastic function with prescribed multimodality characteristics would be of great interest for global optimization. A well researched stochastic function, whose properties are consistent with the general information on a targeted optimization problem, is normally chosen for the statistical model if it is sufficiently simple for algorithmic implementation. Since the Wiener process was the first statistical model used for global optimization, many general discussions on advantages and disadvantages of the approach were related to its properties, e.g. the Wiener process seems to be an acceptable model as long as its local properties can be ignored. In this chapter we consider the problems of choice/construction of statistical models for global optimization in detail. The applicability of the traditional statistical models is discussed. In Sect. 4.1.4 the models which are computationally simpler that traditional models are constructed using the ideas of rational decision theory.

The case of algorithms which can be justified by both, stochastic and deterministic, approaches are of special interest. In Sect. 4.3.3 it is shown that the P-algorithm based on the Wiener process model coincides with the one-dimensional algorithm based on a radial function model.

4.1.2 Random Processes

Let us start with the first statistical model applied in global optimization, namely the Wiener process. It is not only used in global optimization but it is also an important model in the average case analysis of various numerical problems, e.g. one-dimensional interpolation, integration and zero finding [201]. The Wiener process $\xi(x)$, $x \geq 0$, is a Gaussian random process with zero mean and covariance function $cov(\xi(x_1), \xi(x_2)) = \sigma_0^2 \min(x_1, x_2)$, where σ_0^2 is a parameter. The increments of the Wiener process $\xi(x + \delta) - \xi(x)$ are Gaussian random variables $N(0, \sigma_0^2\delta)$, where the mean value is zero, and the variance is equal to $\sigma_0^2\delta$; the increments corresponding to disjoint time intervals are independent. If a target problem is expected to be very complicated, with many local minima, then the assumption of independence of increments of function values for sufficiently distant disjoint subintervals seems acceptable. This is the main argument in favor of the global adequacy of the Wiener process to model complicated multimodal functions. In Fig. 4.1 two sampling functions of a standard Wiener process are presented to illustrate their general behavior. The Wiener process is favorable also from the computational point of view because it is Markovian. However, the sampling functions of the Wiener process are not differentiable with probability one, almost everywhere. This feature draws criticism for the use of the Wiener process as a model because objective functions with such severe local behavior do not arise in applications. Summarizing the advantages and disadvantages, the Wiener process often seems an acceptable model for a global description of objective

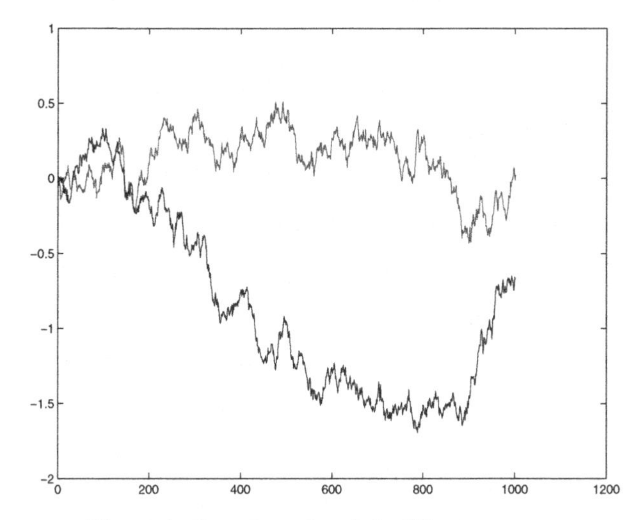

Fig. 4.1. Sample paths of the Wiener process.

functions but not as a local model. The latter conclusion has motivated the introduction of a dual statistical/local model where the global behavior of an objective function is described by the Wiener process, and its behavior in small subintervals of the main local minimizers is described by a quadratic function. An algorithm based on the dual model is considered in Sect. 4.3.6.

An important advantage of the Wiener process is the availability of an analytical formula for the probability distribution of the minimum [224]. The conditional distributions of the minima of $\xi(x)$ over distinct subintervals (x_{i-1}, x_i), $x_{i-1} < x_i$, $i = 1, ..., n$, $x_0 = \xi(0) = 0$, are independent and given by

$$P\left(\min_{x_{i-1} \leq x \leq x_i} \xi(x) \leq y \,|\, \xi(x_1), ..., \xi(x_n)\right) = \exp\left(-2\frac{(f(x_{i-1}) - y)\,(f(x_i) - y)}{x_i - x_{i-1}}\right), \tag{4.1}$$

for $y \leq \min\{f(x_{i-1}), f(x_i)\}$; see [135]. A stopping condition of a global minimization algorithm can be constructively defined via the probability of finding the global minimum with predefined tolerance calculated using (4.1).

We discuss and summarize the application of the Wiener process model later in this chapter. For the original results concerning global optimization as well as approximation and integration we refer to [5], [38], [140], [141], [144], [176], [192], [200], [201], [280], [283], [285], [286].

The Wiener process has attractive advantages and serious disadvantages as a statistical model of the objective function. It seems reasonable to derive the Wiener process based models eliminating its most serious disadvantages. The most severe disadvantage of the Wiener process is nondifferentiability of

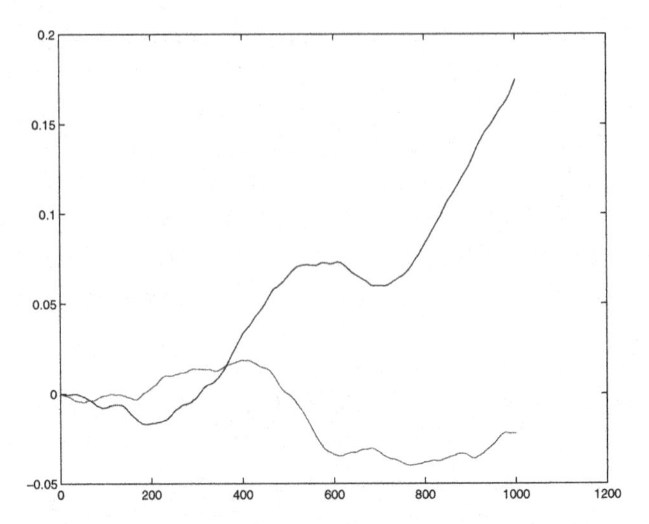

Fig. 4.2. Sample paths of the integrated Wiener process.

sample functions almost everywhere with probability 1. However, the sample functions of the integrated Wiener process

$$\zeta(x) = \int_0^x \xi(\tau)d\tau, \tag{4.2}$$

where $\xi(\tau)$ is the Wiener process, are differentiable, as it follows from the differentiability of the integral (4.2) with respect to x. Correspondingly, the derivative of $\zeta(x)$ is a continuous but nowhere differentiable function. This process is Markovian with respect to the observation of two dimensional vectors $(\xi(x), \zeta(x))$ composed of values of the process and of the process' derivative.

The vector process $\omega(x) = (\xi(x), \zeta(x))$ is a Markov process. If we observe the values of $\xi(x)$ and $\zeta(x)$ at the points $x_1, x_2, \ldots, x_n$, then the distribution of $\omega(x)$ at any point depends only on the values observed at the two nearest observation points to the left and to the right of the point

$$P\left(\xi(x) \leq z, \zeta(x) \leq y \,|\xi(x_j^n), \zeta(x_j^n), j = 1, ..., n\right)$$
$$= P\left(\xi(x) \leq z, \zeta(x) \leq y \,|\xi(x_j^n), \zeta(x_j^n), j = i-1, i\right).$$

where $x_{i-1}^n < x < x_i^n$, and x_i^n denote the ordered observation points x_i.

The corresponding conditional densities are Gaussian. Their parameters are defined by rather long expressions presented in [42]. We omit these expressions since they are not used in this book. The graphs in Fig. 4.2 illustrate the more regular behavior of sampling functions of the integrated Wiener process (4.2) than that of the Wiener process.

An alternative choice of a statistical model with smooth sampling functions is a stationary random process. The choice of Gaussian processes is dictated by implementation considerations. Let us further specialize to the class of stationary Gaussian processes. A process in this class is defined by the mean, variance, and correlation function. The desired smoothness of the sample functions can be granted by choosing an appropriate correlation function. The Ornstein-Uhlenbeck process is the only Markov process in the class of stationary processes. It is well known that the smoothness of the sample functions of a stationary random process is determined by the behavior of the correlation function in the neighborhood of zero. The correlation function of the Ornstein-Uhlenbeck process is $\exp(-c\,|t|)$, where $c > 0$; it is not differentiable at $t = 0$ implying non differentiability of sample functions. Therefore, to have smooth sampling functions we must give up the Markov property. Since there are no other specific arguments for the choice of a correlation function, it may be chosen rather arbitrarily. Let $\xi(x)$ be a stationary Gaussian process with zero mean, unit variance and the correlation function $r(\cdot)$, which we assume to be of the form

$$r(t) = 1 - \frac{1}{2}\lambda_2 t^2 + \frac{1}{4!}\lambda_4 t^4 + o(t^4) \tag{4.3}$$

as $t \to 0$, for finite λ_2, λ_4. We further assume that

$$|d^4 r(t)/dt^4 - \lambda_4| = O(|t|), \quad -r''(t) = \lambda_2 + O(|\log^{-a}|t||)$$

for some $a > 1$ as $t \to 0$, and also that $r(t)\log(t) \to 0$ as $t \to \infty$. These assumptions allow us to choose a version of $\xi(x)$ that has twice continuously differentiable sample functions [143].

Smooth stationary random processes were outside of the attention of researchers in global optimization because of implementation difficulties in the non-Markov case. However, an approximated version of the P-algorithm for a smooth function model can be constructed whose implementation is of similar complexity to that of the P-algorithm based on the Wiener model. This result disagrees with the recent opinion that stationary random processes are not prospective models for global optimization; see Sect. 4.3.3 for the further discussion.

Let us mention a random process specially constructed as a model for global optimization. This random process is proposed in [172] as a model for an information-based approach to global optimization. The discrete time random process ξ_i is defined by means of independent random increments, where the feasible interval is discretized by the points i/N, $i = 0, ..., N$,. The increments $\xi_i - \xi_{i-1}$ are supposed to be independent Gaussian random variables with variance σ^2, and average equal to $-L$ for $i < \alpha$, and equal to L for $i \geq \alpha$. The parameter α is supposed to be a random variable with known (normally uniform) distribution on $\{0,\ 1/N,\ 2/N, ..., 1\}$. The a posteriori distribution of α can be calculated with respect to observed function values. The most likely value of α is expected to be close to the minimizer of the sample function, e.g. this is true for $L \gg \sigma$. The closeness of α to the global minimizer suggested the idea of an algorithm in [172] based on a procedure of maximization of information about the location of α; a further development of this idea is presented in detail in [233].

4.1.3 Random Fields

Natural candidates for the statistical models of multimodal objective functions of many variables are random fields. The theory of random fields is a generalization of the theory of random processes. The former is the theory of stochastic functions of one variable, and the latter is the theory of stochastic functions of many variables. There are many close analogies between properties of random processes and properties of random fields. For example, homogeneous isotropic random fields are generalizations of stationary processes. A Gaussian homogeneous isotropic random field is defined by its mean, variance and correlation function $\varrho(x, y) = r(||x - y||)$, $x, y \in R^n$; the functions $r(t) = \exp(-c|t|)$, $r(t) = \exp(-ct^2)$, $c > 0$, provide two examples of correlation functions used in different applications [166], [231]. However, not all properties of random processes can be easily generalized to random fields. As it was mentioned above, the stationary Gaussian process with correlation function $\exp(-c|t|)$ has Markov property. However, the Gaussian homogeneous isotropic random field with exponential correlation function is not Markovian.

A significant simplification of computations implied by the Markov property in the one-dimensional case can not be maintained in the multidimensional case. The inherent computational complexity of random fields restricts their application to only the optimization of exceptionally expensive (consuming a huge amount of computing time) functions. For examples of the application of Gaussian homogeneous isotropic fields as statistical models for global optimization we refer to [164], [220], [221], [301]. Computational difficulties related to random fields motivated search for simpler statistical models. An axiomatic approach to the construction of statistical models is considered in Sect. 4.1.4. For the simplified models used in the Bayesian approach we refer to [164]. The recent results concerning the approximate one-dimensional P-algorithm constructed for non Markov random process (see Sect. 4.3.3) may attract researchers to reconsider the prospective of random field models in global optimization.

The use of statistical models in optimization by some authors is called "kriging"; see e.g. [130]. This term is borrowed from geo-statistics where random field models are used to predict values of functions of several variables. Methodologically the use of random field models for prediction and for optimization is similar. The main subject of kriging is linear predictors. We cite [231], page vii:

"Kriging is superficially just special case optimal linear prediction applied to random processes in a space or random fields. However, optimal linear prediction requires knowing the covariance structure of the random field. When, as is generally the case in practice, the covariance structure is unknown, what is usually done is to estimate this covariance structure using the same data that will be used for interpolation".

Extrapolation and interpolation of random sequences is a classical part of the theory of stochastic processes. A development of classical ideas towards applications using random field models to prediction (interpolation/extrapolation) problems, where the covariance function of a random field is not precisely known and should be estimated, is called kriging. Among the most important topics of the kriging theory are the fixed-domain asymptotic behavior of predictors corresponding to different covariance functions, and the experimental investigation of prediction errors in various practical problems [231]. The impact of deviations of the model covariance function from the ideal one to the optimization results would be of great interest to global optimization. However, to the best knowledge of the authors of this book, there are no publications generalizing the results of kriging theory to global optimization. This is not surprising since the availability of a good predictor for function values at given points is not sufficient to construct a good global optimization algorithm.

4.1.4 Axiomatic Definition of a Statistical Model

Random fields are useful theoretical models in global optimization, especially to analyze optimal algorithms with respect to average error. However, there are two difficulties in the practical use of random fields. First, the available information on the objective functions is not always sufficient to define a probability measure on the set of functions, i.e. frequently it is difficult to select a random field corresponding to available information about the considered objective function. Second, the computational resources required for the calculations of the conditional mean and variance of a random field grow very fast with the observations number. In the present section we consider alternative statistical models aiming to avoid, or at least to reduce, the difficulties mentioned above.

Several algorithms have been constructed postulating some statistical properties of the considered problems without reference to a stochastic function model. For example, the minimizer of a one-dimensional objective function is assumed to be a random variable with uniform distribution over an interval of interest [246]; the constructed one-dimensional global minimization algorithm combines a Lipshitzmodel and the assumption about randomness of global minimizer. The Lipshitz underestimate is used to reject the subintervals not containing the global minimizer. The choice of a point for current observation is justified by the hypothesis on the uniform distribution of the minimizer over the set of uncertainty. Similar statistical assumptions are combined with various deterministic assumptions to construct algorithms of local minimization and to search for roots of one variable functions in [49], [74], [117], [170]. A simple statistical model is used in [222] to recognize separable functions. The problem is important for applications since the minimization of a function of many variables might be significantly simplified if the function is expressible as a sum of functions of one variable. To test the hypothesis of separability an analysis of variances (ANOVA) based method is proposed in [222] where the function values are interpreted as random variables without the involvement of the underlying random field.

In the cited examples the probability distribution of interest is not derived from an underlying stochastic function model but ad hoc postulated. Such a statistical assumption can further be combined with deterministic assumptions about the objective function. Such a bottom up approach to the construction of a statistical model integrating heuristic information available about the considered problem is advantageous because of the controllability of the computational complexity of the resulting model. The disadvantage of this approach is a possible incompatibility of the accepted assumptions. The top down approach starts with the choice of a stochastic function for a model. The properties of the sample functions of the chosen stochastic model are mathematically provable. The disadvantage of this approach is the computational complexity of the corresponding models. It can also be difficult to find a model with desirable properties, e.g. the Wiener process well suits the global description of complicated multimodal one-dimensional optimization problems but it is not good to to represent the local properties of most practical objective functions.

The bottom up approach seems prospective to construct computationally simple statistical models of mutimodal objective functions. To avoid the disadvantages mentioned above we apply the bottom up approach in axiomatic framework widely used in rational decision theory. We construct a statistical model for global optimization step by step starting with very general rationality assumptions on the uncertainty about the objective function value at a point where it is not yet calculated/observed.

An objective function intended to minimize normally is not an ideal black box since some information on $f(\cdot)$ is available including function values at some points. Let the value $f(x)$ be not yet observed. The weakest assumption on available information seems to be the comparability of the likelihood of inclusions $f(x) \in Y_1$ and $f(x) \in Y_2$ where Y_1, Y_2 are arbitrary intervals. If it seems more likely that $f(x) \in Y_1$ than $f(x) \in Y_2$, such a subjective probability assessment will be denoted $Y_1 \succ_x Y_2$; the index x may be omitted if it is clear from the context. The symbols $\succeq_x$ and $\sim_x$ denote 'not less likely' and 'equally likely' correspondingly. The paradigm of rationality normally does not contradict the following axioms concerning the comparative likelihood of intervals of possible values of the objective function value $f(x)$:

- A1. For any Y_1, Y_2, either $Y_1 \succeq Y_2$, or $Y_1 \preceq Y_2$.
- A2. If $Y_1 \succeq Y_2$ and $Y_2 \succeq Y_3$ then $Y_1 \succeq Y_3$.
- A3. $Y \succ \emptyset$ if and only if $\mu(Y \cap U) > 0$, where U is an interval of all possible values $f(x)$, and $\mu(\cdot)$ is the Lebesque measure.
- A4. Let Y_2 and Y_3 have common end points with Y_1, but $\mu(Y_1 \cap Y_2) = \mu(Y_1 \cap Y_3) = 0$; $Y_2 \succeq Y_3$ iff $\bar{Y}_1 \cup Y_2 \succeq \bar{Y}_1 \cup Y_3$, where bar denotes closure.

- A5. If $Y_1 \succeq Y_2$ then there exist points z_1, z_2 in the interval $Y_1 = (y_1^-, y_1^+)$ such that $Y_2 \sim (y_1^-, z_1)$, $Y_2 \sim (z_2, y_1^+)$ where .

The axiom A1 is a rather week assumption of comparability of the intervals of possible values of the objective function with respect to their likelihood. The axioms A2 - A5 express complexity in the objective function and difficulties in prediction of its values. Let us note, that in the general axiomatics of rational decision theory the comparability of the elements of an algebra is assumed [76], [208]. In our axiomatics, comparative probability is first defined for intervals, and it is further extended to an algebra of unions of intervals. Assume that comparative probability is defined for intervals, and that it satisfies A1-A5. It can be proved that for a set of disjoint intervals $Y_1, Y_2, ..., Y_k$ more likely than $\emptyset$ there exists the increasing sequence of real numbers $b_1, b_2, ..., b_k$ such that $(-\infty, b_1) \sim Y_1$, $(b_{i-1}, b_i) \sim Y_i$, $i = 2, ..., k$, and b_k does not depend on the numeration of Y_i. This result justifies the extension of comparative probability by means of the following axiom

- A6. $(-\infty, b_k) \sim \bigcup_{i=1}^{k} Y_i$.

For the technical questions of extension of the comparative likelihood we refer to [287], [289].

The axioms A1 - A6 imply the existence of a probability density compatible with the axioms, i.e. the existence of a probability density $p_x(t)$ such that

$$Y_1 \succeq_x Y_2 \Leftrightarrow \int_{Y_1} p_x(t)dt \geq \int_{Y_1} p_x(t)dt.$$

We do not formulate a rigorous representation theorem here concerning the existence of the compatible probability density; such a theorem and its proof can be found in [287], [289]. If we agree that the information about the objective function enables us to compare the likelihood of different intervals of possible values of the objective function, then the axioms A1 - A5 are similar to the standard axioms of rational decision theory. Their acceptability is based on the same concept of rationality. It can be proved that the comparative probability extended to unions of intervals by means of A6 has similar properties as those expressed by the axioms A1-A5. We will not go into the details here but summarize the results of [282], [287], [289], [293]. The existence of the probability density compatible with the comparative probability enables us to interpret an unknown value of the objective function $f(x)$ as a random variable $\xi(x)$ with probability density $p_x(\cdot)$.

Let the a priori information on $f(x), x \in A \subseteq \mathbb{R}^d$ and the known function values $f(x_i)$, $i = 1, ..., n$ induce the comparative probability (likelihood) of function values $f(x)$ compatible with the assumptions of rationality A1 – A6. The conclusion above implies the acceptability of a random variable $\xi(x)$ as a model of the unknown function value. Correspondingly, the family of random variables $\xi(x), x \in A$, is acceptable as a statistical model of a class of objective functions. The distribution of $\xi(x)$ is normally accepted to be

Gaussian because of computational reasons. To specify the parameters of the family of random variables we maintain the bottom - up approach defining a predictor (extrapolator/interpolator) of unknown value of the objective function axiomatically. For a methodological basis of an axiomatic approach to extrapolation under uncertainty we refer to [75], [76], [97], [284].

Maintaining the bottom - up approach we define the predictor of the unknown value of the objective function axiomatically. The mean value of $\xi(x)$ is defined to be equal to the predicted value, and it is denoted by $m_n(x|\cdot)$. Taking into account uncertainty about the behavior of the objective function the following assumptions concerning $m_n(x|\cdot)$ seem reasonable

- A7. $m_n(x|x_i, cy_i, i = 1, ..., n) = cm_n(x|x_i, y_i, i = 1, ..., n)$,
- A8. $m_n(x|x_i, y_i + c, i = 1, ..., n) = m_n(x|x_i, y_i, i = 1, ..., n) + c$,
- A9. for any permutation of indices $\{j(i),\ i = 1, ..., n\} = \{1, 2, ..., n\}$ the equality $m_n(x|x_{j(i)}, y_{j(i)}, i = 1, ..., n) = m_n(x|x_i, y_i, i = 1, ..., n)$ holds,
- A10. $m_n(x_j|x_i, y_i, i = 1, ..., n) = y_j,\ j = 1, 2, ..., n$,
- A11. there exists a function $v_n(\cdot)$: $A \times (\mathbb{R} \times A)^{n-1} \to \mathbb{R}$ such that $m_n(x|x_i, y_i, i = 1, ..., n) = m_n(x|x_i, z_i, i = 1, ..., n)$ where $z_j = z = v_n(x, x_i, y_i,\ i = 1, 2, ..., n-1),\ j = 1, 2, ..., n-1,\ z_n = y_n$.

The axioms A7 - A9 postulate invariance of the predicted value with respect to affine transformations of scales of measuring function values, and with respect to the numbering of observation results. The axiom A10 postulates that the observations give precise function values. Finally, A11 restricts the complexity of the predictor assuming that $m_n(x|\cdot)$ can be expressed as a superposition of functions of $n-1$ and of two (aggregated) variables. The only function compatible with axioms A7 - A11 is

$$m_n(x|x_i, y_i, i = 1, ..., n) = \sum_{j=1}^{n} y_j w_j^n(x, x_i, i = 1, ..., n), \tag{4.4}$$

where $w_j^n(x, x_i, i = 1, ..., n)$ are weights satisfying the equalities below

$$\sum_{j=1}^{n} w_j^n(x, x_i, i = 1, ..., n) = 1,$$

$$w_j^n(x, x_i, i = 1, ..., n) =$$

$$w_p^n(x, x_1, ..., x_{j-1}, x_p, x_{j-1}, ..., x_{p-1}, x_j, x_{p+1}, ..., n),$$

$$w_j^n(x_l, x_i, i = 1, ..., n) = \begin{cases} 1, & l = j, \\ 0, & l \neq j. \end{cases}$$

The variance of $\xi(x)$ can be defined by similar axioms; see e.g. [289]. The expressions of weights should be chosen based on heuristical arguments and on results of experimental investigation; for example, in [289] the following formulas for the parameters of $\xi(x)$ are justified

$$m_n(x|x_i, y_i, i = 1, ..., n) = \sum_{j=1}^{n} y_j \cdot w_j^n(x, x_i, i = 1, ..., n), \tag{4.5}$$

$$s_n^2(x|x_i, y_i, i = 1, ..., n) = \tau \sum_{j=1}^{n} ||x - x_j|| \cdot w_j^n(x, x_i, i = 1, ..., n), \ \tau > 0,$$

where $m_n(x|\cdot)$ is mean value of $\xi(x)$, $s_n^2(x|\cdot)$ is variance of $\xi(x)$, and $w_i^n(\cdot)$ are weights, e.g.

$$w_i^n(x, x_i, y_i, i = 1, ..., n) = \delta(||x - x_i||) / \sum_{j=1}^{n} \delta(||x - x_j||),$$
$$\delta(z) = \exp(-cz^2)/z.$$

The methodology of axiomatic construction of statistical models for global optimization is similar to the axiomatic approach of decision making under uncertainty. The results of a psychological experiment in [282] show the acceptability of the axioms to experts routinely solving real world problems.

The class of constructed models includes the models which are computationally simpler than random fields. It is shown in [284] that, using axioms A1 - A6 together with two more specific axioms, a predictor corresponding to the conditional mean of a Gaussian random field can be specified. It is interesting to note that by means of adding a few informal assumptions to the general postulates on uncertainty, a statistical model corresponding, e.g. to the Wiener process, can be constructed [284]. In [300] the sets of function values at several points are considered; it is proved that similar axioms on the rationality of comparison of likelihood imply the existence of a random field compatible with the axioms of comparative likelihood. In the case where the statistical model corresponds to a random field, the characteristics $m_n(x|\cdot)$ and $s_n^2(x|\cdot)$ are the conditional mean and conditional variance of the random field correspondingly [284], [289].

4.1.5 Response Surfaces

Classical local descent methods are constructed using simple polynomial models of an objective function, e.g. the steepest descent method is based on a linear model, and the Newton method is based on a quadratic model. More complicated multivariate approximation functions, frequently called response surfaces, are used to justify choice of observation points in multidimensional global search. To be suitable for global optimization, the approximation should be applicable to scattered data points. Multivariate approximation with scattered data points is a challenging problem. The radial basis function method is one of the most prospective methods in this field [27]. The radial basis interpolant has the form

$$m_n(x) = \sum_{i=1}^{n} \lambda_i \phi(||x - x_i||),$$

where $\phi(\cdot)$ is a basic function, and the coefficients λ_i are chosen to satisfy the equalities $m_n(x_i) = y_i$. The following basis functions are used most frequently: $\phi(r) = \sqrt{r^2 + c^2}$, $\phi(r) = r^2 \log(r)$, $\phi(r) = \exp(-ar^2)$. Similar to the case of the statistical models considered in the previous subsections, such an interpolant predicts unknown function values conditionally to the observed function values. However, it is not only a function value that is predicted by the statistical model but also a measure of uncertainty of the prediction, e.g. in the form of a conditional variance.

It seems attractive to construct a global optimization algorithm that would search for the minimum where expected function values are small. Such an idea is approved in local minimization. Of course, in global optimization a local model of the function would be changed by a global response surface. However, searching in the neighborhood of the global minimizer of a response surface does not guarantee finding a global minimum of the objective function, since the latter can be located in a not yet researched subregion which is qualified as not prospective because of the few observed large function values; see e.g. Fig. 4.5. A measure of uncertainty of prediction enables us to balance globality and locality of search: uncertainty of prediction should be decreased everywhere to decrease the probability of missing the global minimum. A response surface alone is not sufficient for the construction of global optimization algorithms. It is reasonable to extend response surface models introducing a characteristic like uncertainty of prediction. For example, in [106] the radial basis function model is extended introducing the characteristic called 'bampiness'; an increase of bampiness makes the prediction of a global minimizer less certain.

The term 'response surface' is used by some authors, e.g. [130], in a very broad sense including here stochastic functions. Let us mention also that somewhere the term 'surrogate function' has similar meaning as 'response surface' [9], [24].

4.1.6 Estimation of Parameters

Assume a random process or a random field is chosen for a model, e.g. a Gaussian stochastic function with some covariance structure. To specify the model, its parameters should be estimated using methods of mathematical statistics. Before minimization is started, some observations are performed to estimate the parameters of the chosen stochastic function. For example, the parameter of the Wiener process σ_0^2 should be estimated from the observations

$y_i^n = \xi(x_i^n)$, $i = 0, ..., n$, where $0 \le x_i^n \le 1$ are ordered observation points, and $x_0^n = y_0^n = 0$. The following two estimates are appropriate

$$\bar{s}^2 = \frac{1}{n} \sum_{1}^{n} \frac{(y_i^n - y_{i-1}^n)^2}{x_i^n - x_{i-1}^n}, \tag{4.6}$$

$$\tilde{s}^2 = \sum_{1}^{n} (y_i^n - y_{i-1}^n)^2. \tag{4.7}$$

Both estimates are unbiased. The variance of (4.6) is equal to $2\sigma_0^4/n$, and the variance of (4.7) is equal to

$$2\sigma_0^4 \sum_{i=1}^{n} (x_i^n - x_{i-1}^n)^2.$$

The first estimate has advantage because it is unbiased and consistent for dependent observations [303], e.g. the estimate can be updated using observations obtained during the optimization process. This result is generalized in [304] for the maximum likelihood estimate of the parameter of a Gaussian random field where the unknown parameter σ_0^2 appears as a multiplier in the expression for the correlation function $r(x, y) = \sigma_0^2 \rho(x, y)$ with a known function $\rho(x, y)$.

It is shown in Sect. 4.3.7 that the asymptotic convergence rate of P-algorithms does not depend on the parameters of underlying statistical models. On the other hand, there is not much known on the influence of the parameters of a model to the performance of algorithms for a finite number of iterations. The theoretical assessment of the performance of numerical algorithms stopped after a finite number of iterations is difficult. Therefore such assessments are normally based on experimental testing. According to the practical experience of one of the co-authors of the book, a sample of 7 to 10 uniformly distributed observations is sufficient for the initial estimation of the parameter of the Wiener process; the estimate is normally updated using observations performed by a minimization algorithm. Generally speaking a rational choice of an estimation method depends on the model, on the constructed algorithm, and on the properties of the objective function. The problems of the estimation of the covariance structure are discussed in kriging theory, see [231].

A family of Gaussian variables may be chosen as an alternative statistical model to a stochastic function as discussed above. Such a model normally includes several parameters. The latter can be estimated using one of standard methods of parameter estimation, e.g. the method of least squares.

4.1.7 Illustrative Comparison

The question of adequacy of a chosen model of objective functions to a real problem is crucial to the success of the method based on that model. An important criterion of adequacy is the precision of prediction of function values. Mathematical criteria, normally considering the asymptotic behavior of errors, are not very helpful here since global search is frequently stopped far before the asymptotic properties are realised.

To present a visual illustration of the predictive power of the considered models, two-dimensional examples should be used, since the visualization of higher dimensional examples is difficult. It is easy to show two-dimensional examples of good fit for all considered models, taking sufficiently large number of known function values at regularly distributed points. However, in a normal situation prediction should be made from scattered data with large inter-point distances. In the figures below the contours of the predicted function values are presented for the well known Branin test function observed at the points shown by circles on the figures. The parameters of random fields are estimated using the method of maximal likelihood, except for a parameter of correlation function which has been tuned.

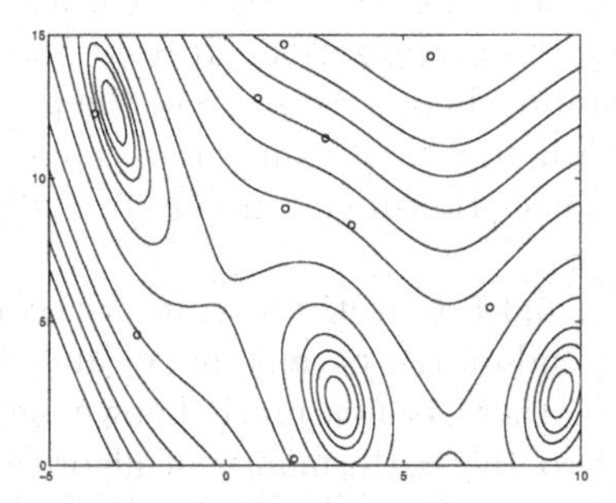

Fig. 4.3. Contour lines of Branin function.

We leave it to the reader to assess which of the presented predictors seems the best in this situation; while predicting function values heuristically, please try not to take into account your full knowledge on the function but only the values at the points denoted by circles.

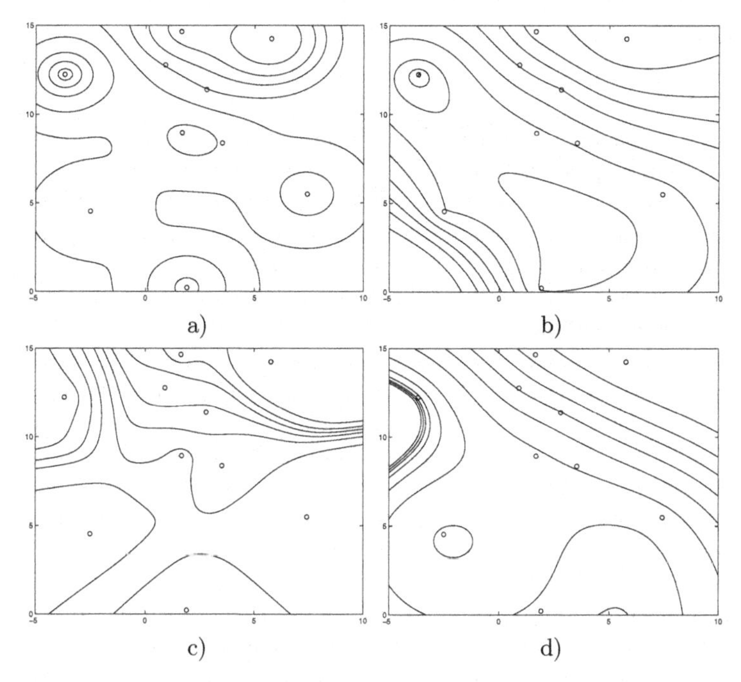

Fig. 4.4. Contour lines of the predictors based on: a) a Gaussian random field with correlation function $\exp\{-0.2r^2\}$, b) a conditional mean of the Wiener field, c) a conditional mean of axiomatic statistic model (4.5), d) a thin plate radial basis function.

4.2 Methods

4.2.1 Introduction

There are several approaches for constructing optimization and other numerical methods [45]. Some optimization methods have been invented which implement heuristic ad hock ideas, e.g. the simplex based Nelder-Mead method. The other approach is based on simulating natural processes where typical examples are simulated annealing and evolutionary search. Theoretical approach to the construction of optimization methods is based on mathematical models of the considered problem. The choice of model is a crucial decision in the development of a method, since the model helps not only to interpret current minimization results but also to enhance the efficiency of further minimization steps. For example, a response surface can be used to predict the function values at the potential observation points conditionally with respect to the observed values $y_i = f(x_i)$, $i = 1, ..., n$. A simple idea is to locate the further observation points where the predicted values are minimal. This idea is fruitful in local minimization. For example, the next observation point by the classical Newton method is chosen at the minimum point of a quadratic model of the objective function based on its first and second derivatives at the current point. However, in global minimization this idea can fail as illustrated in Fig. 4.5. The graph of the response surface after five observations of the objective function values is shown at Fig. 4.5a. Subsequent observations at the minimum point of the response surface improve the fit of the objective functions and the estimate of local minimizer, but the search will stick at a local minimizer, and the global minimum will not be found as shown by Fig. 4.5b.

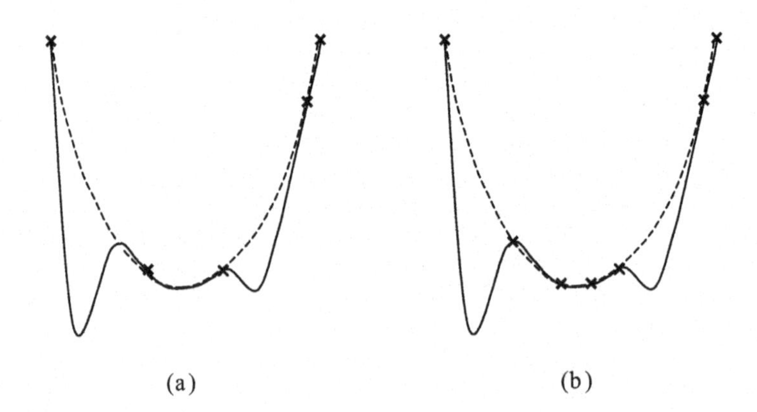

Fig. 4.5. Localization of search based on minimization of an interpolant

To construct a method of rational search for global minimum not only the predicted values but also the uncertainty of prediction is important. Later in this section a global minimization method is defined by assumptions on the rationality of choice of the next observation point, where the predicted function value and the uncertainty of prediction are taken into account via a statistical model of the objective function.

Let us start with a standard definition of a method, optimal with respect to a model defined as a class of objective functions. The efficiency criterion related to deterministic models normally takes into account the worst case situation. Methods optimal with respect to such criteria are called minimax or worst case optimal. To define the minimax method we have to define a class of methods $\mathcal{P}$, and to choose a class of objective functions $\mathcal{F}$. Let a method $\pi \in \mathcal{P}$ be applied to minimize an arbitrary function $f \in \mathcal{F}$. Assume the number of observations is fixed in advance and is equal to n. In this case π is a vector function $\pi(n) = (\pi_1, \pi_2, ..., \pi_{n+1})$, where $x_1 = \pi_1$, $x_{i+1} = \pi_{i+1}(x_j, y_j, j = 1, ..., i), i = 2, ..., n-1$, and $x_{on} = \pi_{n+1}(x_j, y_j, j = 1, ..., i)$ is the estimate of the global minimizer of $f(\cdot)$ obtained by means of π. After n observations the minimization error is equal to

$$\Delta(\pi, f, n) = f(x_{on}) - \min_{x \in A} f(x),$$

where A is the feasible region. The worst case error for the method π is equal to

$$\Delta(\pi, n) = \max_{f \in \mathcal{F}} \Delta(\pi, f, n).$$

After n observations the error smaller than

$$\Delta(n) = \min_{\pi \in \mathcal{P}} \Delta(\pi, n) = \min_{\pi \in \mathcal{P}} \max_{f \in \mathcal{F}} \Delta(\pi, f, n), \tag{4.8}$$

can not be guaranteed, and the method $\bar{\pi}(n)$, delivering minimal guaranteed error, is called minimax or worst case optimal; we assume that all minima and maxima exist. In Sect. 1.1.4 several deficiencies of the minimax approach have been mentioned. Most serious are: 'adaptation does not help', and 'constant is the hardest function'.

An advantage of the statistical approach is the possibility to assess the *average error* of a method with respect to the considered statistical model $\delta(\pi, n) = \mathbf{E}\{\Delta(\pi, f, n)\}$. It seems more rational to construct a method oriented to average conditions than to the worst case conditions which rarely occur in practice. A method $\hat{\pi}$ delivering minimum to $\epsilon(\pi, n)$ is optimal in average with respect to the statistical model. The definition of optimal methods emphasizes the orientation of methods based on deterministic models to the worst case conditions, and orientation of methods based on statistical models to the average conditions.

Implementation of methods optimal with respect to guaranteed as well as to average error is difficult. The global optimization methods considered below have been constructed as optimal in average procedures with respect to different statistical models but simpler criteria than the average error are used there.

The performance of the constructed methods is interesting not only with respect to the statistical model. From the practical point of view even more interesting is the performance of a method under broad conditions of general interest not involving probabilistic assumptions. Therefore we will investigate the convergence of methods, constructed using statistical models, under broader assumptions than those assuring the optimality of the method. The reader should keep in mind that a statistical model in this book serves only to motivate the method; the method is also justified by its convergence properties under the weak assumptions of continuity of the objective function and non degeneracy of the global minimizer.

Statistical models are important not only to aid construction of methods but also for the theoretical research on average complexity of global optimization problems. For example, in [36] it is shown that there does not exist an exponentially fast method where average error is defined with respect to the Wiener measure, i.e. the average error of any method using n function evaluations is not $O(\exp(-cn))$ for any $c > 0$. However, the theory of average complexity of global optimization is more complicated than the well developed theory of of average complexity of linear problems presented e.g. in [201]. In Sect. 4.3.9 we cite several results in average complexity of global optimization to illustrate differences from the worst case complexity.

4.2.2 Passive Methods

A model of objective functions is needed not only to develop a method from theoretical optimality assumptions but also to investigate mathematical properties of the methods constructed without such a model. For example, deterministic and stochastic uniform grids can be used as passive minimization methods. There are heuristical arguments for their attractiveness for optimization, however the theoretical assessment of such grids as optimization methods can not avoid the involvement of a model of the objective function.

Since the subsequent observation points in passive search do not depend on the values of the objective function observed during previous minimization steps, passive methods are relatively simple. Uniform grids heuristically seem adequate to absolute uncertainty with respect to the location of a global minimizer. Therefore they seem attractive as a 'zero' in the scale of optimization methods for the assessment of their empirical efficiency. On the other hand, theoretical properties of passive methods are also interesting and are briefly reviewed in this section.

The most frequently considered passive methods are uniform grids. Their performance can be evaluated rather easily in different theoretical frames. For example, the deterministic grid defined as $0,\ 1/n, 2/n, ..., 1$ is uniform over the interval $[0,\ 1]$ assuming that two of the $n+1$ observations should be made at the end points of the interval. It is easy to estimate the worst case error of the uniform search with respect to the class of Lipshitz functions with Lipshitz constant L: it is equal to $\frac{L}{2n}$. Moreover, it is well known that uniform grids define optimal global optimization methods with respect to a class of Lipshitz functions [234].

For comparison the properties of uniform grids with respect to a statistical model are interesting. Let us consider average errors with respect to the Wiener process which is a popular one-dimensional statistical model. Assume that the feasible region is [0, 1]. It is shown in [200] that the uniform grid is order optimal since its error is

$$\epsilon(n) = \Theta(n^{-1/2}),$$

while for any passive method the error is $\Omega(n^{-1/2})$; $\Theta(h(n))$ denotes a function of the same order as $h(n)$, $n \to \infty$, $\Omega(h(n))$ denotes the asymptotic upper bound for $h(n)$. However, for a fixed number of observations the uniform grid is not optimal. For example, in the case where $n = 3$, the optimal observation points are $x_1 = 0.3$, $x_2 = 0.7$, $x_3 = 1$.

The generation of deterministic uniform grids in multidimensional regions is difficult. Frequently such grids are replaced with random uniform grids generated by Monte Carlo methods. An advantage of random grids is the possibility to modify the distribution of random points to take into account information on the location of global minimizer, e.g. to increase the probability density of observation points in subregions where the location of a global minimizer is most probable. On the other hand, the probability density should be positive over the whole feasible region to guarantee the convergence of such a nonuniform search in probability, i.e. to guarantee the convergence of the probability of finding the global minimum to 1 when the number of observation increases to infinity.

Heuristically it seems rational to choose the density of observation points as close as possible to the density of the distribution of minimizers of the stochastic function used as a statistical model of objective functions. However, the distribution of minimizers is not normally known. Moreover, the optimal distribution of observation points does not need to coincide with the distribution of minimizers. For example, the density of global minimizers of the Wiener process in the interval [0, 1] is equal to

$$g(t) = \frac{1}{\pi\sqrt{t(1-t)}},$$

but, as shown in [3], the optimal distribution of observation points is defined by the following formula

$$h(t) = \beta(2/3,\, 2/3)^{-1}[t(1-t)]^{-1/3},\ 0 < t < 1,$$
$$\beta(x, y) = \int_{t=0}^{1} t^{x-1}(1-t)^{y-1}dt.$$

Theoretically it is interesting to know how much better a considered adaptive (sequential) method is compared to the best passive method. However, the example above shows that construction of the optimal passive method is difficult even with respect to asymptotic criteria. Therefore, uniform grids are normally used as representatives of passive methods in experimental and theoretical investigation of efficiency not only in a worst case but also in an average case setting.

4.2.3 Bayesian Methods

The problem of optimal in average methods of global optimization was formulated by J. Mockus in [163]. A brief description of the problem is given in the introductory section. Here we will present a more precise statement of the problem. Let a random field $\xi(x)$, $x \in A \subset \mathbb{R}^d$, be chosen as a model of objective functions. The accepted stochastic model well represents a situation where many functions with similar characteristics should be minimized. Therefore, the construction of an average case optimal method is of great importance. The notion of average optimality is related to the average error of the method.

Let us define a method (with the a priori fixed number of observations N) by means of a vector function $\pi(N) = (\pi_1, ..., \pi_{N+1})$. A point of current observation is defined depending on points of the previous observations x_i, and the function values at these points y_i, i.e.

$$x_{n+1} = \pi_{n+1}((x_i, y_i), \; i = 1, ...n),$$

where the function $\pi_{N+1}(\cdot)$ defines the estimate of the global minimizer x_{oN}.

Very general assumptions guarantee the existence of the average error $\delta(\pi, N)$ of the estimates of the global minima by a method π for the sampling functions of $\xi(x)$

$$\delta(\pi, N) = \mathbf{E}\{\Delta(\pi, \xi, N)\} = \mathbf{E}\{(\xi(x_{oN}) - \min_{x \in A} \xi(x)\}.$$

where $\mathbf{E}$ denotes the expectation operator, and x_{on} denotes the estimate of the minimizer of the sample function obtained by means of π. Let the methods in $\mathcal{P}$ define the current observation point taking into account all function values and observation points at previous minimization steps. The optimal method $\hat{\pi}(N)$

$$\delta(\hat{\pi}(N), N) = \min_{\pi \in \mathcal{P}} \delta(\pi, N), \tag{4.9}$$

is called the Bayesian method; it is defined by the solution of the following system of recurrent equations, as shown in [163]

$$\begin{aligned}
&u_{N+1}((x_i, y_i),\ i = 1, ..., N) = \\
&\min\nolimits_{x \in A} \mathbf{E}\left[\xi(x) | \xi(x_i) = y_i, i = 1, ..., N\right], \\
&\hat{\pi}_{N+1}((x_i, y_i),\ i = 1, ..., N) = \\
&\arg\min\nolimits_{x \in A} \mathbf{E}\left[\xi(x) | \xi(x_i) = y_i, i = 1, ..., N\right], \\
&u_k((x_i, y_i),\ i = 1, ..., k-1) = \\
&\min\nolimits_{x \in A} \mathbf{E}[u_{k+1}((x_i, y_i), (x, \xi(x))) | \xi(x_i) = y_i, i = 1, ..., k-1], \\
&\hat{\pi}_k((x_i, y_i),\ i = 1, ..., k-1) = \\
&\arg\min\nolimits_{x \in A} \mathbf{E}[u_{k+1}((x_i, y_i), (x, \xi(x))) | \xi(x_i) = y_i, i = 1, ..., k-1], \\
&k = N, N-1, ..., 2, \\
&u_1 = \min\nolimits_{x \in A} \mathbf{E}[u_2(x, \xi(x))], \\
&\hat{\pi}_1 = \min\nolimits_{x \in A} \mathbf{E}[u_2(x, \xi(x))].
\end{aligned} \tag{4.10}$$

The solution of the system of equations above exists under rather weak assumptions with respect to the random field chosen as a model. However, because of the complexity of system (4.10), theoretical as well as numerical investigation of Bayesian methods is difficult. Therefore semi-optimal methods are investigated. A one-step Bayesian method $\tilde{\pi}$ was proposed in [163] as a simplified version of the optimal method. $\tilde{\pi}$ is defined as the repeated optimal planning of the last observation, i.e. the current $(n+1)$th observation is always planned as the last one. Therefore $\tilde{\pi}$ is defined by the first two equations of the system (4.10):

$$\begin{aligned}
&u_{n+2}((x_i, y_i),\ i = 1, ..., n+1) = \\
&\min\nolimits_{x \in A} \mathbf{E}\left[\xi(x) | \xi(x_i) = y_i, i = 1, ..., n+1\right], \\
&\tilde{\pi}_{n+1}((x_i, y_i),\ i = 1, ..., n) = \\
&\arg\min\nolimits_{x \in A} \mathbf{E}[u_{n+2}((x_i, y_i), (x, \xi(x))) | \xi(x_i) = y_i, i = 1, ..., n], \\
&k = 2, 3, ..., \\
&x_1 \text{ is arbitrary.}
\end{aligned} \tag{4.11}$$

If the minimum point of u_{n+2} coincides with the point of the minimal observation then the one-step Bayesian method can be defined by the following equation

$$\begin{aligned}
&\tilde{\pi}_{n+1}((x_i, y_i),\ i = 1, ..., n) = \\
&\arg\min\nolimits_{x \in A} \mathbf{E}[\min(\xi(x), y_n, y_{n-1}, ..., y_1) | \xi(x_i) = y_i, i = 1, ..., n], \\
&n = 2, 3, ..., \\
&x_1 \text{ is arbitrary.}
\end{aligned} \tag{4.12}$$

The one step Bayesian method is reduced to the repeated solution of (4.12) with increasing n. However, implementation complexity is still a problem since the computation of the conditional mean in (4.12) is difficult for large n. A further simplification of one step Bayesian concerns the computational complexity of conditional distributions of random fields. We will not discuss the details of implementation of Bayesian methods here since two monographs by J.Mockus [164], and by J.Mockus with co authors [165] cover the topic.

An interesting idea of optimal methods with restricted memory was proposed in [163]. From the computational point of view the implementation of a Bayesian method with restricted memory can be much simpler than the solution of (4.10). However, only a few attempts were made to implement this idea. For example, a method proposed in [301] chooses the current observation point randomly with uniform distribution on a sphere with the center at the current estimate of the global minimizer. The radius of sphere depends on the current estimate of global minimum and on the number of observations left until termination defined by the fixed in advance number of observations.

4.2.4 P-algorithm

As it was discussed in the previous section, a family of random variables can frequently be accepted as a statistical model. Such a model can be chosen, for example, in the case when a choice of a stochastic function can not be justified by the available information on an objective function. The other reason to chose a statistical model (4.13) is its computational advantages. Let a family of Gaussian random variables

$$\xi(x),\, x \in A \subset \mathbb{R}^d \tag{4.13}$$

be accepted as a model of an objective function; a Gaussian random function can be considered as a special case of the statistical model. Since the implementation of the optimal method (4.9) is difficult, let us define the method as a rational decision at the current minimization step. To define a method we apply here the methodology of decision making under uncertainty. The choice of a point to evaluate the objective function at the current minimization step is indeed a decision under uncertainty. To justify a choice we refer to the axioms of rational decisions [77], [208].

The choice of the next point $x_{n+1} \in A$, where to calculate/observe $f(\cdot)$, may be interpreted as choosing of a distribution function from the distribution functions $F_x(\cdot)$ of the random variables $\xi(x)$. If the preference of choice satisfies some rationality requirements then there exists a unique (to within linear transformation) utility function $u(\cdot)$ compatible with the preferences of this choice: $F_x(\cdot) \succeq F_z(\cdot)$ iff $\int_{-\infty}^{\infty} u(t)dF_x(t) \geq \int_{-\infty}^{\infty} u(t)dF_z(t)$ [77]. To construct the utility function corresponding to a rational search for the global minimum let us consider the preferences between $F_x(\cdot)$. Since $F_x(\cdot)$ are Gaussian, these preferences are equivalent to the preferences between vectors (m_x, s_x) where m_x denotes the mean value, and s_x^2 denotes the variance of $\xi(x)$. Let $\tilde{y}_{on}$ be an aspiration level desirable to reach at the considered minimization step, e.g. $\tilde{y}_{on} = y_{on} - \epsilon$, $y_{on} = \min(y_1, ...y_n)$. The requirements of rationality of search are formulated by the following axioms, where the choice between two Gaussian distributions with the parameters (m_i, s_i), $i = 1, 2$ is considered.

- A1. For arbitrary $m_1 < m_2$, $s_1 > 0$, $m_2 > \tilde{y}_{on}$ there exists $s > 0$ such that $(m_1, s_1) \succ (m_2, s_2)$ if $s_2 < s$.
- A2. For arbitrary m_1, $s_1 > 0$, $m_2 > \tilde{y}_{on}$ it is true that $(m_1, s_1) \succ (m_2, 0)$.
- A3. For arbitrary $s_1 > 0$, $m_2 > \tilde{y}_{on}$, $s_2 > 0$ there exists m_1 $(m_2 > m_1 > \tilde{y}_{on})$ such that $(m_1, s_1) \succ (m_2, s_2)$.
- A4. $u(\cdot)$ is continuous from the left.

Let us comment on the axioms above. For any $\delta > 0$, $p < 1$ there exists such a small s_2 that the inequality $P(|\eta(m_2, s_2) - m_2| < \delta) > p$ holds, where $\eta(m_2, s_2)$ denotes the Gaussian random variable with mean value m_2 and standard deviation s_2 correspondingly. In this case it is almost guaranteed that $\eta(m_2, s_2)$ is larger than m_1. The axiom A1 postulates the irrationality of observations of $f(\cdot)$ at the points at which comparatively large values of $f(\cdot)$ are expected with probability close to 1. Let us emphasize that the axiom A1 does not deny the rationality of evaluating $f(\cdot)$ at the points where large values of $f(\cdot)$ are expected but uncertainty with respect to these values is great.

The observation of $f(\cdot)$ at the points z where the variance of $\xi(z)$ is equal to zero does not add to the knowledge on $f(\cdot)$. The axiom A2 postulates that such observations are irrational.

Within search of the global minimum it is important to obtain a small value of $f(\cdot)$ at the next step. It is also important to reduce the uncertainty with respect to yhe behavior of $f(\cdot)$ in order to arrange for more efficient performance in future steps. Therefore the most preferable observation points may be characterized by a small expected value m_1 and by a large variance s_1^2. This intuitively clear statement is formalized by the axioms A1 and A2. Some priority to the first component of the vector (m, s) is postulated by axiom A3. Axiom A4 postulates some regularity of the utility function.

Theorem 4.1. *The unique (to within linear transformation) utility function satisfying the axioms A1-A4 is*

$$u(t) = 1,\ t \leq \tilde{y}_{on},\ u(t) = 0,\ t > \tilde{y}_{on}.$$

Corollary A method compatible with the rationality axioms performs a current observation at the point

$$x_{n+1} = \arg\max_{x \in A} \mathbf{P}\{\xi(x) \leq \tilde{y}_{on} | \xi(x_1) = y_1, ..., \xi(x_n) = y_n\}, \tag{4.14}$$
$$\tilde{y}_{on} < y_{on} \text{ e.g. } \tilde{y}_{on} = y_{on} - \varepsilon_n,\ y_{on} = \min\{y_1, ..., y_n\},\ \varepsilon_n > 0;$$

for the statistical models with Gaussian distribution

$$\mathbf{P}\{\xi(x) \leq \tilde{y}_{on} | \xi(x_1) = y_1, ..., \xi(x_n) = y_n\} =$$
$$\frac{1}{\sqrt{2\pi}} \int_{-\infty}^{\frac{\tilde{y}_{on} - m_n(x,\cdot)}{s_n(x,\cdot)}} \frac{1}{\sqrt{2\pi}} \exp(-t^2/2) dt, \quad (4.15)$$

and the current observation point is defined as the maximum point of the criterion

$$\frac{(\tilde{y}_{on} - m_n(x|\xi(x_i) = y_i,\, i = 1, ..., n)}{s_n(x|\xi(x_i) = y_i,\, i = 1, ..., n)}. \quad (4.16)$$

The method implementing (4.14) with the Wiener process model was *ad hoc* proposed in [140]. The axioms A1-A4 were suggested and the method (4.14) was named the P-algorithm in [288].

The axioms above justify rationality of the P-algorithm. Nevertheless the question about its convergence should be answered; this question is considered in Sects. 4.3.7, 4.3.8, 4.4.2.

The properties of the P-algorithm depend on the properties of the chosen statistical model, e.g. on the weights in (4.5), and on the only parameter of the method ε_n in (4.14). To understand the influence of the latter we have to analyze the properties of the maximum and the maximizer in (4.14), i.e. the properties of $\max_{x \in A} \phi(x, y\cdot)$, where

$$\phi(x, y) = \mathbf{P}\{\xi(x) \leq y | \xi(x_1) = y_1, ..., \xi(x_n) = y_n\} =$$
$$\frac{1}{\sqrt{2\pi}} \int_{-\infty}^{\frac{y - m_n(x,\cdot)}{s_n(x,\cdot)}} \frac{1}{\sqrt{2\pi}} \exp(-t^2/2) dt. \quad (4.17)$$

The conditional mean and the conditional variance of a random function with probability 1 satisfy the following equalities

$$m_n(x_i|x_j, y_j, j = 1, ..., n) = y_i,\; s_n^2(x_i|x_j, y_j, j = 1, ..., n) = 0. \quad (4.18)$$

In the case of the statistical model $\xi(x)$ defined by (4.5) we assume that for the weights in the formulas (4.5) the equalities

$$w_i^n(x_i, x_j, y_j, j = 1, ..., n) = 1, w_i^n(x_m, x_j, y_j, j = 1, ..., n) = 0, m \neq i,$$

are valid implying that the parameters of $\xi(x)$ satisfy (4.18). We assume also that the weights are chosen in such a way that

$$\min_{x \in A} m_n(x|x_j, y_j, j = 1, ..., n) = \min_{1 \leq i \leq n} \{y_i\},$$
$$s_n(x|x_j, y_j, j = 1, ..., n) > 0,\; x \neq x_j,\; j = 1, ..., n.$$

Theorem 4.2. *If*

$$\tilde{y}_{on} = \min_{x \in A} m_n(x|x_i, y_i, i = 1, ..., n),$$

and

$$\frac{\tilde{y}_{on} - m_n(x|x_j, y_j, j = 1, ..., k))}{s_n(x|x_j, y_j, j = 1, ..., k)} \to 0,$$

when $x \to \arg\min_{x \in A} m_n(x|\cdot)$, *then*

$$\arg\max_{x \in A} \phi(x, \tilde{y}_{on}) = \arg\min_{x \in A} m_n(x|\cdot).$$

If $\tilde{y}_{on} \to -\infty$, *then*

$$|| \arg\max_{x \in A} \phi(x, \tilde{y}_{on}) - \arg\max_{x \in A} s_n(x|x_j, y_j, j = 1, ..., k)|| \to 0.$$

Theorem 4.3. *Assume that* $\phi(x, y)$ *has an unique maximum point in the interior of* A *for all* $y \in \mathbb{R}$, *and let the functions* $m_n(x|\cdot)$, $s_n(x|\cdot)$ *be twice differentiable. Then the inequality*

$$Dx_{n+1} \cdot \nabla_x s_n(x|x_i, y_i, i = 1, ..., k)|_{x = x_{n+1}} < 0$$

holds, where Dx_{n+1} *is the vector of derivatives of the components of the vector* $x_{n+1} = \arg\max_{x \in A} \phi(x, y)$ *with respect to* y, *and* ∇_x *denotes the operator of gradient.*

For the proof of theorem 4.2, and theorem 4.3 see Sect. 4.2.8.

Corollary. Normally it is assumed $\tilde{y}_{on} = y_{on} - \varepsilon_n$. The character of globality/locality of the P-algorithm can be regulated by means of choice of the parameter ε_n: for large ε_n the search is more global than for small ε_n. The P-algorithm with $\varepsilon_n = 0$ degenerates to repeated observations of the objective function values at the best found point. The P-algorithm performs observations at the point of maximal uncertainty when $\varepsilon_n \to \infty$.

Although the P-algorithm is motivated by a stochastic model it is a deterministic algorithm. Its convergence, e.g. for a continuous function, can be analyzed by standard methods of optimization theory. To analyze the convergence with respect to the stochastic function we assume that the objective functions are randomly generated sample functions of the underlying stochastic function. Since the objective functions are random, the optimization results are also random, and the convergence should be considered in a probabilistic sense. Let us emphasize, that the results of minimization by deterministic algorithm are random because of the random nature of the selection of objective functions. In the subsequent chapters convergence of different versions of the P-algorithm is considered under different assumptions about the objective functions.

The probabilistic convergence has mainly theoretical interest since it does not have sense with respect to a unique objective function. However, a user is normally interested in the convergence of the considered algorithm for his problem. Therefore we consider the convergence of an optimization algorithm for a specific problem as a fundamental requisite of the algorithm, and concentrate our attention on the convergence of the P-algorithm for continuous objective functions. The convergence of a global optimization algorithm is desirable under as broad conditions as possible. Rather frequently an objective function of a practical global optimization problem is given by means of a code implementing a complicated algorithm. The theoretical investigation of properties of such functions is difficult. It seems that continuity of the objective function is the weakest reasonable assumption in such a case. In addition to continuity some weak regularity conditions will be assumed where appropriate, e.g. while analyzing the convergence rate. To guarantee convergence for any continuous function the algorithm should generate a dense sequence of trial points as stated by Theorem 1.1. The same necessary and sufficient convergence condition is also valid for narrower classes of objective functions, e.g. for the class of Lipshitz functions with *a priori* unknown Lipshitz constant. Some subintervals of the feasible interval can be excluded from consideration but only in the case where the objective function satisfies sufficiently strong assumptions. For example, a subinterval not containing global minimizers can be indicated by the guaranteed lower bounds for function values defined via a known Lipshitz/Hölder constant. Similarly subintervals not containing global minimizers can be excluded using interval arithmetic based computations. Sometimes in the convergence theorems for the algorithms with not dense trial sequences a known lower bound is assumed implicitly, e.g. assuming that a parameter of the algorithm is sufficiently large. A typical statement of this type is, that the estimate of the global minimum obtained by the considered algorithm with parameter λ larger than $\lambda*$ converges to the true global minimum for any Lipshitz function. In this case the explicit assumption that the Lipshitz constant is known, is replaced by a similar assumption implicitly related to the Lipshitz constant. Such an assumption can be advantageous if an appropriate value of $\lambda*$ can be more easily selected than a good estimate of the Lipshitz constant. A similar remark may be made concerning the convergence of adaptive algorithms: different Lipshitz constants can be assumed for different subsets of the feasible region, but they should overestimate true constants to guarantee global convergence. For details of convergence proofs for algorithms with dense and not dense trial sequences we refer to [39], [40], [41], [91], [111], [112], [189], [214], [216], [233], [279], [288], [290], [291].

4.2.5 Information Theory Based Approach

The construction stochastic processes of well researched in probability theory, is motivated mainly by the problems of physics and technology. Their disadvantage as models of objective functions is their complicated probabilistic characterization of the minima of the sample functions. To avoid such difficulties in this information approach to global optimization a special random process is constructed explicitly including a parameter related to the minimizers of sample functions. Consider a stochastic process $\xi(t)$, $t = 0, 1/N, 2/N..., 1$ with discrete time $0 \leq t \leq 1$ and independent increments; see Sect. 4.1.2. The variance of Gaussian increments $\xi(t) - \xi(t-1)$ is constant, and its mean values are (negative) equal to $-L$ for all time moments $t = 0, ..., \alpha$ and are (positive) equal to L for $t = \alpha + 1/N, ..., 1$. These assumptions mean on average decrease in the objective function values in the interval $0 \leq t < \alpha$ and on average increase in the objective function values in the interval $\alpha < t \leq 1$. Therefore the parameter α indicates the probable neighborhood of the global minimizers of the sample functions of the process $\xi(t)$. Assuming reliable information on the location of the global minimizer of an objective function is absent, α is assumed random with uniform distribution over feasible region $0 \leq t \leq 1$. The sample functions of such a process in average decrease in the interval $[0, \alpha]$ and increase in the interval $[\alpha, 1]$. Since the value of the parameter α defines a point of change in the global behavior of a sample function, this value may be expected to be closely located to a global minimizer. Therefore, it seems rational to select the observation points in such a way that information on α is maximized. The information method was originally constructed in [172] where a current observation of a function value is performed at the point of maximum Shannon information about α where information is calculated conditionally with respect to the function values observed at previous optimization steps. Some modifications of this method are proposed in [232] where, the criterion of information is substituted for the criterion of average loss with a step-wise loss function

$$l(t) = \begin{cases} 1, \ t \neq \alpha, \\ 0, \ t = \alpha. \end{cases}$$

The loss function $l(\cdot)$ expresses the desire to hit the unknown parameter α at the current step of search. The method, that is optimal with respect to the average loss, performs the current observation of the objective function at the maximally probable location of α

$$x_{n+1} = \arg \max_{0 \leq t \leq 1} \ p_n(t), \qquad (4.19)$$
$$p_n(t) = \mathbf{P}\{\alpha = t \ |\xi(x_1), ..., \xi(x_n)\},$$

where the aposteriori probability is calculated using Bayes formula. The advantage of the considered model is the analytic solvability of (4.20). The point of current observation is defined by the following formulae:

$$x_{n+1} = \frac{x_j^n - x_{j-1}^n}{2} - \frac{y_j^n - y_{j-1}^n}{2L}, \qquad (4.20)$$

where j, is an index corresponding to the maximum of

$$R_i = L(x_i^n - x_{i-1}^n) + \frac{(y_i^n - y_{i-1}^n)^2}{L(x_i^n - x_{i-1}^n)} - 2(x_i^n + x_{i-1}^n),$$

and x_i^n, y_i^n denote increasingly ordered observation points and corresponding objective function values. The selection of the point of current observation can be interpreted as a two stage decision: the subinterval j is chosen at the first stage, and the point according to formula (4.20) is chosen at the second stage. It is interesting to note, that the second stage decision coincides with the decision of the well known Pijavskij method based on a Lipshitzian model of the objective functions where a current observation is performed at the minimum point of the piecewise linear underestimate of the function values. The parameter of the statistical model L is similar to Lipshitz constant of a class of Lipshitz functions. To guarantee correctness of (4.20) the inequality

$$L \geq \max_{i=2,\ldots,n} \frac{y_i^n - y_{i-1}^n}{x_i^n - x_{i-1}^n},$$

should be satisfied. Similarly to global optimization methods based on Lipshitz models, an overestimate of the Lipshitz constant should be known and used as the parameter m. In practical applications an adaptive estimate of the Lipshitz constant frequently is used. The method (4.20) with an adaptive estimate of the Lipshitz constant normally outperforms the method with a precise value of the Lipshitz constant, however, it is not guaranteed that the adaptive method will find the global minimum.

We will not go into the details of the information based approach here since this approach is thoroughly presented in the recently published book [233], including theoretical results as well as methods and some applications.

4.2.6 Response Surface Based Methods

Several local optimization algorithms are constructed using a model of an objective function, e.g. quadratic approximation is obtained from a truncated Taylor expansion. The next observation is planned at the minimum point of the model function. Newton's method is a well known example of the implementation of this idea. In local optimization the idea works well therefore it seems attractive to generalize it to global optimization. For such a generalization to be made an appropriate global model of an objective function is needed.

Given previous observation points and objective function values an interpolating function can be constructed and used as a global model for the planning of the next observation. The terms, response surface, surrogate function, and kriging model define a function used in optimization to approximate the objective function. Although these terms are used in the slightly different context of constructing optimization algorithms they have similar meaning. The term 'response surface' is chosen for the title of this section because it seems most context neutral. For example, the conditional mean of a random field can be considered as a response surface.

The idea of choosing the minimizer of the interpolating function for the next observation point can not guarantee rationality of search for the global minimum. Fig. 4.5 illustrates the case where the search is concentrated in a neighborhood of a local minimizer. Since the observations are not performed in the subregion of the global minimizer, the approximation errors there are not decreasing, and the global minimum remains hidden. To ensure global convergence, observations should be planned not only where expected function values are small but also in non-researched subregions. The latter can be indicated by large values of the conditional variance of the statistical model. To define the rational search strategy prediction of function values is not sufficient. Possible deviations in the objective function from the predicted values should also be taken into account. Therefore, to construct a rational global optimization algorithm response surface should be complemented by a measure of approximation uncertainty. The response surface with a measure of approximation uncertainty is nothing more than a statistical model of an objective function.

One of the first response surface methods with a random field model is proposed in [219]. At the current $n + 1$-th step the conditional mean of a random field $\xi(x)$

$$m_n(x|x_i, y_i, i = 1, ..., n) = \mathbf{E}\left[\xi(x)|\xi(x_i) = y_i, i = 1, ..., n\right].$$

is used as a response surface, and the objective function value is calculated at the point

$$x_{n+1} = \arg\min_{x \in A} m_n(x|x_i, y_i, i = 1, ..., n). \tag{4.21}$$

This method is a naive extension of kriging to global optimization. A Gaussian random field with a constant mean value and variance have been used in the implementation as well as a correlation function depending on the distance between the points (calculated taking into account the different scales of the variables) $r(t) = \exp(-ct^2)$, $c > 0$, where $t = ||x - y||$. Some experimental results presented in [219] demonstrate that the method performs rationally. However, the next paper by the same author [220] concludes that the method (4.21) is too local. The localization would be fatal if (4.21) would be based on the stationary Gaussian model with the correlation function $r(t) = \exp(-ct)$, and minimization would be sufficiently precise. The method would degenerate due to the fact that in this case the minimum of the conditional mean coincides with the minimum known value, causing repeated calculation of the objective function value at the best found point. This case presents the strongest form of the localization of search mentioned in the beginning of the section. To improve the performance in [220] a modification of the original kriging method is suggested combining several iterations according to algorithm (4.21) with several subsequent iterations according to P-algorithm.

The disadvantages of the naive kriging global optimization (4.21) are also stated in recent reviews on response surface methods [130], [132]. These reviews include methods based on statistical models, e.g. one step Bayesian and the P-algorithm, presenting them in terms of response surfaces. An advantage of such a presentation is the minimum use of probability theory. Although using the concept of a response surface the method can be explained without the theory of stochastic processes, theoretical background of the considered methods presented in such a light is missed.

An interesting radial basis function method is proposed by Gutmann [106]. The method well represents the idea of response surface based global optimization. Values of an objective function $f(x)$, $x \in \mathbb{R}^d$ are predicted by the radial basis function (RBF)

$$m_n(x|x_i, y_i, i = 1, ..., n) = \sum_{i=1}^{n} \lambda_i \phi(||x - x_i||), \tag{4.22}$$

that interpolates the data $(x_i, y_i = f(x_i))$; we use a form of RBF without the extra polynomial summands. Different basis functions $\phi(\cdot)$ can be chosen, e.g. the Gaussian function $\phi(r) = \exp(-\gamma r^2)$, $r \geq 0$, $\gamma > 0$. The coefficients λ_i are defined by the system of linear equations $m_n(x_i|\cdot) = y_i, i = 1, ..., n$ whose solution is guaranteed by the positive definiteness of the matrix $\Phi = (\phi(||x_i - x_j||))$. Let $\tilde{y}_{on}$ be a target value of the objective function that it is desired to reach at the current minimization step, e.g. $\tilde{y}_{on} = y_{on} - \varepsilon_n$, $y_{on} = \min_{i=1,...,n} y_i$.

Using a statistical model the P-algorithm can be constructed where the most probable point to exceed the target level is chosen for the next observation. In the RBF based algorithm the current observation of $f(\cdot)$ is performed at he point x_{n+1} where the value of $f(\cdot)$ equal to $\tilde{y}_{on}$ is most likely. Although the radial basis function (4.22) does not directly contain such information, a conclusion about the reasonable site for x_{n+1} can be derived from the idea of a minimal norm interpolator.

Let x_i, $i = 1, ..., n$ and y_i, $i = 1, ..., n + 1$, $y_{n+1} = \tilde{y}_{on}$. Let us find a point x_{i+1} such that the norm of $m_{n+1}(x|x_i, y_i, i = 1, ..., n + 1)$ is minimal. Such a point for value y_{n+1} is most 'likely' from the point of view of interpolation, and according to the terminology of [106] it minimizes 'bumpiness' of the response surface. The point x_{n+1} is chosen as the point of current observation. Formally, the algorithm is constructed sequentially tuning the radial function interpolator by means of minimization of a norm (in fact of a semi-norm) with respect to the forecasted global minimum value y_{n+1}.

In the consideration below we use the shorthand $m_n(x) = m_n(x|\cdot)$, and the notations

$$\Lambda = (\lambda_1, ..., \lambda_n)^T, \ \Phi(x) = (\phi(||x - x_1||), ..., \phi(||x - x_n||))^T.$$

The formula (4.22) using these notations can be rewritten in the form $m_n(x) = \Lambda^T \Phi(x)$. Similarly the RBF interpolator using an extended set of data $(x_i, y_i = f(x_i))$, $i = 1, ..., n + 1$ is defined by the formulas

$$m_{n+1}(x) = \sum_{i=1}^{n+1} \mu_i \phi(||x - x_i||) = \mathcal{M}^T \cdot \begin{pmatrix} \Phi(x) \\ \phi(||x - x_{n+1}||) \end{pmatrix}.$$

The vectors of coefficients Λ^T and $\mathcal{M}^T$ can be calculated as solutions of systems of linear equations corresponding to the condition of interpolation

$$\Lambda = \Phi^{-1} \cdot Y,\ Y = \begin{pmatrix} y_1 \\ \dots \\ y_n \end{pmatrix},\ \mathcal{M} = \Psi^{-1} \cdot \begin{pmatrix} Y \\ \tilde{y}_{on} \end{pmatrix}, \tag{4.23}$$

$$\Phi = \begin{pmatrix} \phi(0) & \dots & \phi(||x_1 - x_n||) \\ \dots & \dots & \dots \\ \phi(||x_n - x_1||) & \dots & \phi(0) \end{pmatrix},$$

$$\Psi = \begin{pmatrix} \Phi & \Phi(x_{n+1}) \\ \Phi(x_{n+1})^T & \phi(0) \end{pmatrix}.$$

The squared semi-norm of m_{n+1} is equal to

$$||m_{n+1}(x)||^2 = \mathcal{M}^T \Psi \mathcal{M} = (Y^T, \tilde{y}_{on}) \Psi^{-1} \begin{pmatrix} Y \\ \tilde{y}_{on} \end{pmatrix}, \tag{4.24}$$

where the expression of $\mathcal{M}$ from (4.23) is taken into account.

To invert matrix Ψ presented as a block matrix in (4.23) the formula by Frobenius can be applied

$$\Psi^{-1} = \begin{pmatrix} \Phi^{-1} + \frac{1}{h}\Phi^{-1}\Phi(x_{n+1})\Phi(x_{n+1})^T\Phi^{-1} & -\frac{1}{h}\Phi^{-1}\Phi(x_{n+1}) \\ -\frac{1}{h}\Phi(x_{n+1})^T\Phi^{-1} & \frac{1}{h} \end{pmatrix},$$

where

$$h = \phi(0) - \Phi(x_{n+1})^T \Phi^{-1} \Phi(x_{n+1}).$$

Calculation of the norm (4.24) using the latter expression of Ψ^{-1} gives the following result

$$||m_{n+1}(x)||^2 = (Y^T, \tilde{y}_{on}) \Psi^{-1} \begin{pmatrix} Y \\ \tilde{y}_{on} \end{pmatrix} = (Y^T, \tilde{y}_{on}) \cdot$$

$$\cdot \begin{pmatrix} \Phi^{-1}Y + \frac{1}{h}\Phi^{-1}\Phi(x_{n+1})\Phi(x_{n+1})^T\Phi^{-1}Y - \frac{\tilde{y}_{on}}{h}\Phi^{-1}\Phi(x_{n+1}) \\ -\frac{1}{h}\Phi(x_{n+1})^T\Phi^{-1}Y + \frac{\tilde{y}_{on}}{h} \end{pmatrix} =$$

$$= \Lambda \Phi \Lambda + \frac{(\tilde{y}_{on} - \Phi(x_{n+1})^T \Phi^{-1} Y)^2}{\phi(0) - \Phi(x_{n+1})^T \Phi^{-1} \Phi(x_{n+1})}. \tag{4.25}$$

From (4.25) the subsequent equality follows

$$||m_{n+1}(x)||^2 = ||m_n(x)||^2 + \frac{(\tilde{y}_{on} - \Phi(x_{n+1})^T \Phi^{-1} Y)^2}{\phi(0) - \Phi(x_{n+1})^T \Phi^{-1} \Phi(x_{n+1})}, \tag{4.26}$$

where the first summand does not depend on x_{n+1}. Therefore for x_{n+1} a minimum point of the second summand is chosen. But minimization of the latter is equivalent to maximization of the criterion

$$\frac{\tilde{y}_{on} - \Phi(x_{n+1})^T \Phi^{-1} Y}{\sqrt{\phi(0) - \Phi(x_{n+1})^T \Phi^{-1} \Phi(x_{n+1})}}. \tag{4.27}$$

Let us consider a homogeneous isotropic Gaussian random field $\xi(x)$, $x \in \mathbb{R}^n$ with zero mean and covariance function $\phi(\cdot)$. The conditional mean and the conditional variance of $\xi(x)$ with respect to $\xi(x_i) = y_i$, $i = 1, ..., n$, is equal to

$$m_n(x|\xi(x_i) = y_i,\ i = 1, ..., n) = \Phi(x)^T \Phi^{-1} Y, \text{ and}$$
$$s_n^2(x|\xi(x_i) = y_i,\ i = 1, ..., n) = \phi(0) - \Phi(x)^T \Phi^{-1} \Phi(x),$$

correspondingly. Therefore (4.27) corresponds to the criterion (4.16)

$$\frac{\tilde{y}_{on} - m_n(x|\xi(x_i) = y_i,\ i = 1, ..., n)}{s_n(x|\xi(x_i) = y_i,\ i = 1, ..., n)},$$

which is used to define the P-algorithm, and is justified by the axiomatic approach in Sect. 4.2.4. Therefore the axiomatic approach based on statistical models, and the approach based on RBF, provide similar algorithms.

4.2.7 Modifications

A solution obtained by means of a global optimization method is frequently calculated more precisely by means of a local optimization method. The efficiency of combining the two methods depends on the transition rule from global to local search. Usually a transition rule is defined heuristically, and is based on the belief that a point in the attraction region of a global minimizer is found. The statistical model of the objective function may be helpful in defining a transition rule justified by statistical methodology. For example, general information on a one-dimensional global optimization problem frequently is compatible with Wiener process as a global statistical model of the objective function. However, normally the objective function is smooth, at least in the vicinities of local minimizers. To improve the efficiency of search in the neighborhoods of the best points, the P-algorithm based on Wiener process model can be combined with an efficient local search algorithm. The transition from global to local search is reasonable to control by means of testing the hypothesis if a local minimum is found. The P*-algorithm, combining in this way the P-algorithm with local search, is considered in detail in Sect. 4.3.6.

In the multidimensional case an equivalent simply to test hypothesis is not known. The implementation of an algorithm, combining global and local search by testing the hypothesis if a local minimum is found, requires sophisticated programming also in the one-dimensional case. However, even a simple combination of local and global search is often advantageous. For example, Theorem 4.12 and Theorem 4.24 show that the convergence rate of the P-algorithm can be improved by simply alternating the choice of observation points according to global and local algorithms.

The global minimization methods based on statistical models can be generalized for interesting and important problems of minimization in the presence of noise, i.e. for minimization of the functions whose values are corrupted by noise. The inclusion of noise in the statistical model does not cause theoretical difficulties since the information on function values is integrated using conditional probability distributions. However, computational difficulties occur even in the case of the Wiener process model since the Markov property with respect to noisy observations is not valid. A special technique to cope with the computational difficulties is developed in [286] and the code of the proposed method is presented in [285]. Similar information theory based algorithms for minimization in the presence of noise is described in [233]. For the recent results on convergence we refer to [43].

4.2.8 Proofs of Theorems

Proof of Theorem 4.1.

First we show that the utility function is not increasing. Suppose that this were not the case, i.e. $u(m_2) - u(m_1) = \Delta > 0$, where $m_1 < m_2$. If $u(\cdot)$ is continuous at m_1, m_2, then there exists small s_0 so that for $s < s_0$ the inequality

$$|U(m, s) - u(m)| < \Delta/2, \; m = m_1, \; m = m_2 \tag{4.28}$$

holds, where

$$U(m, s) = \int_{-\infty}^{\infty} u(t) \frac{1}{s\sqrt{2\pi}} exp\left(-\frac{(t-m)^2}{2s^2}\right) dt.$$

From (4.28) it follows, that $s_1 < s_0$, $s_2 < s_0$ implying the inequality

$$U(m_2, s_2) - U(m_1, s_1) > u(m_2) - \Delta/2 - u(m_1) - \Delta/2 > 0,$$

which also implies a contradiction to axiom A1: $U(m_2, s_2) > U(m_1, s_1)$. Therefore, for all points of continuity of $u(\cdot)$ the inequality $m_1 < m_2$ implies $u(m_1) \geq u(m_2)$. This is also true for the points of discontinuity since the function $u(\cdot)$ is continuous from the left.

From axiom A2 it follows that the utility function is constant, $u(t) = c$ for $t > \tilde{y}_{on}$. If this were not the case then for some m_1, m_2, $\tilde{y}_{on} < m_2 < m_1$, the inequality $u(m_2) - u(m_1) = \Delta > 0$ would hold. Then there would exist small s_1 implying the following relations $U(m_1, s_1) < u(m_1) + \Delta = u(m_2) = U(m_2, 0)$ which contradicts axiom A2 requiring $U(m_1, s_1) > U(m_2, 0)$. Thus $u(t)$ is constant for $t \geq \tilde{y}_{on}$, and without loss of generality we can assume $u(t) = c = 0$, $t \geq \tilde{y}_{on}$.

From axiom A1 it follows that $u(\cdot)$ is not identical to a constant. Since $u(\cdot)$ is not increasing there exists $y < \tilde{y}_{on}$ such that $u(t) \geq \delta > 0$ for $t < y$. We will show that $u(\cdot)$ is discontinuous at the point $\tilde{y}_{on}$. Let $m_2 > \tilde{y}_{on}$, $s_2 = (y - m_2)/T_p$ where T_p is the $p-$quantile of the standard Gaussian distribution, $p < 0.5$. Then $U(m_2, s_2) > \delta p = \Delta$. If $u(\cdot)$ were continuous at the point $\tilde{y}_{on}$ then, because $u(t) = 0$, $\tilde{y}_{on}$, there would exist such a small s_1 that $U(\tilde{y}_{on}, s_1) < \Delta$.

Since $U(m, s_1)$ is not increasing in m, then for the chosen m_2, s_2, s_1, there does not exist m_1, $\tilde{y}_{on} < m_1 < m_2$, such that $U(m_1, s_1) > U(m_2, s_2)$. The last statement contradicts to axiom A3. Therefore, $u(\cdot)$ is discontinuous at the point $\tilde{y}_{on}$. Without loss of generality we assume $u(\tilde{y}_{on}) = 1$.

To show that $u(t) = 1$ for $t < \tilde{y}_{on}$ assume the contrary $u(t) > 1+\Delta$, $\Delta > 0$ for $t \leq z_1 < \tilde{y}_{on}$. Then for $m_2 > \tilde{y}_{on}$, $s_2 = ((z_1-m_2)/T_p$, $p = 0.5/(1+\epsilon_1)$, $\Delta = 3\epsilon_1 + 2\epsilon_1^2$, $\epsilon_1 > 0$, the inequality $U(m_2, s_2) > 0.5 + \epsilon_1$ holds. There exists such a small s_1 that $U(\tilde{y}_{on}, s_1) < 0.5 + \epsilon_1$. Therefore, there does not exist $\tilde{y}_{on} < m_1 < m_2$ such that $U(m_1, s_1) > U(m_2, s_2)$, which contradicts axiom A3.

Proof of Theorem 4.2.

If $x_{on} = \arg\min_{x\in A} m_n(x|x_i, y_i, i = 1, ..., k)$ and $\tilde{y}_{on} = m_n(x_{on}|x_i, y_i, i = 1, ..., k)$, then $\Phi(x_{on}, \tilde{y}_{on}) \geq 0.5$, but $\Phi(x, \tilde{y}_{on}) < 0.5$, $x \neq x_{on}$. These simple facts prove the first part of the theorem.

The second part of the theorem concerns the case $\tilde{y}_{on} \to -\infty$. Since $\Phi(x, \tilde{y}_{on})$ is a monotonically increasing function of $\frac{\tilde{y}_{on}-m_n(x|\cdot)}{s_n(x|\cdot)}$ then it is increasing with respect to $s_n(x|\cdot)$, and decreasing with respect to $m_n(x|\cdot)$. Let m^+ and m^- denote the maximum and minimum of $m_n(x|\cdot)$, respectively, and $B_\varepsilon = \{x : x \in X,\ s_n(x|\cdot) \geq s^+ - \varepsilon\}$ where $s^+ = \max_{x\in A} s_n(x|\cdot)$.

Let $\tilde{y}_{on}$ be defined as follows

$$\tilde{y}_{on} = m^- - \frac{(s^+ - \varepsilon)(m_+ - m^-)}{\varepsilon},\ 0 < \varepsilon < s^+, \tag{4.29}$$

then for $x \notin B_\Delta$

$$\frac{\tilde{y}_{on} - m_n(x|\cdot)}{s_n(x|\cdot)} < \frac{\tilde{y}_{on} - m^-}{s^+ - \varepsilon} = -\frac{m^+ - m^-}{\varepsilon}.$$

At the maximum point of $s_n(x|\cdot)$ which is denoted by x^+ the inequality

$$\frac{\tilde{y}_{on} - m_n(x^+|\cdot)}{s_n(x^+|\cdot)} \geq \frac{\tilde{y}_{on} - m^+}{s^+} = -\frac{m^+ - m^-}{\varepsilon}$$

holds implying $\arg\max_{x\in X} \Phi(x, \tilde{y}_{on}) \in B_\varepsilon$. Since $s_n(x|\cdot)$ is continuous then for sufficiently small ε the distance

$$||\arg\max_{x\in A} \Phi(x, \tilde{y}_{on}) - \arg\min_{x\in A} m_n(x|\cdot)||$$

is arbitrarily small. Let $\tilde{y}_{on} \to -\infty$, then ε defined from (4.29) tends to zero, implying

$$||\arg\max_{x\in A} \Phi(x, \tilde{y}_{on}) - \arg\min_{x\in A} m_n(x|\cdot)|| \to 0.$$

Proof of Theorem 4.3.

To shorten the formulae, in this proof we use the notation m for $m_n(x|\cdot)$, and s for $s_n(x|\cdot)$ correspondingly. Since $\Phi(x, \tilde{y}_{on})$ is a monotonically increasing function of $(\tilde{y}_{on} - m)/s$ then the necessary maximum conditions for $\Phi(x, \tilde{y}_{on})$ with respect to x are

$$\nabla_x \frac{\tilde{y}_{on} - m}{s} = \frac{-s \cdot \nabla_x m - (\tilde{y}_{on} - m) \cdot \nabla_x s}{s^2} = \frac{G}{s^2} = 0. \tag{4.30}$$

The negative definiteness of the Hessian

$$H = \nabla_x^2 \frac{\tilde{y}_{on} - m}{s},$$

and the equality(4.30) are sufficient maximum conditions. Taking into account (4.30) it is easy to show that

$$H = \frac{-s \cdot \nabla_x^2 m - \nabla_x m \cdot (\nabla_x s)^T + \nabla_x s \cdot (\nabla_x m)^T - (\tilde{y}_{on} - m) \cdot \nabla_x^2 s}{s^2}$$

where T denotes transposition. The maximum point of $\Phi(x, \tilde{y}_{on})$ is defined as the implicit function of $\tilde{y}_{on}$ by the equation $G = 0$. The differentiation of this equation with respect to $\tilde{y}_{on}$ taking into account $x = x(\tilde{y}_{on})$ gives

$$\sum_{j=1}^{n} \frac{\partial g^i}{\partial x_j} \cdot \frac{\partial x_j}{\partial \tilde{y}_{on}} - \frac{\partial g^i}{\partial \tilde{y}_{on}} = 0, \ i = 1, ..., n, \tag{4.31}$$

where $G = (g^1, ..., g^n)^T$, $x = (x_1, ..., x_n)^T$. Since the equalities

$$\frac{\partial g^i}{\partial x_j} = -\frac{\partial^2 m}{\partial x_i \partial x_j} \cdot s - \frac{\partial m}{\partial x_i} \cdot \frac{\partial s}{\partial x_j} +$$
$$\frac{\partial m}{\partial x_j} \cdot \frac{\partial s}{\partial x_i} - (\tilde{y}_{on} - m) \cdot \frac{\partial^2 s}{\partial x_i \cdot \partial x_j},$$
$$\frac{\partial g^i}{\partial \tilde{y}_{on}} = -\frac{\partial s}{\partial x_i}, \ i, j = 1, ..., n,$$

hold, then taking into account (4.31) they can be rewritten as

$$\frac{\partial s}{\partial x_i} = -\frac{\partial g^i}{\partial \tilde{y}_{on}} = \sum_{j=1}^{n} \frac{\partial g^i}{\partial x_j} \cdot \frac{\partial x_j}{\partial \tilde{y}_{on}} = \sum_{j=1}^{n} s^2 H_{ij} D x_j,$$

and finally reduced to the matrix form $s^2 \cdot H \cdot Dx = \nabla_x s$. Since the Hessian is negatively definite then at the maximum point the inequality

$$Dx \cdot \nabla_x s = s^2 \cdot Dx \cdot H \cdot Dx < 0$$

holds, and this conclusion completes the proof.

4.3 One-dimensional Algorithms

4.3.1 Introduction

In the one-dimensional case an algorithmic implementation of a method is normally rather close to the theoretical method, and the theoretical investigation is often not overly complicated. Therefore in the one-dimensional case the empirical testing results can be predicted and interpreted theoretically. Theoretical analysis of one-dimensional algorithms is important not only for one-dimensional optimization, since some conclusions concerning one-dimensional algorithms can be generalized to the multidimensional case. We concentrate in this section on the P-algorithm based on different random process models. We briefly discuss the one step Bayesian algorithm to help in its conceptual comparison with the P-algorithm.

The algorithms in this section are aimed to minimize a continuous objective function $f(x)$ over the interval $x \in [0, 1]$ (by rescaling we can treat the general interval). We are mainly interested in the case where $f(\cdot)$ does not satisfy stronger regularity conditions, such as convexity, unimodality, or Lipshitz continuity.

4.3.2 One-step Bayesian Algorithm

A version of a one-dimensional one-step Bayesian algorithm based on the Wiener process model was proposed and investigated in [279]. It was shown there that the calculation of a current observation point, defined by the equations

$$x_{n+1} = \arg\min_{0 \le x \le 1} \mathbf{E}\{\min(\xi(x), y_{on}) | \xi(x_1) = y_1, ..., \xi(x_n) = y_n\},$$
$$y_{on} = \min\{y_1, ..., y_n\},$$

may be reduced to the solution of $n-1$ unimodal optimization problems. It was shown that this algorithm asymptotically generates an everywhere dense sequence of points, and it therefore converges for any continuous function. However, the experimental investigation has shown that for a modest number of trials the one-step algorithm performs a very local search concentrating the observations in close vicinities of the best found points [279]. This property may well be explained as a consequence of one-step optimality: if the very last observation is planned then it does not seem rational to perform the observation with an uncertain result; it seems more rational to make an observation near to the best point, where an improvement (maybe only small) is likely. However, such an explanation is not very helpful for modifications of the one-step Bayesian algorithm to increase the globality of the search. In [279] the efficiency of the search was improved heuristically by means of artificially increasing the parameter of the statistical model. Such a modification works well but it is not justified theoretically. For some improvements of the one-step Bayesian algorithm and its further theoretical investigation we refer to [144].

4.3.3 P-algorithm

The stochastic process $\{\xi(x) : 0 \leq x \leq 1\}$ is accepted as a statistical model of the objective functions; the parameters of the model can be estimated from an initial sampling of function values at points uniformly distributed in $[0, 1]$. Fix $\epsilon > 0$. The $n+1$-th observation of the function value is performed by the P-algorithm at the point

$$x_{n+1} = \arg \max_{0 \leq x \leq 1} P\{\xi(x) < y_{on} - \epsilon \,|\, \xi(x_i) = y_i, i = 1, \ldots, n\}, \qquad (4.32)$$

where $x_i, y_i = f(x_i)$ are the results of observations at previous minimization steps and $y_{on} = \min_{1 \leq i \leq n} y_i$. Let us denote the ordered observation points by

$$0 = x_0^n < x_1^n < \cdots < x_n^n = 1,$$

and the corresponding function values by $y_i^n = f(x_i^n)$, $i \leq n$.

Let us start with the P-algorithm based on the standard Wiener process model. The Markov property implies the reduction of (4.32) to the selection of the best of the following minimizers

$$\arg \max_{x_{i-1}^n \leq x \leq x_i^n} P\{\xi(x) < y_{on} - \epsilon | \xi(x_{i-1}^n) = y_{i-1}^n,\ \xi(x_i^n) = y_i^n\}. \qquad (4.33)$$

Applying the known formulas of conditional mean and conditional variance of the Wiener process

$$m_n(x|\xi(x_{i-1}^n) = y_{i-1}^n,\ \xi(x_i^n) = y_i^n) = \frac{y_{i-1}^n(x_i^n - x) + y_i^n(x - x_{i-1}^n)}{x_i^n - x_{i-1}^n},$$

$$s_n^2(x|\xi(x_{i-1}^n) = y_{i-1}^n,\ \xi(x_i^n) = y_i^n) = \frac{(x_i^n - x)(x - x_{i-1}^n)}{x_i^n - x_{i-1}^n}, \qquad (4.34)$$

it is easy to show that the maximum of (4.33) is achieved at the point

$$t_i = x_{i-1}^n + \tau_i \cdot (x_i^n - x_{i-1}^n),\ \tau_i = \frac{1}{1 + \frac{y_i^n - y_{on} + \epsilon}{y_{i-1}^n - y_{on} + \epsilon}}, \qquad (4.35)$$

and is equal to

$$P_n(t_i) = \Phi\left(\frac{y_{on} - \epsilon - \frac{y_{i-1}^n \cdot (x_i^n - t_i) + y_i^n \cdot (t_i - x_{i-1}^n)}{x_i^n - x_{i-1}^n}}{\sqrt{\frac{(x_i^n - t_i) \cdot (t_i - x_{i-1}^n)}{x_i^n - x_{i-1}^n}}}\right), \qquad (4.36)$$

$$\Phi(t) = \frac{1}{\sqrt{2\pi}} \int_{-\infty}^{t} \exp(-z^2/2) dz.$$

To find $\max \Phi(z_i)$, $i = 1, ..., n-1$, it is sufficient to compare the values z_i, $i = 1, ..., n-1$, since $\Phi(t)$ is a monotonically increasing function of t. Applying simple algebra the maximal probability (4.37) may be shown to be monotonically related to the interval criterion value

$$\gamma_i^n = \frac{x_i^n - x_{i-1}^n}{(y_i^n - y_{on} + \epsilon)(y_{i-1}^n - y_{on} + \epsilon)}.$$

Finally, the algorithm is reduced to the selection of the subinterval with maximal criterion value γ_i^n, and the site of the observation at this interval is chosen to be equal to t_i defined in (4.35). The details of the implementation of the Wiener process version of the algorithm (4.33) may be found in [286].

Now we turn to the P-algorithm for the smooth function model (4.3). The implementation of the algorithm according to (4.33) is not practical in the non-Markov case because of difficulties implied by the inversion of the covariance matrix during the computation of the conditional mean and conditional variance of the process. Let us simplify the calculations by taking into account only the two neighboring points while computing the conditional mean and variance [39]. With such a simplification, computing the coordinate of the current trial point x_{n+1} according to (4.33) is reduced to the following procedure. Similarly to the P-algorithm based on the Wiener process model for each subinterval $[x_{i-1}^n, x_i^n]$, $i = 1, ..., n$, calculate

$$\max_{x_{i-1}^n \leq x \leq x_i^n} P\{\xi(x) < y_{on} - \epsilon | \xi(x_{i-1}^n) = y_{i-1}^n,\ \xi(x_i^n) = y_i^n\}, \tag{4.37}$$

and for the interval with the largest probability, calculate the point that maximizes the probability in (4.37); the maximum point is the new observation point x_{n+1}. For $x_{i-1}^n \leq x \leq x_i^n$ the probability in (4.37) is calculated according to the formula

$$P\{\xi(x) < y_{on} - \epsilon | \xi(x_{i-1}^n) = y_{i-1}^n,\ \xi(x_i^n) = y_i^n\} =$$
$$= \Phi\left(\frac{y_{on} - \epsilon - m_n(x|\xi(x_{i-1}^n) = y_{i-1}^n,\ \xi(x_i^n) = y_i^n)}{s_n(x|\xi(x_{i-1}^n) = y_{i-1}^n,\ \xi(x_i^n) = y_i^n)}\right), \tag{4.38}$$

where

$$m_n(x|\xi(x_{i-1}^n) = y_{i-1}^n,\ \xi(x_i^n) = y_i^n) = (y_{i-1}^n, y_i^n) \cdot R^{-1} \cdot (r(x - x_{i-1}^n), r(x - x_i^n))^T,$$

and

$$s_n^2(x|\xi(x_{i-1}^n) = y_{i-1}^n,\ \xi(x_i^n) = y_i^n) =$$
$$= 1 - (r(x - x_{i-1}^n), r(x - x_i^n)) \cdot R^{-1} \cdot (r(x - x_{i-1}^n), r(x - x_i^n))^T,$$

are the conditional mean and variance, respectively. Here we use the notation

$$R = \begin{pmatrix} 1 & r(x_i^n - x_{i-1}^n) \\ r(x_i^n - x_{i-1}^n) & 1 \end{pmatrix}. \tag{4.39}$$

The computations according to the modified algorithm (4.37) are much simpler than the computations according to the original algorithm (4.33). The formulae for the conditional mean and conditional variance can be rewritten as

$$m_n(x|\xi(x_{i-1}^n) = y_{i-1}^n,\ \xi(x_i^n) = y_i^n) =$$
$$\frac{r(x - x_{i-1}^n) - r(x_i^n - x_{i-1}^n)r(x_i^n - x)}{1 - r(x_i^n - x_{i-1}^n)^2} y_{i-1}^n +$$
$$+ \frac{r(x_i^n - x) - r(x_i^n - x_{i-1}^n)r(x - x_{i-1}^n)}{1 - r(x_i^n - x_{i-1}^n)^2} y_i^n,$$

and

$$s_n^2(x|\xi(x_{i-1}^n) = y_{i-1}^n,\ \xi(x_i^n) = y_i^n) = 1 -$$
$$- \frac{r(x-x_{i-1}^n)^2 + r(x_i^n - x)^2 - 2r(x - x_{i-1}^n)r(x_i^n - x)r(x_i^n - x_{i-1}^n)}{1 - r(x_i^n - x_{i-1}^n)^2}.$$

For further simplification of these formulae let us apply the expansion (4.3). The first order expansion gives the following formula for the conditional mean:

$$m_n(x|\xi(x_{i-1}^n) = y_{i-1}^n,\ \xi(x_i^n) = y_i^n)$$
$$= \frac{y_{i-1}^n \cdot (x_i^n - x) + y_i^n \cdot (x - x_{i-1}^n)}{x_i^n - x_{i-1}^n} + o(x_i^n - x_{i-1}^n) \tag{4.40}$$

as $x_i^n - x_{i-1}^n \to 0$; i.e., the conditional mean is approximated by linear interpolation. With the third order expansion of the numerator and the first order expansion of the denominator the conditional variance is approximated :

$$s_n^2(x|\xi(x_{i-1}^n) = y_{i-1}^n,\ \xi(x_i^n) = y_i^n) =$$
$$= \tfrac{1}{4}(\lambda_4 - \lambda_2^2) \cdot (x_i^n - x)^2 \cdot (x - x_{i-1}^n)^2 + o(x_i^n - x_{i-1}^n)^4 \tag{4.41}$$

as $x_i^n - x_{i-1}^n \to 0$. From now on we incorporate approximations (4.40) and (4.41) (ignoring the remainder terms) into (4.38). Thus our algorithm is as follows: After n steps, choose the next point x_{n+1} to maximize the function $P_n(\cdot)$ defined by

$$P_n(x) = \Phi\left(\frac{y_{on} - \epsilon - \frac{y_{i-1}^n \cdot (x_i^n - x) + y_i^n \cdot (x - x_{i-1}^n)}{x_i^n - x_{i-1}^n}}{\frac{1}{2}\sqrt{\lambda_4 - \lambda_2^2} \cdot (x_i^n - x) \cdot (x - x_{i-1}^n)} \right) \tag{4.42}$$

for $x \in [x_{i-1}^n, x_i^n]$, $i \le n$. We can think of $P_n(x)$ heuristically as the conditional probability that $\xi(x)$ falls below $y_{on} - \epsilon$, though we do not assume the existence of a stochastic process with the corresponding marginal distribution given by $P_n(\cdot)$.

Theorem 4.4. *The function $P_n(x)$ has a unique local maximum in the interval $x_{i-1}^n \le x \le x_i^n$, which is attained at the point*

$$t_i = x_{i-1}^n + \tau_i \cdot (x_i^n - x_{i-1}^n), \tag{4.43}$$

where

$$\tau_i = \frac{\Delta_{ni}}{\Delta_{n\,i-1} + \sqrt{\Delta^2_{n\,i-1} + (y^n_i - y^n_{i-1}) \cdot \Delta_{n\,i-1}}},$$

$$\Delta_{ni} = y^n_{i-1} - y_{on} + \epsilon. \tag{4.44}$$

Corollary 4.5. *To implement the P-algorithm the subinterval (x^n_{i-1}, x^n_i) should be selected with the maximal value in (4.42). Since $\Phi(\cdot)$ is monotonically increasing, the subinterval with maximal value of*

$$\gamma^n_i = \frac{(x^n_i - x^n_{i-1})^2 \tau_i (1 - \tau_i)}{y^n_i \tau_i + y^n_{i-1}(1 - \tau_i) - y_{on} + \epsilon}$$

should be chosen. The next observation of the objective function is taken in the chosen subinterval at the point t_i given in (4.43).

The criterion of selection of a subinterval may be reduced to the following simpler form

$$\gamma^n_i = \frac{x^n_i - x^n_{i-1}}{\sqrt{y^n_i - y_{on} + \epsilon} + \sqrt{y^n_{i-1} - y_{on} + \epsilon}}. \tag{4.45}$$

4.3.4 P-algorithm for a Model with Derivatives

The integrated Wiener process is an intermediate between the Wiener process model and the smooth function model, since its sample functions are only once differentiable. The P-algorithm for the integrated Wiener process $\zeta(x)$, $0 \leq x \leq 1$, is defined in the usual way, the $(n+1)$-th point is chosen to maximize the probability

$$P\left(\zeta(x) < y_{on} - \epsilon | \xi(x^n_i), \zeta(x^n_i), i = 0, ..., n\right), \tag{4.46}$$

where $\xi(\cdot)$ denotes the derivative of $\zeta(\cdot)$.

In light of the Markov property maximization of $P(\cdot)$ in (4.46) over the interval [0, 1] can be replaced by maximization over n subintervals

$$\max_{x^n_{i-1} \leq x \leq x^n_i} P\{\zeta(x) < y_{on} - \epsilon | \zeta(x^n_j) = y^n_j,\ \xi(x^n_j) = z^n_j, j = i-1, i\},$$

where y^n_i, z^n_i denote the values of the objective function and its derivative at the point x^n_i correspondingly. The best of n maximizers is chosen for the site of the $(n+1)$-th observation. The objective function can be expressed via $\Phi(\cdot)$, the conditional mean and the conditional variance of the statistical model similarly as in (4.38); for the formulae of conditional mean and variance we refer to Sects. 4.1.2 and [42]. Since the general form of these formulae is cumbersome the maximization problem is analyzed for the standardized subinterval $[0, t]$, with endpoint values $\xi(0) = z_1, \zeta(0) = y_1, \xi(t) = z_2, \zeta(t) = y_2$. Consider the maximization of

$$P(\zeta(s) < c|\xi(0) = z_1, \zeta(0) = y_1, \xi(t) = z_2, \zeta(t) = y_2) = \\ = \Phi\left(-\frac{a_1 z_1 - a_2 z_2 + b_1 y_1 + b_2 y_2 - ct^3}{t^3\sqrt{s^3(t-s)^3/3t^3}}\right),$$

$$a_1 = st(t-s)^2,\ a_2 = s^2 t(t-s),\ b_1 = 2s^3 - 3s^2 t + t^3,\ b_2 = 3s^2 t - 2s^3,$$

over the interval $s \in (0, t)$, where $c = y_{on} - \epsilon$. This is equivalent to choosing $s \in (0, t)$ to maximize

$$\frac{t^3\sqrt{s^3(t-s)^3/3t^3}}{st(t-s)^2 z_1 - s^2 t(t-s) z_2 + (2s^3 - 3s^2 t + t^3) y_1 + (3s^2 t - 2s^3) y_2 - ct^3}, \tag{4.47}$$

or, after the substitution $r = s/t$, to choosing $r \in (0, 1)$ to maximize

$$\frac{t^{3/2}\sqrt{r^3(1-r)^3}}{t[r(1-r)][(1-r)z_1 - rz_2] + (2r^3 - 3r^2 + 1)y_1 + (3r^2 - 2r^3)y_2 - c}. \tag{4.48}$$

Setting the derivative of this last expression to 0 implies that

$$p(r) = t(z_1 - z_2)r^3 + (tz_2 - 2tz_1 + 3y_1 - 3y_2)r^2 + \\ +(6c + tz_1 - 6y_1)r - 3c + 3y_1 \ = \ 0.$$

The maximizer r_0 can be found by comparing the value in (4.48) at the roots of p. (If the maximizer is not unique, choose the maximizer nearest the midpoint of the interval.) We call the maximum of (4.48) the *criterion value* of the interval $[0, t]$.

After these preliminary calculations we can precisely state the general algorithm. For each subinterval $[x_{i-1}^n, x_i^n]$, $i = 1, 2, \ldots, n$, find the r_0^i that maximizes (4.48), and the maximum criterion value; to apply formula (4.48) to the subinterval $[x_{i-1}^n, x_i^n]$ the following obvious replacements are needed: c with $y_{on} - \epsilon$, t with $x_i^n - x_{i-1}^n$, and y_1, y_2, z_1, z_2 with $y_{i-1}^n, y_i^n, z_{i-1}^n, z_i^n$. Choose the interval with the largest criterion value, e.g. $[x_{j-1}^n, x_j^n]$, and set $x_{n+1} = x_{j-1}^n + r_0^j(x_j^n - x_{j-1}^n)$.

As $t \downarrow 0$, the factor of t in the denominator of (4.48) approaches 0. Ignoring that term the criterion becomes

$$\frac{t^{3/2}\sqrt{r^3(1-r)^3}}{(2r^3 - 3r^2 + 1)y_1 + (3r^2 - 2r^3)y_2 - y_{on} + \epsilon}. \tag{4.49}$$

The limit of the criterion value does not depend on the derivatives at the end points of the interval. Indeed, it is not surprising, since both derivatives converge to the same value equal to the limit of $(y_i^n - y_{i-1}^n)/(x_i^n - x_{i-1}^n)$, i.e. the function values at the ends of the interval contain information needed to calculate the criterion value for sufficiently short intervals. To maximize the criterion (4.49) we solve the the equation

$$3(y_1 - y_2)r^2 - 6(y_1 - y_{on} + \epsilon_n)r + 3(y_1 - y_{on} + \epsilon_n) = 0.$$

The only root in $(0, 1)$ is equal to

$$\frac{1}{1 + \sqrt{\frac{y_2 - y_{on} + \epsilon}{y_1 - y_{on} + \epsilon}}}.$$

Substituting this root into the expression for criterion (4.49) yields

$$\frac{t^{3/2}}{\left((y_1 - y_{on} + \epsilon)(y_2 - y_{on} + \epsilon)\right)^{1/4}\left(\sqrt{y_1 - y_{on} + \epsilon} + \sqrt{y_2 - y_{on} + \epsilon}\right)}.$$

Since we are only concerned with the order of the criterion values in choosing new subintervals, we can apply an increasing function to the criterion values. Raising the values to the power $2/3$ gives

$$\frac{t}{\left((y_1 - y_{on} + \epsilon)(y_2 - y_{on} + \epsilon)\right)^{1/6}\left(\sqrt{y_1 - y_{on} + \epsilon} + \sqrt{y_2 - y_{on} + \epsilon}\right)^{2/3}}.$$

Thus the criterion that we maximize at each step of the algorithm is

$$\gamma_i^n = \frac{(x_i^n - x_{i-1}^n)\left(\sqrt{y_{i-1}^n - y_{on} + \epsilon} + \sqrt{y_i^n - y_{on} + \epsilon}\right)^{-2/3}}{(y_{i-1}^n - y_{on} + \epsilon)^{1/6}(y_i^n - y_{on} + \epsilon)^{1/6}}. \quad (4.50)$$

Let $\gamma^n = \max_{i=1,\ldots,n} \gamma_i^n$. The algorithm works by choosing the subinterval $[x_{j-1}^n, x_j^n]$ with the largest criterion value and then choosing the next observation at the point $x_{j-1}^n + \tau_j(x_j^n - x_{j-1}^n)$, where

$$\tau_j = \frac{1}{1 + \sqrt{\frac{y_j^n - y_{on} + \epsilon}{y_{j-1}^n - y_{on} + \epsilon}}}.$$

4.3.5 Comparison of P-algorithms Based on Different Models

The versions of the P-algorithm constructed for three different stochastic processes in the previous section have a common structure which may be represented by the following pseudo code.

P-algorithm(n,$\{\epsilon_n\}$, Y, X)

1 $\quad y_0 \leftarrow Y(0)$, $x_0 \leftarrow X(0)$

```
2      min ← y_0
3      for k ← 0 to n−1 do
4          t_{k+1} ← argmax P(Y(t) < min −ε|x_i, y_i, i ≤ k)
5          y_{k+1} ← Y(t_{k+1}), x_{k+1} ← Y'(t_{k+1})
6          if y_{k+1} < min then min ← y_{k+1}
7      return min
```

The maximization at line 4 is in fact reduced to the selection of a subinterval j corresponding to the maximal criterion value $j = \text{argmax}_{1 \leq i \leq n} \gamma_i^n$, where

- γ_i^n for the smooth function model is defined by

$$\gamma_i^n = \frac{x_i^n - x_{i-1}^n}{\sqrt{y_i^n - y_{on} + \epsilon} + \sqrt{y_{i-1}^n - y_{on} + \epsilon}}, \tag{4.51}$$

- γ_i^n for the integrated Wiener process is defined by

$$\gamma_i^n = \frac{(x_i^n - x_{i-1}^n)\left(\sqrt{y_{i-1}^n - y_{on} + \epsilon} + \sqrt{y_i^n - y_{on} + \epsilon}\right)^{-2/3}}{(y_{i-1}^n - y_{on} + \epsilon)^{1/6}(y_i^n - y_{on} + \epsilon)^{1/6}}, \tag{4.52}$$

- γ_i^n for the Wiener process is defined by

$$\gamma_i^n = \frac{x_i^n - x_{i-1}^n}{(y_i^n - y_{on} + \epsilon)(y_{i-1}^n - y_{on} + \epsilon)}. \tag{4.53}$$

The point in the selected interval is defined by the formula

$$x = x_{j-1}^n + \tau_j \cdot (x_j^n - x_{j-1}^n),$$

where

- τ_j for the smooth function model is defined by

$$\tau_j = \frac{1}{1 + \frac{y_j^n - y_{on} + \epsilon}{y_{j-1}^n - y_{on} + \epsilon}}, \tag{4.54}$$

- τ_j for the integrated Wiener process is defined by

$$\tau_j = \frac{1}{1 + \sqrt{\frac{y_j^n - y_{on} + \epsilon}{y_{j-1}^n - y_{on} + \epsilon}}}, \tag{4.55}$$

- τ_j for the Wiener process is defined by

$$\tau_j = \frac{1}{1 + \sqrt{\frac{y_j^n - y_{on} + \epsilon}{y_{j-1}^n - y_{on} + \epsilon}}}. \tag{4.56}$$

The interval selection criteria are different for all different versions of the algorithm, but the formula for τ_j is the same for the integrated Wiener process model and for the smooth function model.

4.3.6 P^* -algorithm

Let us assume that general information on an optimization problem is compatible with the Wiener process as a global statistical model of the objective function. On the other hand, as it is normally the case, assume that the objective function is smooth in the vicinities of local minimizers. To improve the efficiency of search in neighborhoods of the best points, the P-algorithm based on the Wiener process can be combined with an efficient local search algorithm. Consider one-dimensional minimization by means of the P-algorithm based on the Wiener process model. To indicate a suitable transition time to local minimization, in parallel with global search a hypothesis is tested if a subinterval of a local minimizer is found. Since the concentration of observation points is relatively high over a subinterval of a low local minimum, then the fulfillment of the following inequalities is expected for a sufficiently large n and an index L

$$y_i^n > y_{i+1}^n,\ i = L-3,\ L-2,\ L-1,$$
$$y_{i+1}^n > y_i^n,\ i = L,\ L+1,\ L+2.$$

The probability of fulfillment of such inequalities for a Wiener process is less than 0.016. On the other hand, these inequalities would be fulfilled if the objective function was unimodal over the subinterval (x_{L-3}^n, x_{L+3}^n). Therefore, if these inequalities are fulfilled, then the Wiener process model is assessed as inadequate over the subinterval (x_{L-3}^n, x_{L+3}^n), and here the unimodal function model is accepted. An example of a dual model with one found local minimum is illustrated in Fig. 4.6.

In the case where the hypothesis of finding a local minimum is accepted, a current local minimizer is calculated with predefined accuracy by means of a local minimization method. For example, the local minimization method based on quadratic interpolation can be used here. The found interval of a current local minimizer is excluded from further global search. Global minimization by the P-algorithm is continued over a the left region where the Wiener process model is used. Global search is continued until the probability the global minimum is found with a prescribed accuracy exceeds a predefined level. This probability can be calculated according to the formula (see 4.1)

$$g_n = \mathbf{P}(\max_{x\in G_n} \xi(x) \geq y_{on} - \varepsilon) =$$
$$= \prod_{i;\,(x_{i-1}^n, x_i^n)\in G_n} \exp\left(-2\frac{(y_{i-1}^n - y_{on}+\varepsilon)(y_i^n - y_{on}+\varepsilon)}{x_i^n - x_{i-1}^n}\right),$$

where G_n is the subset of the feasible region where the statistical model is valid, i.e. the subset of the feasible region complementing the union of indicated subintervals of local minima. Minimization terminates when the probability g_n of finding the global minimum exceeds the predefined level, e.g. 0.99.

Such a combination of local and global minimization strategies complementing their strengths was originally proposed in [280] and named P*-algorithm.

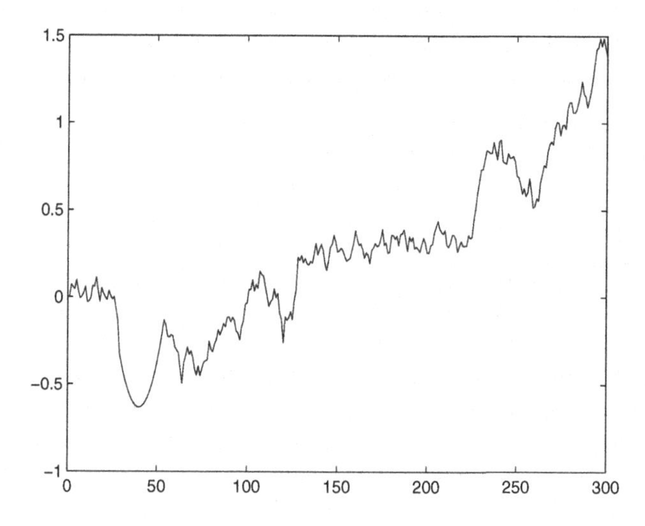

Fig. 4.6. Example of a dual model

4.3.7 Convergence

The convergence of algorithms based on a stochastic function model has two aspects. We may consider convergence with respect to the stochastic function, assuming that randomly generated sample functions should be minimized; such type of convergence is considered in Sect. 4.3.9. However, from the point of view of a user the convergence of the algorithm for an arbitrary objective function satisfying some generally acceptable conditions is even more important. To analyze the convergence we assume only the continuity of objective functions.

The proof of guaranteed convergence for any continuous function is equivalent to the proof that trial points form an everywhere dense sequence; see Theorem 1.1. To prove the latter it is, for example, sufficient to prove that every subinterval will be eventually chosen and subdivided by the new trial point with a bounded ratio. Convergence is an asymptotic property and can not guarantee small minimization error after an appropriate number of observations. The convergence rate is frequently considered an efficiency criterion, but it is also an asymptotic characteristic. In view of this some users rely entirely on the assessment of the efficiency of optimization algorithms by means of testing experiments. Nevertheless, the analysis of convergence is the most important subject in optimization theory. For the proper assessment of the advantages/disadvantages of an algorithm the analysis of convergence proof is frequently very helpful.

Let $m = \min_{0 \leq x \leq 1} f(x)$, and let x_* denote a global minimizer; if the global minimizer is not unique we will assume $x_* = \inf\{x : f(x) = m\}$. First we consider the convergence of the P-algorithm with fixed threshold value $\epsilon > 0$.

Theorem 4.6. *All three versions of the P-algorithm converge for any continuous function in the sense that* $y_{on} \downarrow m$.

Corollary 4.7. *Under the operation of the P-algorithm, the subintervals are eventually bisected.*

The statement of the corollary shows similar asymptotic behavior of the P-algorithm for all considered statistical models. To prove the corollary the assumption of continuity of objective functions is essential. The proofs of the theorem and of the corollary suggest the conclusion that for the asymptotic behavior of the P-algorithm, the properties of the objective function are more important than the properties of the underlying statistical model.

The threshold value ϵ defines the globality of search as shown in Sect. 4.4.2: for larger values of ϵ the search is more global. Therefore, it seems reasonable to start with a larger values of ϵ and reduce it during the search. One may hope to improve in this way the convergence of the algorithm. The original idea was proposed in [140] although without detailed convergence analysis. It appears, however, that the algorithm may lose the global convergence property if the sequence $\{\epsilon_n\}$ vanishes too fast. The following theorem shows an appropriate rate of decrease of a sequence ϵ_n.

Theorem 4.8. *Let* $\epsilon_n = n^{-1+\delta}$, *for some small positive* $\delta < 1$. *Then the P-algorithm based on the smooth function model will converge for any continuous objective function.*

4.3.8 Convergence Rates

The convergence rate defines how fast an error bound approaches zero. We start with the analysis of minimization of arbitrary continuous functions satisfying weak regularity conditions. A guaranteed error bound is considered, i.e. it is an upper bound for the absolute value of difference between the estimate of the global minimum and its true value: $\Delta_n \geq y_{on} - m$. The magnitude of the error depends on the limit of quantities associated with the objective function $f(\cdot)$. These quantities characterize an objective function with respect to the 'hardness' criteria defined below; we assume that the objective functions satisfy conditions of existence of these characteristics. The *global minimum weight* criterion $\Gamma_\beta(\epsilon)$, depending on the parameters $\epsilon > 0$ and $\beta > 0$, is defined by

$$\Gamma_f(\beta, \epsilon) \triangleq \int_{x=0}^{1} \left(1 + \frac{f(x) - m}{\epsilon}\right)^{-\beta} dx, \tag{4.57}$$

and it shows the size of the subset of the feasible region where the objective function values are close to the global minimum. Note that $\Gamma_f(\beta, \epsilon)$ is decreasing in β and increasing in ϵ. Larger values of $0 < \Gamma_\beta(\epsilon) \leq 1$ correspond to more difficult minimization problems due to the necessity to search over larger subsets where the function values are close to the global minimum. As the second criterion of hardness we consider

$$\Lambda_f = \lim_{n \to \infty} n^\alpha \sup_{|x| \le 1/n} [f(x_* + x) - m] \ge 0, \tag{4.58}$$

where it is assumed that the limit exists for some $\alpha > 0$. The criterion $\Lambda(f)$ may be interpreted as the *sharpness of the global minimum.*

The meaning of criteria $\Gamma_f(\beta, \epsilon)$ and Λ_f is illustrated by the following examples. With respect to the criterion $\Gamma_f(\beta, \epsilon)$ a constant is the hardest objective function since it has maximal global minimum weight $\Gamma_f(\beta, \epsilon) = 1$. The hardness of a constant function with respect to this criterion should not be surprising since all function values are equal (i.e. as close as possible) to the global minimum. A constant is also the worst case objective function for global optimization algorithms based on Lipshitz model; see Sect. 1.1.4. However, it would be quite unnatural to accept a constant as the hardest function. Indeed, with respect to the criterion Λ_f a constant $f(x) = c$ is the simplest function since $\Lambda_c = 0$. For the two next examples of simple functions the global minimum weight and minimum sharpness criteria are calculated analytically.

Example 1. Let $f_1(x) = (x - x_*)^2$, where $x_* = 1/2$. Then for $\beta = 1/2$ we have

$$\Gamma_{f_1}(1/2, \epsilon) = \int_{x=-1/2}^{1/2} \left(1 + x^2/\epsilon\right)^{-1/2} dx =$$

$$= \sqrt{\epsilon} \ln\left(1 + \frac{1}{2\epsilon} + \frac{1}{2\epsilon}\sqrt{1 + 4\epsilon}\right) \sim \sqrt{\epsilon} \ln(1 + 1/\epsilon) \tag{4.59}$$

as $\epsilon \downarrow 0$. Taking $\beta = 2$, we have

$$\Gamma_{f_1}(2, \epsilon) = \int_{x=-1/2}^{1/2} \left(1 + x^2/\epsilon\right)^{-2} dx =$$

$$= \frac{1}{2\left(1 + \frac{1}{4\epsilon}\right)} + \sqrt{\epsilon} \arctan\left(\frac{1}{2\sqrt{\epsilon}}\right) \sim \frac{\pi}{2}\sqrt{\epsilon} \tag{4.60}$$

as $\epsilon \downarrow 0$.

If $f(\cdot)$ is smooth with a positive second derivative at x_*, then $\alpha = 2$ and $\Lambda_f = f''(x_*)/2$; therefore for this example $\Lambda_{f_1} = 1$.

Example 2. Let $f_2(x) = |x - x_*|$, where $x_* = 1/2$. Then

$$\Gamma_{f_2}(1/2, \epsilon) = 2 \int_{x=0}^{1/2} (1 + x/\epsilon)^{-1/2} dx = 4\epsilon\left(\sqrt{1 + \frac{1}{2\epsilon}} - 1\right) \tag{4.61}$$

and

$$\Gamma_{f_2}(2,\epsilon) = 2\int_{x=0}^{1/2} (1 + x/\epsilon)^{-2}\, dx = \frac{2\epsilon}{1+2\epsilon}. \tag{4.62}$$

The criterion Λ_f can easily be calculated for the slightly generalized function of this example, i.e. $f(x) = b|x - x_*|$: the parameter α is equal to one, and the local characteristic Λ_f is equal to b.

Theorem 4.9. *Let an objective function $f(x)$ be continuous over the minimization interval and have local minimum sharpness characteristic Λ_f with exponent α at the unique global minimizer x_*, and global minimum weight characteristic $\Gamma_f(\beta,\epsilon)$. Then the P-algorithm with fixed threshold value ϵ based on one of the following models*

- *random process with smooth sample functions,*
- *integrated Wiener process*
- *Wiener process*

converges to the global minimum with the following convergence rate:

$$\limsup_{n\to\infty} n^\alpha \Delta_n \le 2^\alpha \Lambda_f \Gamma_f^\alpha(1/2,\epsilon) \tag{4.63}$$

for the algorithm based on a random process with smooth sample functions,

$$\limsup_{n\to\infty} n^\alpha \Delta_n \le 2^\alpha \Lambda_f \Gamma_f^\alpha(3/2,\epsilon) \tag{4.64}$$

for the algorithm based on the integrated Wiener process model, and

$$\limsup_{n\to\infty} n^\alpha \Delta_n \le 2^\alpha \Lambda_f \Gamma_f^\alpha(2,\epsilon) \tag{4.65}$$

for the algorithm based on the Wiener process model.

The conclusion was slightly surprising since the rate of convergence of the P-algorithm depends mainly on the properties of the objective function, and not on the stochastic process used in the construction of the algorithm as a model of the objective functions. The dependence on the stochastic model only enters in the constant factor.

Since the same form of conditional mean is used in the cases of the Wiener process and smooth function models, the way the observations are allocated by the P-algorithm depends on the conditional variance. The algorithm based on the smooth model for the same objective function spreads observations more uniformly over the interval than the Wiener process based algorithm with the same ϵ.

Examination of (4.57) shows that $\Gamma_f(\beta,\epsilon)$ is large when $f(\cdot)$ spends a lot of time near the global minimum m. In this case the search effort is not concentrated close to x_* since the promising region is large. In contrast, if $f(\cdot)$ has a narrow "spike" at x_*, then the algorithm can concentrate the search effort there (in this case $\Gamma_f(\beta, \epsilon)$ is relatively small). However, a small $\Gamma_f(\beta, \epsilon)$ does not necessarily mean that a problem is easy. The bound also depends on Λ_f. The latter increases the bound for a "spiky" global minimum. Such a balance seems reasonable, since the sharpness of the global minimum implies the necessity of greater density of points in the basin of the global minimizer to achieve the appropriate precision.

The error bound for the P-algorithm based on the integrated Wiener process model (4.64) is intermediate between the bounds of the two other algorithms (4.63) and (4.65). Therefore further discussion refers to the algorithms based on the Wiener process and smooth function models. As previously noted, $\Gamma_f(1/2, \epsilon) > \Gamma_f(2, \epsilon)$. However, this does not necessary imply that the Wiener process based algorithm is more efficient for a smooth objective function: the reason is that a good choice of ϵ may be different for the two cases. While choosing ϵ a trade-off between its influence on the convergence rate and on the global distribution of trial points should be taken into account. This balance is analyzed below.

If the same value of ϵ was chosen for both versions of the P-algorithm then the algorithm based on the Wiener process model would converge faster than the algorithm based on smooth function models. However, the choice of ϵ should normally be larger for the first than for the second version of the algorithm as shown by the theorem below.

Theorem 4.10. *Let us denote the smallest subinterval containing point x after n minimization steps by $\delta(n, x)$. For the Wiener process model based P-algorithm the ratio of lengths of subintervals satisfies the inequality*

$$\frac{1}{2}\left(\frac{f(x) - m + \epsilon}{f(z) - m + \epsilon}\right)^2 \leq \liminf_{n\to\infty} \frac{\delta(n, x)}{\delta(n, z)} \leq$$
$$\limsup_{n\to\infty} \frac{\delta(n, x)}{\delta(n, z)} \leq 2\left(\frac{f(x) - m + \epsilon}{f(z) - m + \epsilon}\right)^2;$$

for smooth function model based P-algorithm the following inequalities are satisfied

$$\frac{1}{2}\sqrt{\frac{f(x) - m + \epsilon}{f(z) - m + \epsilon}} \leq \liminf_{n\to\infty} \frac{\delta(n, x)}{\delta(n, z)} \leq$$
$$\limsup_{n\to\infty} \frac{\delta(n, x)}{\delta(n, z)} \leq 2\sqrt{\frac{f(x) - m + \epsilon}{f(z) - m + \epsilon}}.$$

The choice of ϵ defines not only the value of the constant $\Gamma_f(\beta, \epsilon)$ in the estimate of convergence rate but also the distribution of observation points. The value of ϵ should be chosen small enough to ensure a high convergence rate. But the algorithm with small ϵ spends much time searching the neighborhoods of the best points, postponing search in the subintervals where expected function values are not so good. It is important from the very beginning of minimization to balance density of observations over the feasible region since relatively large unsearched subintervals would cause a large probability of missing the global minimum if minimization was stopped after a modest number of observations. The value of ϵ should be chosen sufficiently large to ensure a not too fast decrease in the density of observation points with departure from the best found points. The ratio R of the length of a subinterval, where the expected function values correspond to a chosen level F, and the length of the interval corresponding to the global minimum can be controlled by the choice of ϵ, since the following inequalities hold

$$\frac{1}{2}\left(\frac{F-m+\epsilon}{\epsilon}\right)^2 \leq R \leq 2\left(\frac{F-m+\epsilon}{\epsilon}\right)^2$$

for the Wiener process model based algorithm, and by the choice

$$\frac{1}{2}\sqrt{\frac{F-m+\epsilon}{\epsilon}} \leq R \leq 2\sqrt{\frac{F-m+\epsilon}{\epsilon}}$$

for smooth function model based algorithm. These inequalities suggest the choice $\epsilon = \frac{F-m}{\sqrt{R}-1}$ for the first case and a choice $\epsilon = \frac{F-m}{R^2-1}$ for the second case. Such a choice of ϵ for the Wiener process model based P-algorithm corresponds to a formula empirically found in [286] for the level F removed by 1/4 of the variation of function values from an estimate of the minimum, and $R = 4$.

Let us consider the minimization problems presented by Example 1 and Example 2, assuming F is equal to the minimum plus 1/4 of the variation of values of the function. The desired ratio R value is chosen $R = 4$.

In the case of the minimization problem of Example 1 the threshold values calculated by the formulae above are $\epsilon = 1/16$ for the algorithm based on the Wiener process model, and $\epsilon = 1/240$ for the algorithm based on a smooth function model, implying $\Gamma_{f_1}(2, 1/12) = 0.3786$ and $\Gamma_{f_1}(1/2, 1/240) = 0.3543$. In this case the second version of the algorithm converges faster than the first version; although the convergence order of both algorithms is the same, the constant is smaller for the algorithm based on a smooth function model.

In the case of the minimization problem of Example 2 the threshold values calculated by the formulae above are $\epsilon = 1/8$ for the algorithm based on the Wiener process model, and $\epsilon = 1/120$ for the smooth function model based algorithm, implying $\Gamma_{f_2}(2, 1/12) = 0.2000$ and $\Gamma_{f_2}(1/2, 1/240) = 0.2270$. In this case the first version of the algorithm converges faster than the second version, i.e. the constant is smaller for the algorithm based on the Wiener process model.

The conclusions about convergence rate are coherent with the smoothness of objective function at the minimum point: the objective function of Example 1 is smooth at the minimum point, and the objective function of Example 2 is non differentiable at minimum point. Let us mention that the local minimum sharpness characteristic Λ_f is not important when comparing different versions of the algorithm since it only depends on the function and not on the statistical model used in algorithm construction.

An interesting possibility to improve the convergence rate is in replacing the constant threshold value ϵ with a decreasing sequence $\{\epsilon_n\}$. The basic idea is proposed by Kushner in [140]. However, in [140] it was not determined for what sequence the algorithm would converge, nor what the convergence rate of the modified algorithms are. It does not seem possible to achieve a high convergence rate without assuming smoothness of an objective function. Correspondingly the version of the P-algorithm based on stationary random process with smooth sample functions will be considered.

Theorem 4.11. *Assume an objective function is twice continuously differentiable, $f(\cdot) \in C^2([0,1])$, and has a unique global minimum point. The P-algorithm based on a smooth function model with threshold sequence $\epsilon_n = n^{-1+\delta}$ converges to the global minimum with convergence rate $O(n^{-3+\delta})$; here $\delta > 0$ is a small fixed constant.*

The local search of the Wiener process model based P-algorithm with fixed ϵ practically is not very efficient. To improve the behavior of the P-algorithm in the vicinities of local minimizers the P*-algorithm has been proposed; see Sect. 4.3.6. The P*-algorithm is based on dual model: the Wiener process for global description of an objective function, and a quadratic interpolant as a local model in the neighborhoods of the detected local minima. The transition from global to local search is based on testing a statistical hypothesis that the subinterval of a local minimum is found. The transition from local to global search is defined by a stopping condition of the local search, meaning that a local minimum is found with acceptable accuracy. The P*-algorithm combines the advantages of efficient global strategy inherited from the P-algorithm, with the efficient local strategy based on quadratic interpolation of the objective function in the neighborhood of a local minimizer.

To investigate convergence, the stopping condition of local minimization should be relaxed. But in this case the switch from local to global search would not happen. Therefore, let us modify the P*-algorithm in the following way. The global search is performed as before but the subintervals of the found local minimizers are not excluded from the search. The procedure of local improvement is incorporated into the global search: each s-th trial is made according to the local optimization algorithm based on quadratic fit for the points x_{l-1}^n, x_l^n, x_{l+1}^n, where x_l is the best found point.

Theorem 4.12. *Let the objective function $f(x)$ be twice continuously differentiable and have the unique global minimum point x_*. Then the (modified) P*-algorithm converges in the sense that $y_{on} \downarrow m$, and the convergence rate is superlinear.*

Corollary 4.13. *Although the convergence of the P*-algorithm is superlinear the order is close to 1 even for modest values of s and high order of convergence of the local algorithm.*

4.3.9 Probabilistic Convergence Rates

In this section we analyze the average error of an algorithm Δ_n where averaging is meant with respect to a probability measure over a class of potential objective functions; see Sect. 4.2.1. Average error is an important criterion for assessment of efficiency of an algorithm in multiple minimizations when impact of errors is uniform over the class of problems.

We start with a review of results on the convergence of passive methods where each point x_i is chosen independently of the function evaluations. Ritter [200] established the best possible convergence rate of any passive optimization algorithm under the Wiener measure: the average error is $\Omega(n^{-1/2})$ meaning that $n^{-1/2}$ is an asymptotic lower bound for the average error. Such a convergence order is attained for equispaced points. That is, if the number of evaluations n is fixed in advance, then $x_i = i/n$ is order-optimal, i.e. the average error is of order $n^{-1/2}$ (this fact is normally denoted as $\Delta_n = \Theta(n^{-1/2})$). There are many passive algorithms with errors $\Delta_n = \Theta(n^{-1/2})$.

The minimizer of the Wiener process is almost surely unique and has the arcsine density

$$p(t) = \frac{1}{\pi\sqrt{t(1-t)}}$$

on $(0,1)$. Thus a natural choice for a randomized algorithm is to choose the x_i independently according to the distribution $p(\cdot)$. With this choice,

$$\sqrt{n}E(\Delta_n) \to \frac{1}{\sqrt{2\pi}}\beta(3/4, 3/4) \approx 0.675978,$$

where $\beta(\cdot)$ is the beta function. This is not the best choice of sampling distribution; if the x_i are chosen independently according to the $\beta(2/3, 2/3)$ distribution, then

$$\sqrt{n}E(\Delta_n) \to \frac{1}{\pi\sqrt{2}}\beta(2/3, 2/3)^{3/2} \approx 0.662281,$$

and a slight improvement is achieved. Both distributions are better than the uniform distribution, for which

$$\sqrt{n}E(\Delta_n) \to \frac{1}{\sqrt{2}} \approx 0.707107.$$

These results are proved in [3].

The deterministic versions of the above passive algorithms have slightly better convergence rates. A sequence of knots $\{x_i^n; 1 \le i \le n\}$ is a *regular sequence* if the knots form quantiles with respect to a given density $g(\cdot)$; i.e.,

$$\int_{t=0}^{x_i^n} g(t)\,dt = \frac{i-1}{n-1}$$

for $1 \leq i \leq n$; see [201]. If we take $g(\cdot)$ to be the uniform distribution, and construct an algorithm based on the corresponding regular sequence, then

$$\sqrt{n}E(\Delta_n) \to c \approx 0.5826.$$

With $g(\cdot)$ being the arcsine density, the constant is improved to approximately $0.956c$, and with $g(\cdot)$ being the $\beta(2/3, 2/3)$ density, this improves to approximately $0.937c$; see [33]. The deterministic algorithms described above are non-composite in that the number of evaluations n must be specified in advance; this can be seen as a disadvantage relative to the randomized algorithms that tends to offset their better convergence rates; see [3].

Some of these results on passive algorithms have been extended to diffusion processes other than Wiener process [37]. Particulary it is shown that the normalized error converges in distribution (for both random and deterministic passive algorithms) when the path is random, but pathwise the normalized error fails to converge in distribution for almost all paths [38].

So far we have discussed passive algorithms. Most algorithms in use, e.g. those considered in previous sections, are adaptive in that each function evaluation depends on the results of previous evaluations. In the worst case setting, adaptation does not help under quite general assumptions. For example, the minimax optimal with respect to a class of Lipshitz functions adaptive global optimization algorithm coincides with the minimax optimal passive algorithm, i.e. with the uniform grid [234]. If F is convex and symmetric (in the sense that $-F = F$), then the maximum error under an adaptive algorithm with n observations is not smaller than the maximum error of a passive method with $n+1$ observations; see [175]. Thus adaptive methods in common use can not be justified by a worst case analysis, which motivates our interest in the average-case setting.

The convergence of the P-algorithm (4.14) with constant $\epsilon_k = \epsilon > 0$ is studied for the case of minimization of a particular function satisfying some assumptions. It is interesting to compare the convergence rate in the mentioned case with the convergence rate of the average error under the Wiener measure. A sequence of stopping times $\{n_k\}$ can be constructed so that

$$\lim_{k\to\infty} P\left(\frac{\sqrt{n_k}}{\epsilon\sqrt{\rho}}\Delta_{n_k} \leq y\right) = G(y),$$

as $k \to \infty$, where

$$\rho = \int_{t=0}^{1} \frac{dt}{(f(t) - f^* + \epsilon)^2}$$

and where $G(\cdot)$ is the distribution function of the minimum of a two-sided three-dimensional Bessel process over a lattice of diameter 1 [35]. The convergence order of the average error of the P-algorithm in this case is the same as the convergence order of the best passive algorithms. Such a result seems a bit disappointing from a theoretical point of view. However, practical objective functions in close neighborhood of the global minimizer behave very differently than the Wiener process. Therefore from the point of view of applications an estimate of the convergence rate in Sect. 4.3.8 is more important than the estimates of the convergence rate of average error under the Wiener measure.

Adaptive algorithms with better convergence rates of average error under the Wiener measure can be constructed. In [34] a class of adaptive algorithms was introduced that operate with memory of cardinality 2. Within this class, for any $\delta > 0$, it is shown that an algorithm can be constructed that converges to the global minimum at rate $n^{-1+\delta}$ in the number of observations n. More precisely,

$$P\left(n^{1-\delta/2}\Delta_n \le x\right) \to \tanh^2(\sqrt{2}\,x), \qquad x > 0.$$

This rate is in contrast to the $n^{-1/2}$ rate that is the best achievable with a passive algorithm. A useful property of this class of algorithm is that they can be implemented in parallel on two processors with minimal communication. It is possible to construct a similar algorithm with sufficiently large memory such that, under the Wiener measure, the average error decreases to zero faster than any fixed polynomial in n^{-1} [34].

Assuming that the smooth function $f(\cdot)$ has a unique global minimizer that has an absolutely continuous distribution on $[0, 1]$, we can derive descriptions of the normalized point processes of observations near x_*. Basically, x_* is asymptotically uniformly distributed in the interval formed by the two nearest observations, and the subintervals are eventually bisected. There exists a sequence of stopping times $\{n_k : k \ge 1\}$ such that the point process of observations near x_* (suitably normalized) converges to uniform spacing. This, together with the smoothness assumption on $f(\cdot)$ at x_*, implies

$$\frac{n_k^2 \Delta_{n_k}}{C} \xrightarrow{\mathcal{D}} \min\{U^2, (1-U)^2\} \tag{4.66}$$

for some normalizing random variable C (which depends on the algorithm), where U is a uniformly distributed random variable on the unit interval, and $\xrightarrow{\mathcal{D}}$ denotes convergence in distribution, i.e., $\xi_n \xrightarrow{\mathcal{D}} \xi$ for random variables ξ_n, ξ if $P(\xi_n \le x) \to P(\xi \le x)$ for all x such that $P(\xi = x) = 0$. For P-algorithms based on Wiener and smooth function models it is proved in [35], [39] that

$$\frac{n_k^2 \Delta_{n_k}}{\frac{1}{2} f''(x_*) \Gamma_f^2(\beta, \epsilon)} \xrightarrow{\mathcal{D}} \min\{U^2, (1-U)^2\}, \tag{4.67}$$

where $\beta = 2$ for the Wiener process case and $\beta = 1/2$ for the smooth process case. The case of the P-algorithm based on the smooth functions model is considered below in detail.

We assume that $f(\cdot)$ is a random element in the space of continuously differentiable functions on $[0, 1]$, that x_* has an absolutely continuous distribution with continuous density $g(\cdot)$, and that $0 < f''(x_*) < \infty$. All probabilistic statements will be with respect to this probability.

Theorem 4.14. *There exist stopping times n_k such that*

$$\frac{n_k^2 \Delta_{n_k}}{\frac{1}{2} f''(x_*) \Gamma_f(1/2, \epsilon)^2} \xrightarrow{\mathcal{D}} \min\{U^2, (1-U)^2\}.$$

as $k \to \infty$.

A large $\Gamma_f(1/2, \epsilon)$ corresponds to the asymptotic error being relatively large compared to the case when $\Gamma_f(1/2, \epsilon)$ is small. Let us recall, that $\Gamma_f(1/2, \epsilon)$ is a measure of how hard f is to minimize: a large $\Gamma_f(1/2, \epsilon)$ means that f spends a lot of time near the minimizer m. For a function f whose values are far from m over much of the domain $\Gamma_f(1/2, \epsilon)$ is small; in this case a large portion of the domain can be discounted early on, allowing the search to concentrate in the relatively small region where f is near to the global minimum m.

It is interesting to compare Theorem 4.14 with the analogous result for the Wiener process [35]. Let $\{\xi(t) : 0 \leq t \leq 1\}$ be a standard Wiener process. There exists an increasing sequence of stopping times $\{n_k\}$ such that

$$P\left(\sqrt{n_k}\, \Delta_{n_k} \left(\int_{t=0}^{1} \left(1 + \frac{\xi(t) - m}{\epsilon}\right)^{-2} dt\right)^{-1/2} \leq y\right) \to F(y), \qquad (4.68)$$

$y > 0$, where F is the limiting distribution function of the normalized error with equi-spaced observations; that is,

$$P\left(\sqrt{n} \min_{o \leq i \leq n} (\xi(i/n) - m) \leq y\right) \to F(y). \qquad (4.69)$$

The convergence rate is $O(n^{-1/2})$ in the Wiener process case, in contrast to the $O(n^{-2})$ rate in the case of a random process with smooth sample functions.

4.3.10 Testing and Applications

A numerical simulation was performed to determine how accurately the limit theorem predicts the performance of the algorithm for small and moderate values of the number of observations. The version of the P-algorithm based on the smooth function model has been tested. Test functions were random samples with a particular distribution on a class of smooth functions $C^2[0, 1]$. The distribution was defined by the distributions of random variables in the formula below. The minimizer x_* is uniformly distributed on the interval $[0.1, 0.9]$, and conditional on x_*, f is given by

$$f(x) = A_1 (1 - \cos(35(x - x_*))) + A_2 (1 - \cos(70(x - x_*))) \\ + A_3 (1 - \cos(120(x - x_*))),$$

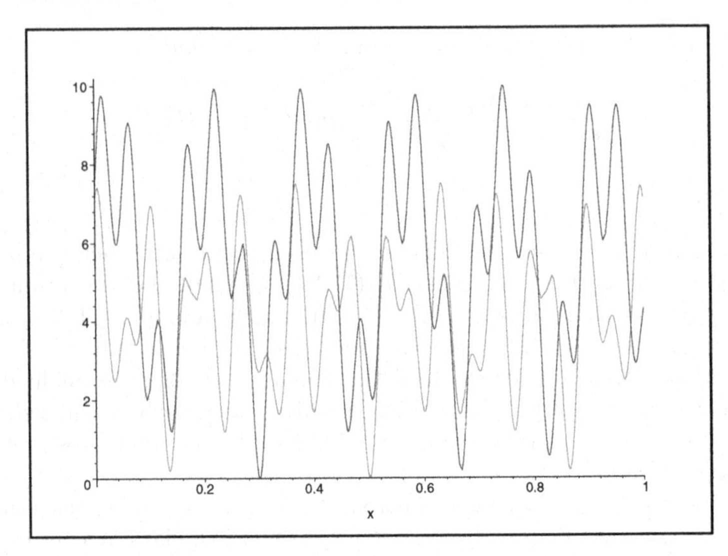

Fig. 4.7. Graphs of two test functions

where the A_i are the absolute values of $N(0, 2)$ random variables. Two sampling functions are presented in Fig. 4.7 to illustrate the complexity of the considered test functions.

The version of the algorithm with a vanishing threshold with parameter $\delta = 0.1$ (i.e. $\epsilon = n^{-0.9}$), was tested. The theoretical convergence rate is defined by Theorem 4.11. To estimate the convergence rate empirically the minimization experiment was repeated independently 1,000 times. The total number of observations was taken to be 1,000 (we did not use the stopping rule described above, but instead simply stopped after n trials). The average normalized error is plotted in Fig. 4.8. That is, we plot the average of the quantities

$$\frac{n^{3-\delta}}{\log(n)^2} \Delta_n$$

for n between 1 and 1,000. As can be seen from Fig. 4.8, the normalized error is fairly stable after about $n = 100$.

Generally speaking, the algorithms based on statistical models compete favorably with the algorithms based on the other approaches; the results of testing experiments with various deterministic and stochastic test functions can be found e.g. in [144], [248], [281].

One-dimensional optimization problems occur in various applications. An example related to statistics is discussed below. The lognormal distribution (LD) is an important probability distribution that has been used in such diverse areas as biology, geology, agriculture, economics, etc. The problem of estimation of three parameters of LD is considered in [264]; it is shown, that numerical computation of the maximum likelihood estimate is difficult. The

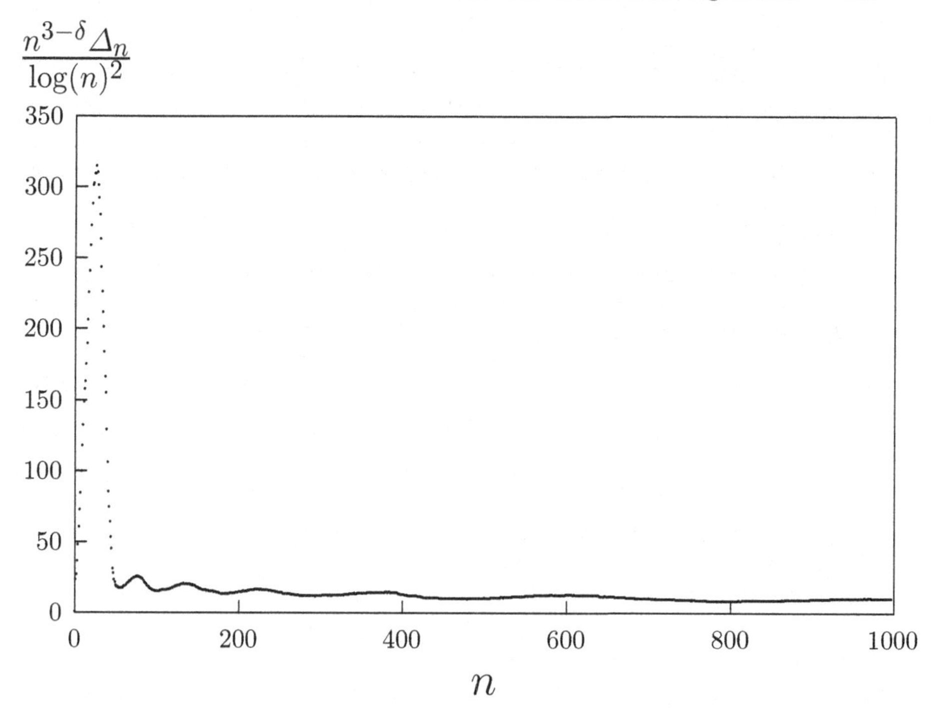

Fig. 4.8. Normalized errors.

probability density function of the LD is

$$p(x) = \frac{1}{\sqrt{2\pi\beta}(x-\gamma)} \exp\left(-\frac{\ln(x-\gamma)-\mu}{2\beta}\right),$$

where γ is threshold parameter, $x > \gamma$, $-\infty < \gamma < \infty$; μ is the mean, and β is the variance of the parent normal population. In [264] the maximization of the logarithmic likelihood is reduced to a one-dimensional global maximization problem

$$\max_{\theta}(-\hat{\mu}(\gamma)) + 0.5\ln(\hat{\beta}(\gamma)),$$
$$\hat{\beta}(\gamma) = \frac{1}{n}\sum(\ln(x_i-\gamma)-\hat{\mu}(\gamma))^2,$$
$$\hat{\mu}(\gamma) = \frac{1}{n}\sum\ln(x_i-\gamma),\ \gamma = x_{min} - \exp(-\theta),$$

where $x_i, i = 1,...,n$ is a sample of independent random observations, and x_{min} is the smallest sample order statistic. The estimates are recalculated using the optimal value of θ. It is shown [264] that the Wiener process model based P*-algorithm is efficient in solving the maximization problem above.

There exist methods of multidimensional optimization based on one-dimensional algorithms. For example, the one-dimensional algorithm can be used by means of cycling the variables. Such a coordinate-wise extension of a one-dimensional algorithm to multi-dimensional problems may be efficient only in the case of weak dependence of the variables. Examples of successful application of the coordinate method to practical multimodal problems are presented in, e.g. [177] (training neural networks), and [222], [248] (optimal design). Let us note that coordinate-wise global optimization is easily parallelized [249].

There are examples of successful practical applications of one-dimensional algorithms extended to multi-dimensional problems by means of dynamic programming and mapping of multi-dimensional sets to one-dimensional intervals [232]. The first type of extension is defined by the following equality

$$\min_{x \in A} f(x) = \min_{x_d^- \le x_d \le x_d^+} (...(\min_{x_1^- \le x_1 \le x_1^+} f(x))...), \tag{4.70}$$

where $A = \{x : x_i^- \le x_i \le x_i^+, i = 1, ..., d\}$. The disadvantage of this method is explicitly exponential growth of the number of function calls with the increase of d. A more efficient extension of one-dimensional methods to a multi-dimensional case is based on Peano type space filling curves [112], [214], [233], [232]. One-dimensional methods have also great impact to multi-dimensional optimization indirectly, via the generalization of their properties to the multidimensional case, see e.g. in [102], [131], [189].

4.3.11 Proofs of Theorems and Lemmas

Proof of Theorem 4.4.

Let us denote

$$\tau = \frac{x - x_{i-1}^n}{x_i^n - x_{i-1}^n}, \; z_1 = y_{i-1}^n, \; z_2 = y_i^n, \; z = y_{on} - \epsilon.$$

Then the maximization of (4.42) reduces to the maximization of

$$\frac{z - z_2\tau - z_1(1-\tau)}{\tau(1-\tau)}$$

over the interval $(0, 1)$. The unique zero of the derivative of the latter expression in the interval $(0, 1)$ is defined by the formula

$$\tau = \frac{-z_1 + z + \sqrt{(z_1 - z)^2 + (z_1 - z)(z_2 - z_1)}}{z_2 - z_1}. \tag{4.71}$$

The latter expression is not well suited for computation since the denominator of (4.71) becomes 0 for $z_1 = z_2$. The multiplication of the numerator and denominator of (4.71) by

$$-z_1 + z + \sqrt{(z_1 - z)^2 + (z_1 - z)(z_2 - z_1)}$$

yields the expression (4.44) in the statement of the theorem.

Proof of Theorem 4.6.

The convergence of the multidimensional P-algorithm for any continuous function is proved in [288] under rather general assumptions on the statistical model of the objective function; see Sect. 4.4.2. The convergence of the one-dimensional P-algorithm based on Wiener process model follows easily from this general result.

Let us consider the convergence of the P-algorithm based on the smooth function model. The general proof of convergence in 4.4.2 can not be applied to this version of the P-algorithm since the conditional variance formula (4.41) does not belong to the class assumed in 4.4.2. Therefore, the general proof should be slightly modified. This version of the P-algorithm involves the maximization of (4.42).

Let us suppose that no global minimum point of $f(x)$ is the cluster point of the sequence $x_1, x_2, \ldots$; i.e. there exists a subinterval $U \subset (0, T)$ of length L containing a global minimum point x_* but no point x_i. Since the feasible interval is bounded, the sequence x_i, $i = 1, 2, \ldots$ has a cluster point w whose arbitrary neighborhood $S(w, \delta)$ contains the points of the considered sequence.

Let

$$\rho = \max_{0 \le x \le 1} f(x) - \min_{0 \le x \le 1} f(x).$$

Since function (4.42) is a decreasing function of y_{i-1}^n, y_i^n,

$$\sup_{x \in U} P_n(x) > \Phi\left(\frac{-\epsilon - \rho}{\frac{1}{16}\sqrt{\lambda_4 - \lambda_2^2}\, L^2}\right), \tag{4.72}$$

$$\sup_{x \in S(w,\delta)} P_n(x) < \Phi\left(\frac{-\epsilon}{\frac{1}{16}\sqrt{\lambda_4 - \lambda_2^2}\, \delta^2}\right). \tag{4.73}$$

But the selection of trial points according to the P-algorithm is impossible in $S(w, \delta)$ for

$$\delta < L\sqrt{\frac{\epsilon}{\rho + \epsilon}},$$

since in such a case the lower bound of (4.72) is larger than the upper bound of (4.73).

Proof of Corollary 4.7.

The point of subdivision of the selected subinterval in the cases of the Wiener and integrated Wiener process models is defined by the coinciding formulae (4.56), (4.55), and for the case of the smooth function model it is defined by formula (4.54). Since the observation sequence is dense as $n \to \infty$, then by continuity of the objective function $y_i^n - y_{i-1}^n \to 0$. Therefore for the cases of the Wiener and integrated models it is obvious that

$$\tau_i = \frac{1}{1 + \frac{y_i^n - y_{on} + \epsilon}{y_{i-1}^n - y_{on} + \epsilon}} \to \frac{1}{2}. \tag{4.74}$$

To prove the corollary for the smooth function model let us write (4.54) in the form

$$\tau_i = \frac{\Delta_{n\,i}}{\Delta_{n\,i-1} + \sqrt{\Delta^2_{n\,i-1} + (y^n_i - y^n_{i-1}) \cdot \Delta_{n\,i-1}}} = \frac{1}{1 + \sqrt{1 + \frac{y^n_i - y^n_{i-1}}{y^n_{i-1} - y_{on} + \epsilon}}}. \tag{4.75}$$

The last expression obviously converges to $1/2$ as $n \to \infty$.

Proof of Theorem 4.8

The error $\Delta_n = \min_{i \le n} f(x_i) - f(x_*)$ converges to zero for all continuous functions if and only if the observation sequence is dense in the unit interval, and this condition is implied by $\liminf \gamma^n = 0$. We will construct a subsequence $\{n_k\}$ with the property that $\gamma^{n_k} \to 0$.

Let ω_n denote the length of the shortest interval formed by the observations 1 through to n (so $\omega_n \le 1/n$), and let n_k be the kth time that a new observation results in a new smallest interval; that is, at time n_k an interval of width $\widetilde{\omega}_{n_k}$ is to be split, with its smallest child then having width ω_{n_k+1}. Then

$$\omega_{n_k+1} = \widetilde{\omega}_{n_k} \min\{\tau_{n_k}, 1 - \tau_{n_k}\}.$$

Now

$$1 - \tau_n = \frac{1}{1 + \sqrt{1 + \frac{y^n_{i-1} - y^n_i}{y^n_i - y_{on} + \epsilon_n}}},$$

and $\tau_n \le 1 - \tau_n$ if and only if $y^n_i \ge y^n_{i-1}$. Therefore,

$$\min\{\tau_{n_k}, 1 - \tau_{n_k}\} = \frac{1}{1 + \sqrt{1 + \frac{\Delta f}{f_s - f_{on_k} + \epsilon_{n_k}}}},$$

where Δf is the absolute difference in function values at the endpoints of the interval $\widetilde{\omega}_{n_k}$ and f_s is the smaller of the two function values. Therefore,

$$\begin{aligned}
\widetilde{\omega}_{n_k} &= \frac{\omega_{n_k+1}}{\min\{\tau_{n_k}, 1 - \tau_{n_k}\}} \\
&\le \frac{1}{(n_k+1)\min\{\tau_{n_k}, 1 - \tau_{n_k}\}} \\
&= \frac{1}{n_k+1}\left(1 + \sqrt{1 + (\Delta f)/(f_s - f_{on_k} + \epsilon_{n_k})}\right) \\
&\le \frac{1}{(n_k+1)\sqrt{\epsilon_{n_k}}}\left(\sqrt{\epsilon_{n_k}} + \sqrt{\epsilon_{n_k} + (\Delta f)}\right) \\
&\le \frac{1}{(n_k+1)\sqrt{\epsilon_{n_k}}}\left(2\sqrt{\epsilon_{n_k}} + \Delta f\right) \\
&= O\left(\frac{1}{(n_k+1)\sqrt{\epsilon_{n_k}}}\right),
\end{aligned}$$

since $\Delta f \leq \max_{0\leq s\leq 1} f(s) - \min_{0\leq s\leq 1} f(s) < \infty$. Therefore,

$$\gamma^{n_k} \leq \frac{\widetilde{\omega}_{n_k}}{2\sqrt{\epsilon_{n_k}}} = O\left(\frac{1}{n_k \epsilon_{n_k}}\right),$$

which converges to 0 if $n_k\epsilon_{n_k} \to \infty$.

Proof of Theorem 4.9.

We begin with the proof of (4.63). Since the observations are dense in $[0,1]$ and f is continuous, $y_{on} \downarrow m$ and

$$\sum_{i=1}^{n} \gamma_i^n \to \frac{1}{2}\int_{x=0}^{1} \frac{dx}{\sqrt{f(x) - m + \epsilon}} = \frac{\Gamma_f(1/2,\epsilon)}{2\sqrt{\epsilon}}. \tag{4.76}$$

Let x_L^n and x_R^n be the observation points to the left and right, respectively, of the minimizer x_*; that is, for some j_n,

$$x_L^n = x_{j_n}^n \leq x_* \leq x_{j_n+1}^n = x_R^n.$$

Let $y_L^n = f(x_L^n)$ and $y_R^n = f(x_R^n)$ be the corresponding function values. Let

$$\gamma_s^n \triangleq \frac{x_R^n - x_L^n}{\sqrt{y_L^n - y_{on} + \epsilon} + \sqrt{y_R^n - y_{on} + \epsilon}} \tag{4.77}$$

and

$$\gamma^n = \max_{i\leq n} \gamma_i^n, \qquad \gamma_n = \min_{i\leq n} \gamma_i^n.$$

Since the subintervals are eventually bisected, $\gamma^n/\gamma_n \to 2$.

Because $y_L^n - y_{on} \to 0$ and $y_R^n - y_{on} \to 0$,

$$\frac{x_R^n - x_L^n}{2\gamma_s^n\sqrt{\epsilon}} = \frac{\sqrt{y_L^n - y_{on} + \epsilon} + \sqrt{y_R^n - y_{on} + \epsilon}}{2\sqrt{\epsilon}} \to 1, \tag{4.78}$$

which, combined with (4.76), implies that

$$\frac{n(x_R^n - x_L^n)}{\gamma_s^n / \frac{1}{n}\sum_{i=1}^n \gamma_i^n} \to 2\sqrt{\epsilon}\,\Gamma_f(1/2,\epsilon). \tag{4.79}$$

Since

$$\frac{\gamma_s^n}{\frac{1}{n}\sum_{i=1}^n \gamma_i^n} \geq \frac{\gamma_n}{\gamma^n},$$

we have that

$$\liminf_{n\to\infty} \frac{\gamma_s^n}{\frac{1}{n}\sum_{i=1}^n \gamma_i^n} \geq \liminf_{n\to\infty} \frac{\gamma_n}{\gamma^n} = \frac{1}{2}. \tag{4.80}$$

Re-writing, we obtain

$$n(x_R^n - x_L^n) = \frac{(x_R^n - x_L^n)}{2\gamma_s^n\sqrt{\epsilon}} \frac{\gamma_s^n}{\frac{1}{n}\sum_{i=1}^n \gamma_i^n} 2\sqrt{\epsilon}\sum_{i=1}^{n} \gamma_i^n, \tag{4.81}$$

and applying (4.102), (4.80), and (4.76) to the three terms on the right-hand side of (4.81), we conclude that

$$\limsup_{n\to\infty} n(x_R^n - x_L^n) \le 4\sqrt{\epsilon}\frac{\Gamma_f(1/2,\epsilon)}{2\sqrt{\epsilon}} = 2\Gamma_f(1/2,\epsilon). \tag{4.82}$$

We now turn our attention from the gap surrounding the global minimizer to the error $\Delta_n = y_{on} - m$. Because of our assumption on the existence of Λ_f,

$$\limsup_{n\to\infty} \frac{\Delta_n}{\Lambda_f(x_R^n - x_L^n)^2} \le 1.$$

Therefore, by (4.88),

$$\limsup_{n\to\infty} \frac{n^\alpha \Delta_n}{\Lambda_f} = \limsup_{n\to\infty} \frac{\Delta_n}{\Lambda_f(x_R^n - x_L^n)^\alpha}[n(x_R^n - x_L^n)]^\alpha \le 2^\alpha \Gamma_f^\alpha(1/2,\epsilon), \tag{4.83}$$

which proves (4.63).

The convergence rate of the P-algorithm based on the differentiable function model is evaluated in similar way. We present the main steps without comments:

$$\sum_{i=1}^{n} \gamma_i^n \to \frac{1}{\sqrt[3]{4\epsilon^2}} \int_{x=0}^{1} \frac{dx}{\sqrt[3]{(f(x) - m + \epsilon)^2}} = \frac{\Gamma_f(2/3,\epsilon)}{\sqrt[3]{4\epsilon^2}}, \tag{4.84}$$

$$\gamma_s^n \triangleq \frac{(x_R^n - x_L^n)(\sqrt{y_L^n - y_{on} + \epsilon} + \sqrt{y_R^n - y_{on} + \epsilon})^{-2/3}}{(y_L^n - y_{on} + \epsilon)^{1/6} + (y_R^n - y_{on} + \epsilon)^{1/6}}, \tag{4.85}$$

$$\frac{x_R^n - x_L^n}{\gamma_s^n \sqrt[3]{4\epsilon^2}} = \frac{(y_L^n - y_{on} + \epsilon)^{1/6} + (y_R^n - y_{on} + \epsilon)^{1/6}}{(x_R^n - x_L^n)(\sqrt{y_L^n - y_{on} + \epsilon} + \sqrt{y_R^n - y_{on} + \epsilon})^{-2/3}\sqrt[3]{4\epsilon^2}} \to 1, \tag{4.86}$$

$$n(x_R^n - x_L^n) = \frac{x_R^n - x_L^n}{\gamma_s^n \sqrt[3]{4\epsilon^2}} \frac{\gamma_s^n}{\frac{1}{n}\sum_{i=1}^n \gamma_i^n} \sqrt[3]{4\epsilon^2} sum_{i=1}^n \gamma_i^n, \tag{4.87}$$

$$\limsup_{n\to\infty} n(x_R^n - x_L^n) \le 2\sqrt[3]{4\epsilon^2}\frac{\Gamma_f(2/3,\epsilon)}{\sqrt[3]{4\epsilon^2}} = 2\Gamma_f(2/3,\epsilon), \tag{4.88}$$

$$\limsup_{n\to\infty} \frac{n^\alpha \Delta_n}{\Lambda_f} = \limsup_{n\to\infty} \frac{\Delta_n}{\Lambda_f(x_R^n - x_L^n)^\alpha}[n(x_R^n - x_L^n)]^\alpha \le 2^\alpha \Gamma_f^\alpha(2/3,\epsilon). \tag{4.89}$$

We now derive a similar upper bound for the Wiener process-based algorithm; for details see [35]. Recall that the quantity to be maximized over all intervals in this case is

$$\gamma_i^n \triangleq \frac{x_i^n - x_{i-1}^n}{(y_{i-1}^n - y_{on} + \epsilon)(y_i^n - y_{on} + \epsilon)}.$$

A similar analysis to that carried out for the smooth case yields

$$\frac{x_R^n - x_L^n}{\epsilon^2 \gamma_s^n} \to 1,$$

and

$$\sum_{i=1}^{n} \gamma_i^n \to \int_{t=0}^{1} \frac{dt}{(f(t) - m + \epsilon)^2} = \frac{1}{\epsilon^2} \Gamma_f(2, \epsilon).$$

Therefore,

$$\limsup_{n\to\infty} n(x_R^n - x_L^n) = \limsup_{n\to\infty} \frac{(x_R^n - x_L^n)}{\gamma_s^n} \frac{\gamma_s^n}{\frac{1}{n}\sum_{i=1}^{n} \gamma_i^n} \sum_{i=1}^{n} \gamma_i^n \le 2\Gamma_f(2, \epsilon), \tag{4.90}$$

and so under the Wiener process-based algorithm,

$$\limsup_{n\to\infty} \frac{n^\alpha \Delta_n}{\Lambda_f} = \limsup_{n\to\infty} \frac{\Delta_n}{\Lambda_f (x_R^n - x_L^n)^\alpha} [n(x_R^n - x_L^n)]^\alpha \le 2^\alpha \Gamma_f^\alpha(2, \epsilon). \tag{4.91}$$

This completes the proof of (4.65).

Proof of Theorem 4.10.

Let us consider the Wiener process model case. The analysis for the smooth function statistical model case is similar. The trial points are dense everywhere in the minimization interval. Therefore in minimization process each subinterval will be chosen for subdivision. For continuously differentiable functions $x_i^n - x_{i-1}^n \to 0$ implies that the point x maximizing (4.32) is equal to

$$\frac{x_i^n + x_{i-1}^n}{2} + o((x_i^n - x_{i-1}^n)^2); \tag{4.92}$$

i.e., the intervals are eventually bisected by both versions of the P-algorithm. For sufficiently large n the criterion value γ_i^n of subinterval (x_{i-1}^n, x_i^n) is no larger than the criterion value of any other subinterval before subdivision, i.e. it is no larger than double the criterion value of any other subinterval. Estimating the limit of the ratio of corresponding criterion values (4.51), (4.53) yields the proof of the theorem.

Proof of Theorem 4.11.

The proof of the theorem is separated into the following lemmas.

Lemma 4.15. *The maximal selection criterion value* $\gamma^n = max_{i=1,\dots,n} \gamma_i^n$ *tends to 0 as* $n \to \infty$.

Lemma 4.16. *The error approaches zero faster than* ϵ_n

$$y_{on} - m = o(\epsilon_n).$$

Lemma 4.17. *The average selection criterion value is majorized by* $\log(n)/n$*:*

$$\frac{1}{n} \sum_{i=1}^{n} \gamma_i^n = O\left(\frac{\log(n)}{n}\right). \tag{4.93}$$

The idea of the proof of Theorem 4.11 is to bound the error Δ_n in terms of the bound for the average of the γ's.

Let us introduce stopping times for the algorithm. Recall that ω_n^s is the length of the interval containing x_* after n observations, and γ_n^s is the γ value corresponding to that interval. We will consider as candidate stopping times for the algorithm only those times n_k when γ_n^s crosses the average $\sum_1^n \gamma_i^n/n$ from below. Then $\gamma_{n_k}^s$ is asymptotically equivalent to the average, so from (4.17),

$$\gamma_{n_k}^s = \frac{\omega_{n_k}^s}{\sqrt{\epsilon_{n_k}} + \sqrt{\epsilon_{n_k} + o(\epsilon_{n_k})}} = \Theta\left(\frac{\log(n_k)}{n_k}\right),$$

which implies that

$$\omega_{n_k}^s = \Theta\left(\sqrt{\epsilon_{n_k}}\frac{\log(n_k)}{n_k}\right) = \Theta\left(\frac{\log(n_k)}{n_k^{(3-\delta)/2}}\right).$$

Since eventually y_{on} will be the minimum of the two function values at either side of x_*, $\Delta_n = \Theta(\omega_n^s)^2$, and

$$\Delta_{n_k} = \Theta\left(\frac{\log(n_k)^2}{n_k^{3-\delta}}\right) = O\left(n^{-3+\delta'}\right),$$

where δ' is any number greater than δ. Since δ is arbitrary, this completes the proof on convergence rate $O\left(n^{-3+\delta}\right)$ of the P-algorithm with decreasing threshold values $\epsilon_n = O\left(n^{-1+\delta}\right)$.

Proof of Lemma 4.15.

The following simple property of continuously differentiable functions will be useful in the following analysis. Let $L_n(\cdot)$ be the linear interpolator of the observed values, defined for $x_{i-1}^n \leq s \leq x_i^n$ by

$$L_n(s) = \frac{x_i^n - s}{x_i^n - x_{i-1}^n} y_{i-1}^n + \frac{s - x_{i-1}^n}{x_i^n - x_{i-1}^n} y_i^n.$$

Because $f \in C^2([0,1])$ and so f'' is bounded on $[0,1]$, a basic result on linear interpolation ([48], p. 53) implies that there exists a number B such that

$$\max_{x_{i-1}^n \leq s \leq x_i^n} |f(s) - L_n(s)| \leq B(x_i^n - x_{i-1}^n)^2 \tag{4.94}$$

for $i \leq n$.

If $\gamma^n = \max_{1 \leq i \leq n} \gamma_i^n$ is small, then the γ values for the two children of a split interval will be close to one-half that of the parent. To see this, suppose that an interval of width T with left and right function values y_L and y_R, respectively, is to be split at time n (so the γ value for the interval is the maximum, γ^n). Let us consider the γ value for, say, the left child, which we denote γ_L^{n+1}. To simplify the expressions, let

$$a_L = y_L - y_{on} + \epsilon_n, \qquad a_R = y_R - y_{on} + \epsilon_n.$$

If the new function value is $\bar{y}$, then

$$\gamma_L^{n+1} = \frac{\sqrt{a_L}}{\sqrt{a_L} + \sqrt{a_R}} \frac{T}{\sqrt{y_L - y_{o\,n+1} + \epsilon_{n+1}} + \sqrt{\bar{y} - y_{o\,n+1} + \epsilon_{n+1}}}.$$

Therefore,

$$\begin{aligned}\frac{\gamma_L^{n+1}}{\gamma^n} &= \frac{\sqrt{a_L}}{\sqrt{y_L - y_{o\,n+1} + \epsilon_{n+1}} + \sqrt{\bar{y} - y_{o\,n+1} + \epsilon_{n+1}}} \\ &= \frac{1}{\sqrt{1 + \frac{y_{on} - y_{o\,n+1}}{a_L} - \frac{\epsilon_n - \epsilon_{n+1}}{a_L}} + \sqrt{1 + \frac{y_{on} - y_{o\,n+1}}{a_L} - \frac{\epsilon_n - \epsilon_{n+1}}{a_L} + \frac{\bar{y} - y_L}{a_L}}} \\ &\leq \frac{1}{\sqrt{1 - \frac{\epsilon_n - \epsilon_{n+1}}{a_L}} + \sqrt{1 - \frac{\epsilon_n - \epsilon_{n+1}}{a_L} + \frac{\bar{y} - y_L}{a_L}}}.\end{aligned}$$

The limit $(\epsilon_n - \epsilon_{n+1})/\epsilon_n \to 0$ implies the limit $(\epsilon_n - \epsilon_{n+1})/a_L \to 0$. From (4.94) it follows

$$\frac{\bar{y} - y_L}{a_L} = \frac{\bar{y} - y_L}{T^2} \frac{T^2}{a_L} \leq B(\gamma^n)^2 \to 0.$$

Therefore, the upper bound on the ratio γ_L^{n+1}/γ^n approaches 1/2.

Fix $\eta > 0$. Since $\liminf \gamma^n = 0$, and $\frac{\gamma_L^{N+1}}{\gamma^N} \leq \frac{1}{2} + \eta$ there is an n such that $\gamma^N \leq \gamma^n 2^{(1-\delta)/2}$. Consider γ_i^n and γ_i^{n+k} with $k \leq n$ and such that $[x_{i-1}^{n+k}, x_i^{n+k}] = [x_{i-1}^n, x_i^n]$; that is, no new observation is placed in the interval between time n and $n+k$. In this case the value of γ_i^n only increases due to ϵ_{n+k} decreasing. Therefore, if

$$\gamma_i^n = \frac{T}{\sqrt{\epsilon_n + y_L - y_{on}} + \sqrt{\epsilon_n + y_R - y_{on}}},$$

then letting

$$b_L = \epsilon_{n+k} + y_L - y_{on}, \qquad b_R = \epsilon_{n+k} + y_R - y_{on}.$$

it follows that

$$\begin{aligned}\gamma_i^{n+k} &\leq \frac{T}{\sqrt{b_L} + \sqrt{b_R}} = \\ \frac{T}{\sqrt{\epsilon_n + y_L - y_{on}} + \sqrt{\epsilon_n + y_R - y_{on}}} & \frac{\sqrt{\epsilon_n + y_L - y_{on}} + \sqrt{\epsilon_n + y_R - y_{on}}}{\sqrt{b_L} + \sqrt{b_R}} = \\ &\leq \gamma_i^n \left(\frac{\epsilon_n}{\epsilon_{n+k}}\right)^{1/2} \leq \gamma_i^n \left(\frac{\epsilon_n}{\epsilon_{2n}}\right)^{1/2} = \quad (\text{since } k \leq n) \\ &\gamma_i^n \left(\frac{n^{-1+\delta}}{(2n)^{-1+\delta}}\right)^{1/2} = \gamma_i^n 2^{(1-\delta)/2}.\end{aligned}$$

Therefore, the γ value for any child of a split interval between time n and $2n$ will be at most $(1/2+\eta)2^{(1-\delta)/2}$ times that of its parent. In particular, we can choose η so that any child will have a γ value at time $2n$ at most $\frac{3}{4}\gamma^n$. Since all n original subintervals at time n can be split by time $2n$, we can conclude that $\gamma^{2n} \leq \frac{3}{4}\gamma^n$, and that $\max_{k\leq n} \gamma^{n+k} \leq \frac{3}{2}\gamma^n$. This implies that $\gamma^n \to 0$, completing the proof.

Proof of Lemma 4.16.

We are mainly interested in the error in terms of the function values, which we denote by

$$\Delta_n = \min_{i\leq n} f(x_i) - m.$$

We now analyze the rate at which the error Δ_n converges to 0. We continue to assume that $f \in C^2([0,1])$, but now we will assume in addition that f has a unique minimizer x_*, and that $f''(x_*) > 0$. These assumptions imply that y_{on} will eventually be the function value at one of the observation points adjacent to x_*, and so by (4.94) we eventually have that

$$\Delta_n = y_{on} - m \leq B\left(\omega_n^s\right)^2, \tag{4.95}$$

where ω_n^s is the distance between the two observations adjacent to x_* (the width of the straddling interval).

By the assumptions on f eventually the γ value of the interval containing x_*, which we denote by γ_s^n, becomes

$$\gamma_s^n = \frac{\omega_n^s}{\sqrt{\epsilon_n} + \sqrt{\epsilon_n + \tilde{y}_{on} - y_{on}}},$$

where y_{on} is the function value at one of the endpoints of this interval and $\tilde{y}_{on}$ is the function value at the other endpoint. Since $\lim \gamma^n = 0$, $\lim \gamma_s^n = 0$. Thus $\omega_n^s = o(\sqrt{\epsilon_n})$ follows if $(\tilde{y}_{on} - m)/(\omega_n^s)^2$ is bounded. To see that this is true, suppose that $x_L \leq x_* \leq x_R$. By Taylor's theorem, there exists a $z \in [x_*, x_R]$ such that

$$f(x_R) - f(x_*) = \frac{1}{2} f''(z)(x_R - x_*)^2.$$

With the similar bound for $f(x_L) - f(x_*)$, we conclude that

$$\frac{\tilde{y}_{on} - y_{on}}{(\omega_n^s)^2} \leq \frac{\tilde{y}_{on} - m}{(\omega_n^s)^2} + \frac{y_{on} - m}{(\omega_n^s)^2} \leq \sup_{z\in[0,1]} f''(z).$$

Since $\omega_n \leq 1/n$, where ω_n is the width of the smallest subinterval after n observations, it follows that $\omega_n/\sqrt{\epsilon_n} \to 0$. Then

$$y_{on} - m \leq B(\omega_s^n)^2 = o(\epsilon_n). \tag{4.96}$$

Proof of Lemma 4.17

The integral version of the γ's may be expressed using the linear interpolator $L_n(\cdot)$ (4.94)

$$\gamma_i^n = \int_{s=x_{i-1}^n}^{x_i^n} \frac{ds}{2\sqrt{L_n(s) - y_{on} + \epsilon_n}}$$

we have that

$$\begin{aligned}
\sum_{i=1}^n \gamma_i^n &= \frac{1}{2}\int_{s=0}^1 \frac{ds}{\sqrt{L_n(s) - y_{on} + \epsilon_n}} \\
&= \frac{1}{2}\int_{s=0}^1 \frac{ds}{\sqrt{L_n(s) - m + \epsilon_n - (y_{on} - m)}} \\
&= \frac{1}{2}\int_{s=0}^1 \frac{ds}{\sqrt{L_n(s) - m + \epsilon_n'}},
\end{aligned}$$

where $\frac{\epsilon_n'}{\epsilon_n} = \frac{\epsilon_n - (y_{on} - m)}{\epsilon_n} \to 1$ by (4.96).

Since $f''(x_*) > 0$ and f'' is continuous, we can choose a subinterval $[\alpha, \beta]$ on which f is convex, where $0 \le \alpha < x_* < \beta \le 1$. On this subinterval f minorizes $L_n(\cdot)$. We can choose a positive number η small enough that $[x_* - \eta, x_* + \eta] \subset [\alpha, \beta]$, and for $s \in [x_* - \eta, x_* + \eta]$,

$$f(s) - m \ge \frac{1}{4} f''(x_*)(s - x_*)^2. \tag{4.97}$$

For large n, $L_n(s) - y_{on}$ will be bounded below by $\min\{f(x_* - \eta), f(x_* + \eta)\} - m$ on $[0,1] \setminus [x_* - \eta, x_* + \eta]$, and so

$$\begin{aligned}
\frac{1}{n}\sum_{i=1}^n \gamma_i^n &= \frac{1}{2n}\int_{s\in[x_*-\eta, x_*+\eta]} \frac{ds}{\sqrt{L_n(s) - m + \epsilon_n'}} \\
&\quad + \frac{1}{2n}\int_{s\in[0,1]\setminus[x_*-\eta, x_*+\eta]} \frac{ds}{\sqrt{L_n(s) - m + \epsilon_n'}} \\
&\le \frac{1}{2n}\int_{s\in[x_*-\eta, x_*+\eta]} \frac{ds}{\sqrt{f(s) - m + \epsilon_n'}} + O(1/n) \\
&\le \frac{1}{2n}\int_{s=x_*-\eta}^{x_*+\eta} \frac{ds}{\sqrt{\frac{1}{4}f''(x_*)(s - x_*)^2 + \epsilon_n'}} + O(1/n) \\
&\le \frac{1}{n}\int_{s=0}^{\eta} \frac{ds}{\sqrt{\frac{1}{4}f''(x_*)s^2 + \epsilon_n'}} + O(1/n) \\
&\le \frac{1}{n}\frac{2}{\sqrt{f''(x_*)}}\int_{s=0}^{\eta} \frac{ds}{\sqrt{s^2 + \epsilon_n''}} + O(1/n),
\end{aligned}$$

where $\epsilon_n'' = 4\epsilon_n'/f''(x_*)$. Since,

$$\frac{1}{n}\sum_{i=1}^n \gamma_i^n \le \frac{1}{n}\frac{2}{\sqrt{f''(x_*)}}\int_{s=0}^{\eta/\sqrt{\epsilon_n''}} \frac{ds}{\sqrt{s^2 + 1}} + O(1/n),$$

and

$$\int_{s=0}^{\eta/\sqrt{\epsilon''_n}} \frac{ds}{\sqrt{s^2+1}} + O(1/n) =$$
$$\left(\log\left(\frac{2}{\sqrt{\epsilon''_n}}\left(\sqrt{\eta^2+\epsilon''_n}+\eta\right)\right) - \log(2)\right) + O(1/n) =$$
$$\left(\log\left(\sqrt{\eta^2+\epsilon''_n}+\eta\right) + \frac{1}{2}\log\left(\frac{1}{\epsilon_n}\right) + \frac{1}{2}\log\left(\frac{\epsilon_n}{\epsilon''_n}\right)\right) + O(1/n) =$$
$$\left(\frac{1}{2}(1-\delta)\log(n) + O(1)\right) + O(1/n),$$

then

$$\frac{1}{n}\sum_{i=1}^{n}\gamma_i^n \leq \Theta\left(\frac{\log(n)}{n}\right).$$

Proof of Theorem 4.12.

The convergence of the P*-algorithm is implied by the convergence of the Wiener process based P-algorithm for any continuous function. After a finite number of steps, three points (the best in between) will be found in the region of attraction of the global minimum dominating all the points not belonging to this region. The further trial points by the local algorithm will be chosen in the detected subinterval.

The local search algorithm based on quadratic fit converges to a local minimizer superlinearly:

$$||x_i - x*|| \leq C \cdot ||x_{i+1} - x*||^p,\ p > 1,\ C > 0, \tag{4.98}$$

with the order of convergence $p = 1.3$ for the considered case [150], where i denotes the iteration number of a local algorithm. However, only the s-th trial in the frame of P*-algorithm is made by the local minimization algorithm. Since the objective function is approximately quadratic in the neighborhood of the global minimum point, and in (4.98) there holds $i = n/s$, the order of convergence of y_{on} to $\min_{0\leq x\leq T} f(x)$ is equal to $(2p)^{1/s}$.

Proof of Theorem 4.14.

The proof of the theorem is subdivided in three lemmas below. Let x_L^n and x_R^n be the observation points to the left and right, respectively, of the minimizer x_*; that is, for some j_n,

$$x_L^n = x_{j_n}^n \leq x_* \leq x_{j_n+1}^n = x_R^n;$$

.

The next Lemma proves that in the limit, x_* is uniformly distributed over the interval $[x_L^n, x_R^n]$. Define

$$\theta_n \triangleq (x_* - x_L^n)/(x_R^n - x_L^n).$$

Lemma 4.18. *For* $0 \leq z \leq 1$, *as* $n \to \infty$

$$P(\theta_n \leq z) \to z.$$

Proof of Lemma 4.18.

Let $\{n_k\}$ be the times at which the interval straddling x_* is split (i.e., $x_{n_k} \in (x_L^{n_k-1}, x_R^{n_k-1})$), and set $Y_k = \theta_{n_k}$. Let

$$s_k = \frac{1}{1+\sqrt{1-(f(x_R^{n_k})-f(x_L^{n_k}))/(f(x_L^{n_k})-y_0\, n_k+\epsilon)}}$$

represent the relative location in the interval $[x_L^{n_k}, x_R^{n_k}]$ where the new observation is placed (denoted τ_i at (4.43)). Recall from Corollary 4.7 that $s_k \to 1/2$. The $\{Y_n\}$ satisfy the recursive relationship

$$Y_{n+1} = \begin{cases} (Y_n - s_n)/(1-s_n) & \text{if } Y_n > s_n, \\ Y_n/s_n & \text{if } Y_n < s_n. \end{cases} \tag{4.99}$$

Therefore,

$$P(Y_n \le z) = P\left(\cup_{i=0}^n [a_i^n \le x_* < a_i^n + z(a_{i+1}^n - a_i^n)]\right)$$

where the $\{a_i^n; i \le n\}$ are the order statistics of the set

$$\left\{ \prod_{k=0}^n s_k^{m_k}(1-s_k)^{1-m_k} : m_k \in \{0,1\}, k \le n \right\}.$$

Clearly the $\{a_i^n; i \le n\}$ partition $[0,1]$ and $\max_{i\le n} |a_{i+1}^n - a_i^n| \to 0$ as $n \to \infty$ since $s_k \to 1/2$. Since g is continuous,

$$P\left(\cup_{i=0}^n [a_i^n \le x_* < a_i^n + z(a_{i+1}^n - a_i^n)]\right) = \sum_{i=0}^n \int_{s=a_i^n}^{a_i^n+z(a_{i+1}^n-a_i^n)} g(s)\,ds$$

$$\to z\int_{s=0}^1 g(s)\,ds = z.$$

The proof is completed.

Define

$$\Delta_n = \min_{1\le i\le n} f(x_i^n) - m, \qquad \hat{\Delta}_n = \min\{f(x_L^n), f(x_R^n)\} - m. \tag{4.100}$$

The error random variable that we are mainly interested in is Δ_n. Let us recall that U denotes a random variable uniformly distributed on $[0,1]$, independent of f.

Since eventually the minimizer will be either x_L^n or x_R^n,

$$n^2(\hat{\Delta}_n - \Delta_n) \to 0,$$

the proof of the theorem follows from Lemma 4.20.

Lemma 4.19. *As* $n \to \infty$,

$$\frac{\hat{\Delta}_n}{\frac{1}{2}f''(x_*)(x_R^n - x_L^n)^2} \xrightarrow{\mathcal{D}} \min\{U^2, (1-U)^2\}, \tag{4.101}$$

where $\xrightarrow{\mathcal{D}}$ *denotes convergence in distribution; that is,*

$$P\left(\frac{\hat{\Delta}_n}{\frac{1}{2}f''(x_*)(x_R^n - x_L^n)^2} \le z\right) \to 2\sqrt{z}$$

for $0 \le z \le 1$.

Proof of Lemma 4.19.
Using the definition of θ_n,

$$\begin{aligned}\hat{\Delta}_n &= \min\{f(x_L^n), f(x_R^n)\} - m \\ &= \min\{f(x_* - \theta_n(x_R^n - x_L^n)), f(x_* + (1-\theta_n)(x_R^n - x_L^n))\} - m.\end{aligned}$$

Because f is assumed to have a positive and finite second derivative at x_*,

$$f(x_* + z) = m + \frac{1}{2}z^2 f''(x_*) + o(z^2)$$

as $z \to 0$. To complete the proof let us note that the equalities

$$\begin{gathered}\frac{\hat{\Delta}_n}{\frac{1}{2}f''(x_*)(x_R^n-x_L^n)^2} = \\ \min\left(\frac{\frac{1}{2}\theta_n^2(x_R^n-x_L^n)^2 f''(x_*)+o(x_R^n-x_L^n)^2}{\frac{1}{2}f''(x_*)(x_R^n-x_L^n)^2},\right. \\ \left.\frac{\frac{1}{2}(1-\theta_n)^2(x_R^n-x_L^n)^2 f''(x_*)+o(x_R^n-x_L^n)^2}{\frac{1}{2}f''(x_*)(x_R^n-x_L^n)^2}\right) = \\ \min\{\theta_n^2 + o(1), (1-\theta_n)^2 + o(1)\} + o(1) \\ \xrightarrow{\mathcal{D}} \min\{U^2, (1-U)^2\},\end{gathered}$$

are valid by Lemma 4.18.

Let $y_L^n = f(x_L^n)$, $y_R^n = f(x_R^n)$, and define

$$\gamma_s^n = \frac{x_R^n - x_L^n}{\sqrt{y_L^n - y_{on} + \epsilon} + \sqrt{y_R^n - y_{on} + \epsilon}},$$

the γ value for the interval $[x_L^n, x_R^n]$ straddling x_*. Because $y_L^n - y_{on} \to 0$ and $y_R^n - y_{on} \to 0$,

$$\frac{x_R^n - x_L^n}{2\gamma_s^n\sqrt{\epsilon}} = \frac{\sqrt{y_L^n - y_{on} + \epsilon} + \sqrt{y_R^n - y_{on} + \epsilon}}{2\sqrt{\epsilon}} \to 1. \tag{4.102}$$

To simplify the limit theorem (actually Theorem 4.14) the normalized error random variables at certain stopping times is considered. The quantity

γ_s^n is not observable by the optimizer by time n, since it requires knowledge of which interval contains x_*. Let us approximate γ_s^n by

$$\gamma_{i_n}^n \triangleq \frac{x_{i_n+1}^n - x_{i_n}^n}{\sqrt{f(x_{i_n}^n) - y_{on} + \epsilon} + \sqrt{f(x_{i_n+1}^n) - y_{on} + \epsilon}},$$

where $x_{i_n}^n$ is the minimizing observation of the first n; i.e., $f(x_{i_n}^n) \leq f(x_j^n)$ for $j \leq n$. Define an increasing sequence of stopping times $\{n_k\}$ by the successive times that the ratio

$$z_n \triangleq \frac{\gamma_{i_n}^n}{\frac{1}{n}\sum_{i=1}^n \gamma_i^n}$$

crosses 1 from below. By the construction of the stopping times $\{n_k\}$, $z_{n_k} \to 1$, since by definition $z_{n_k} \geq 1$ and $z_{n_k-1} < 1$, and $z_{n_k} - z_{n_k-1} \to 0$.

Along the sequence n_k, $\gamma_s^{n_k}$ and $\gamma_{i_{n_k}}^{n_k}$ do not differ by much; in particular,

$$\frac{\gamma_s^{n_k}}{\gamma_{i_{n_k}}^{n_k}} \to 1. \tag{4.103}$$

This is because eventually $x_{i_{n_k}}^{n_k}$ will be one of $x_L^{n_k}$ or $x_R^{n_k}$. Therefore, if $\gamma_s^{n_k}/\gamma_{i_{n_k}}^{n_k}$ had a limit point other than 1, it would need to be 1/2 or 2 (since by Corollary 4.7 the intervals are eventually bisected). But at the times n_k, $\gamma_{i_{n_k}}^{n_k}$ is approximately the average of the $\gamma_i^{n_k}$'s, which have a range of 2 in the limit (as the intervals are eventually bisected, the γ_i^n's are eventually 'cut in two' and the largest is twice the smallest).

Lemma 4.20. *Under the P-algorithm as $k \to \infty$,*

$$\frac{n_k^2 \hat{\Delta}_{n_k}}{\frac{1}{2}f''(x_*)\Gamma_f(1/2,\epsilon)^2} \xrightarrow{\mathcal{D}} \min\{U^2, (1-U)^2\}. \tag{4.104}$$

Proof of Lemma 4.20.

$$\frac{{n_k}^2 \hat{\Delta}_{n_k}}{\frac{1}{2}f''(x_*)\Gamma_f(1/2,\epsilon)^2} = \frac{\hat{\Delta}_{n_k}}{\frac{1}{2}f''(x_*)(x_R^{n_k} - x_L^{n_k})^2} \frac{(x_R^{n_k} - x_L^{n_k})^2}{4(\gamma_s^{n_k})^2\epsilon} \times$$
$$\times \left(\frac{\gamma_s^{n_k}}{\gamma_{i_{n_k}}^{n_k}}\right)^2 \left(\frac{\gamma_{i_{n_k}}^{n_k}}{\frac{1}{n_k}\sum_{i=1}^{n_k}\gamma_i^{n_k}}\right)^2 \frac{(\sum_{i=1}^{n_k}\gamma_i^{n_k})^2}{\Gamma_f(1/2,\epsilon)^2/4\epsilon}.$$

By Lemma 4.19, the first term in the last line converges in distribution to $\min\{U^2, (1-U)^2\}$. The second term converges to 1 by (4.102), and the third term converges to 1 by (4.103). The fourth term converges to 1 since $z_{n_k} \to 1$, and since

$$2\sqrt{\epsilon}\sum_{i=1}^{n}\gamma_i^n \to \Gamma_f(1/2,\epsilon),$$

the last term also converges to 1. Therefore,

$$\frac{{n_k}^2\hat{\Delta}_{n_k}}{\frac{1}{2}f''(x_*)\Gamma_f(1/2,\epsilon)^2} \xrightarrow{\mathcal{D}} \min\{U^2,(1-U)^2\},$$

as was to be shown.

4.4 Multidimensional Algorithms

4.4.1 Introduction

The structure of multidimensional algorithms based on statistical models is similar to the structure of corresponding one-dimensional algorithms. However, the problems of implementation in the multidimensional case are more complicated because of complicated auxiliary subproblems. The impact of the implementation on the efficiency of the multidimensional algorithms is much stronger than on the efficiency of one-dimensional algorithms where theoretical method can be implemented almost precisely, and the underlying theory predetermines the efficiency of the algorithm. In this section we mainly consider the multidimensional implementations of the P-algorithm. Slightly different ideas have been applied to develop multidimensional one-step Bayesian algorithms; these algorithms are not considered here since they are presented in detail in [164], [165]. Methods of multidimensional extension of one-dimensional algorithms by means of space filling curves are well presented in [233].

4.4.2 P-algorithm

The standard probabilistic generalization of stochastic processes to the multidimensional case are random fields, i.e. stochastic functions of several variables. As discussed in Sect. 4.1.3 let the random field $\xi(x)$, $x \in A \subset \mathbb{R}^d$ be accepted as a model of the objective functions. The P-algorithm at $(n+1)$-th minimization step is defined as follows

$$x_{n+1} = \arg\max_{x \in A} \mathbf{P}\{\xi(x) \leq y_{on} - \varepsilon) | \xi(x_1) = y_1, ..., \xi(x_n) = y_n\}, \tag{4.105}$$
$$y_{on} = \min\{y_1, ..., y_n\}.$$

Maximization in (4.105) is equivalent (see e.g. (4.17)) to the maximization of

$$\frac{y_{on} - \varepsilon - m_n(x|x_j, y_j, j = 1, ..., k)}{s_n(x|x_j, y_j, j = 1, ..., k)}, \tag{4.106}$$

where $m_n(x|\cdot)$ and $s_n^2(x|\cdot)$ are the conditional mean and conditional variance of the random field $\xi(x)$ correspondingly. This maximization problem is difficult since the objective function is not concave, and the computational complexity of evaluation of the objective function increases with n. Indeed, the complexity of computation of the conditional mean and variance for Gaussian random fields is $O(n^3)$ since the formulae

$$m_n(x|x_j, y_j, j = 1, ..., k) = m(x) +$$
$$(\sigma(x, x_1), ..., \sigma(x, x_n)) \cdot \Sigma_n^{-1} \cdot (y_1 - m(x_1), ..., y_n - m(x_n))^T,$$
$$s_n(x|x_j, y_j, j = 1, ..., k) = \sigma^2(x) -$$
$$(\sigma(x, x_1), ..., \sigma(x, x_n)) \cdot \Sigma_n^{-1} \cdot (\sigma(x, x_1), ..., \sigma(x, x_n))^T,$$
$$\sigma(x_i, x_j) \text{ is thecovariance between } \xi(\mathrm{x_i}) \text{ and } \xi(\mathrm{x_j}),$$
$$\Sigma_n = \begin{pmatrix} \sigma(x_1, x_1) & ... & \sigma(x_1, x_n) \\ ... & ... & ... \\ \sigma(x_n, x_1) & ... & \sigma(x_n, x_n) \end{pmatrix},$$

include the inversion of a covariance matrix of size $n \times n$, where $m(x)$ and $\sigma^2(x)$ are the apriori mean and variance of $\xi(x)$. On the other hand, during the maximization of (4.105) the objective function is calculated many times with different values of x but with the same matrix Σ_n. If the number of the objective function evaluations is larger than n then the amortized complexity of one function evaluation may be considered as $O(n^2)$. Nevertheless such complexity is high, and methods based on random field models can be prospective only for very expensive (needing much computer time for evaluation) objective functions.

Summarizing we emphasize two main difficulties in constructing *multidimensional* (not only the P-algorithm) algorithms based on random fields:

FIRST: the current trial point is defined by means of optimization of a merit function (e.g., average improvement or probability of improvement), which again is multimodal optimization problem.

SECOND: the computation of a merit function is hard because it implies inversion of correlation matrixes whose dimensionality is $n \times n$ at the $(n+1)^{th}$ minimization step.

The second difficulty can be weakened by applying axiomatically based simple statistical models (see Sect. 4.1.4). Let us assume for the model of objective functions the family of Gaussian random variables instead of a random field, i.e. let the family of Gaussian random variables ξ_x, $x \in A$ be accepted for a model, and the parameters of ξ_x be defined below as:

$$m_n(x|x_i, y_i, i = 1, ..., n) = \sum_{i=1}^{n} y_i w_i^n(x, x_i, y_i, i = 1, ..., n),$$
$$s_n^2(x|x_i, y_i, i = 1, ..., n) = \tau \sum_{i=1}^{n} ||x - x_i|| w_i^n(x, x_i, y_i, i = 1, ..., n),$$
$$\tau > 0, \tag{4.107}$$

where $m_n(x|\cdot)$ is mean value of ξ_x, $s_n^2(x|\cdot)$ is variance of ξ_x, and $w_i^n(\cdot)$ are weights.

In this case the complexity of computation of $m_n(x|\cdot)$ and $s_n(x|\cdot)$ is $O(n)$. For large n the computational advantage of simple statistical models over random fields can be significant. General properties of $m_n(x|\cdot)$ and $s_n(x|\cdot)$ in both cases are similar; for visual comparison we refer to Figure 4.4, where the objective function values modelled by means of simple statistical model (4.107) with weights

$$w_i^n(x, x_i, y_i, i = 1, ..., n) = \beta(||x - x_i||)/\sum_{j=1}^{n}\beta(||x - x_j||),$$
$$\beta(z) = \exp(-cz^2)/z,$$

are shown in Figure 4.4c.

Although the first difficulty is not avoided in the case of simple statistical models, the lower complexity of computation of (4.105) in this enables to find better approximations to global minimum of (4.105) using heuristic methods. Normally a combination of a local descent algorithm with global evolutionary search is used.

Since the auxiliary computations require much computing time the P-algorithm (and other methods based on statistical models) are primarily aimed for the minimization of expensive objective functions. In such a case the search for the global minimum should be stopped after not very many evaluations of the objective function. It is difficult to analyze or even define the efficiency of global optimization algorithms aiming to find a rough approximation of the global minimum. Frequently the approximation found by means of a global optimization algorithm is used as a starting point for local search. Our main argument to substantiate the P-algorithm is the axiomatic definition of typical information about the objective function and of rational search strategy. Nevertheless the standards of optimization theory require to supplement these arguments by the convergence analysis. The theorem below states that the P-algorithm converges under rather broad assumptions.

Theorem 4.21. *Let the following conditions be satisfied:*

- *a) optimization problem: the objective function $f(x)$, $x \in A \subset \mathbb{R}^d$ is continuous and the set A is compact;*
- *b) statistical model: a family of Gaussian random variables ξ_x is accepted as a statistical model with parameters (4.107), where $0 < \tau^- < \tau < \tau^+ < \infty$;*
- *c) the weights $w_i^n(x, \cdot)$ are continuous with respect to x, and for a sequence $z_i \in A$ with a cluster point z, and for arbitrary $0 < \gamma < 1, \delta > 0$ there exists N such that*

$$\sum_{i \in I(z,\delta)} w_i^n(z, z_j, j = 1, ..., n) > 1 - \gamma,$$

where $n \geq N$, $I(z, \delta) = \{i : i \leq n, ||z - z_i|| < \delta\}$.

Then the P-algorithm converges to the global minimum: $y_{on} \rightarrow \min_{x \in A} f(x)$.

4.4.3 Two Dimensional Select and Clone

The implementation of the one-dimensional P-algorithm can be greatly simplified by choosing a Markov random process for a model of the objective functions: comparison of the equations (4.32) and (4.33) clearly shows the computational advantage of the Markov process model. Random fields (generalization of random processes to multidimensional case) with such favorable properties are not known. However a simple statistical model (a family of Gaussian random variables ξ_x) with a generalized Markov property is possible. Let us start with the two-dimensional case $d = 2$ and aim to construct a simple statistical model ξ_x whose mean and variance would be generalized versions of the conditional mean and conditional variance of the Wiener process. The choice of the latter as a prototype is partly justified by the following arguments:

- the functional structure of conditional characteristics (conditional mean and conditional variance) is the simplest possible,
- in the one-dimensional case the (deterministic) convergence order of the P-algorithm based on the Wiener process model and of the P-algorithm based on the smooth function model are the same with respect to continuous functions whose global minimizers are unique and regular; see Theorem 4.9,
- it is supposed that the constructed P-algorithm is used in combination with a local descent algorithm, i.e. the high precision of local search by the P-algorithm is not a concern.

Assume that A is an equilateral triangle. Only those subsets of A are considered which can be obtained by repeated application of the cloning procedure, where cloning is the subdivision of a triangle into four triangles with half length sides shown in Fig. 4.9. Calculations of objective function values are allowed only on the vertices of feasible triangles. A global optimization algorithm consisting of selection and cloning procedures can be constructed. A triangle with the maximal criterion value is selected where criterion should be defined according to the chosen optimization paradigm, e.g. using Lipshitz [47] or statistical models of objective functions [295], [305]. The selected triangle is cloned bisecting its sides. The structure of such an algorithm is similar to the structure of the one-dimensional P-algorithm: instead of intervals the triangles are considered. In the two-dimensional case cloning subdivides the chosen triangle into four equal subtriangles while in the one-dimensional case the selected interval is subdivided (see Corollary4.7) into two asymptotically equal subintervals.

Let some objective function values be known: $y_i = f(x_i), i = 1, ..., n$, where x_i are vertices of equilateral subtriangles covering the feasible region. We accept a family of Gaussian random variables ξ_x as a simple statistical model of the objective functions. Generalizing the Markov property we assume that the mean value and variance of ξ_x depend only on x_i, y_i corresponding to the triangle T_x where $x \in T_x$. Extending the formula of the conditional

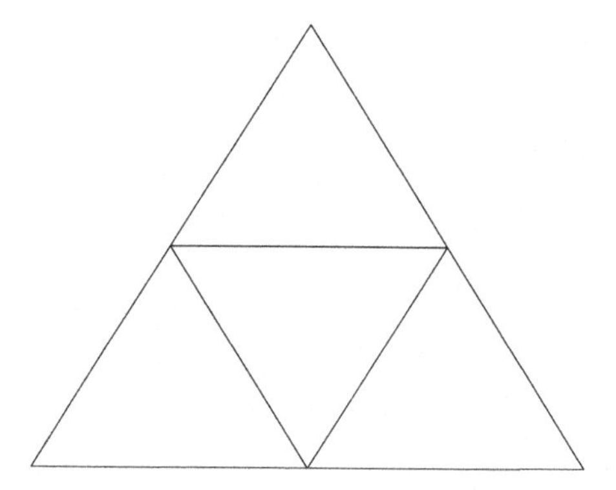

Fig. 4.9. Cloning of an equilateral triangle

mean of the Wiener process (4.34) to the two-dimensional case, we obtain the piecewise linear in x formula for the mean of ξ_x

$$m_n(x|x_i, y_i, i = 1, ..., n) = \sum_{x_i \in T_x} y_i \cdot \nu_i(x, x_j,\ j = 1, ..., n), \qquad (4.108)$$

where the baricentric coordinates $\nu_i(x, x_j,\ j = 1, ..., n)$ are defined by the equalities

$$\nu_i(x, x_j,\ j = 1, ..., n) = 0, \text{ for indices } i \text{ such that } x_i \notin T_x,$$
$$\text{and } x = \sum_{x_i \in T_x} x_i \cdot \nu_i(x, x_j,\ j = 1, ..., n).$$

The variance of ξ_x is defined similarly, generalizing the formula of the condition variance of the Wiener process. Assuming the length of the side of T_x is equal to h we define the variance of ξ_x as the piecewise quadratic in x function attending maximum at the center of the triangle c_x and equal to zero at the vertices:

$$s_n^2(x|x_i, y_i, i = 1, ..., n) = \sigma_0^2(\frac{h^2}{3} - ||c_x - x||^2). \qquad (4.109)$$

To implement the P-algorithm the probability (4.105) should be maximized. Since the mean and variance of ξ_x (defined by (4.108), (4.109)) depend only on the vertices of T_x and the corresponding objective function values, we have to solve a series of maximization problems with triangle feasible regions.

Let us analyze a standardized problem whose feasible region is the (equilateral) triangle with side length equal to h, and with the vertices $\omega_1 = (0, 0)$, $\omega_2 = (0, h)$, $\omega_3 = (h/2, h\sqrt{3}/2)$. The objective function values at the vertices are denoted by φ_i, $i = 1, 2, 3$, and without loss of generality it is assumed that $\varphi_1 \leq \varphi_2 \leq \varphi_3$. Since a more general case ($d \geq 2$) of maximization of $\mathbf{P}\{\xi_x \leq \tilde{y}_{on}|\cdot\}$ over general simplices is considered in Sect. 4.4.4 we present

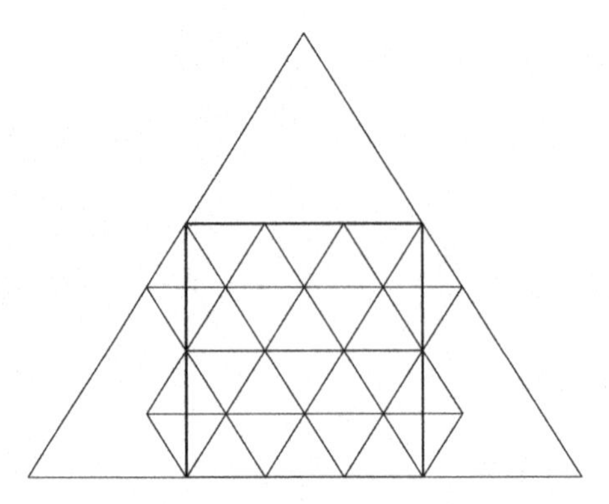

Fig. 4.10. Two examples of initial cover of a rectangle

here the result without proof: the maximum of $\mathbf{P}\{\xi_x \leq \tilde{y}_{on}|\cdot\}$, $\tilde{y}_{on} = y_{on} - \varepsilon$ over the equilateral triangle with side length h is attained at the point

$$x_{max} = h\left(\frac{1}{2} - \frac{\varphi_2}{\varphi_2 + \varphi_3 - 3\tilde{y}_{on}}, \ \frac{\sqrt{3}}{3}\left(\frac{1}{2} - \frac{2\varphi_3 - \varphi_2}{\varphi_2 + \varphi_3 - 3\tilde{y}_{on}}\right)\right),$$

if $\varphi_2 - \varphi_3 - \tilde{y}_{on} > 0$. In this case x_{max} is an interior point of the triangle, and the maximal probability is equal to

$$P_{max} = \Pi\left(-\frac{1}{h\sigma_0}\sqrt{2(\varphi_2 - \varphi_3 - \tilde{y}_{on})(\varphi_3 - \tilde{y}_{on}) + (\tilde{y}_{on} - \varphi_3)^2 - \varphi_2^2}\right). \tag{4.110}$$

In the case $\varphi_2 - \varphi_3 - \tilde{y}_{on} \leq 0$, the constraint $x_{\cdot 2} = 0$ should be taken into account, and the maximum is defined by the following formulae

$$x_{max} = h\left(\frac{1}{2} - \frac{\varphi_2}{2(\varphi_2 - 2\tilde{y}_{on})}, \ 0\right), P_{max} = \Pi\left(-\frac{2}{h\sigma_0}\sqrt{-(\varphi_2 - \tilde{y}_{on})\tilde{y}_{on}}\right). \tag{4.111}$$

Normally general multimodal functions are minimized over rectangle feasible regions. Thus suppose A to be a rectangle. To apply the select and clone algorithm, A should be covered by equilateral triangles with disjoint interiors. An initial cover can be constructed rather arbitrarily. For example, a rectangle can be re-scaled to the standard region and embedded into equilateral triangle as shown in the Fig. 4.10 where the standard region is defined as: $0 \leq x_{\cdot 1} \leq 1$, $0 \leq x_{\cdot 2} \leq 2\sqrt{3}/3$. However, overcovering is not desirable. The cover of the standard region by the set of 28 triangles shown in the Fig. 4.10 has an obvious advantage. The edge length of a triangle in the cover consisting of 28 triangles is equal to 1/3, and the vertices ω_i, $i = 1, \ldots, 22$ have the coordinates $\omega_{i1} \in \{j/6, \ j = -1, \ldots, 7\}$, $\omega_{i2} \in \{k \cdot \frac{\sqrt{3}}{6}, k = 1, 3,$ if j is uneven and $k = 0, 2, 4,$ if j is even$\}$.

The proof of the convergence of the two-dimensional select and clone algorithm is a special case of the convergence proof in Sect. 4.4.4.

The only parameter of the statistical model σ_0^2 is estimated from the data obtained during the optimization. The initial estimate can be obtained by the assumption that s_0^2, i.e. the variance of values of $f(\cdot)$ at the points of the initial cover, is equal to the variance (4.109) at the center of a simplex of the initial cover, i.e. in the case of the initial cover consisting of 28 triangles shown in Fig. 4.10 it is equal to $27s_0^2$. New function values calculated during the optimization can be used to update the estimate of σ_0^2. The deviation Δ_f (of the new function value at the middle point of the edge of the selected simplex from the mean value of the model (4.108)) and the variance of the statistical model at the trial point $\sigma_0^2 h^2/4$ define the new entry for updating: $4\Delta_f^2/h^2$. Let us note, that the proof of convergence of the algorithm also remains valid for the case of the adaptive σ_0^2 under the weak assumption $\sigma_0^2 \geq \sigma_{min}^2 > 0$.

Rather frequently the general behavior of $f(\cdot)$ over A severely differs from that over a small subregion of A. For example, the $f(\cdot)$ values grow very fast when approaching the boundary of A. In such a case some entries for updating the estimate of σ_0^2 should be interpreted as outliers. To reduce the influence of such excesses a robust statistical procedure should be applied, e.g. omitting a small percentage of extremal elements of the sample.

The parameter of the algorithm ε defines the aspiration level to reach in the current iteration. The globality of search increases (i.e. the trial points are generated more uniformly) with increasing ε. The choice of ε in the one-dimensional case is discussed in detail in Sect. 4.3.8. The value of ε is chosen similarly in the two dimensional algorithm: $\varepsilon = (\max_{i=1,\dots,k} y_i - \min_{i=1,\dots,k} y_i)/4$.

In the one-dimensional case the statistical hypothesis of finding of a subinterval of a local minimum is tested during the global search. A similar procedure is not known for the two-dimensional case. The heuristic idea is implemented: if the variation of function values at the vertexes of a simplex are less than a prescribed threshold then such a simplex is discarded from further consideration. The stopping condition is defined by the probability 0.99 of no improvement during the next 100 evaluations of $f(\cdot)$.

The algorithm has been implemented in C. To facilitate the search of the triangle with maximal probability value all (not discarded) triangles are arranged into the priority queue in the array of structures containing all the necessary information on the triangles. Testing results of the algorithm are presented in Sect. 4.4.5.

4.4.4 P-algorithm With Simplicial Partitioning

In the previous section a simple statistical model of objective functions is discussed generalizing properties of the conditional mean and conditional variance of the Wiener process to the two-dimensional case. The P-algorithm constructed using this model inherits many advantages of the one-dimensional

P-algorithm based on the Wiener process model, e.g. the selection criterion (maximum of probability (4.105) over an equilateral triangle) is defined by an analytical formula. In this way a multimodal maximization problem is reduced to a series of rather simple unimodal maximization problems. In the present section possibilities to develop similar global optimization algorithms for problems of higher dimensionality are considered.

For the definition of the select and clone algorithm the essential element was a triangle. To extend the algorithm to general multidimensional case we have to select a d dimensional element similar to a triangle in the two-dimensional case. Such an element is a simplex, i.e. a polyhedron in $\mathbb{R}^d$ with $d+1$ vertices. Special case of simplices is the standard simplex whose vertices are $\omega_0, \omega_1, ..., \omega_d$,

$$\omega_0 = (-a, ..., -a),\ \omega_i = e_i,\ i = 1, ..., d, \tag{4.112}$$

$$a = (\sqrt{d+1} - 1)/d,\ e_i \text{ is } i - \text{th unit vector.} \tag{4.113}$$

The center of (4.112) is

$$C = (b, ..., b),\ b = \left(\sqrt{d+1} - 1\right) / \left(d\sqrt{d+1}\right), \tag{4.114}$$

and the distance between center and vertices is equal to $D = \sqrt{\frac{d}{d+1}}$. A regular simplex is obtained from the standard simplex by means of an orthogonal mapping and extension/contraction of $\mathbb{R}^d$ equally in all coordinates.

The two-dimensional statistical model of the previous section can easily be extended to a higher dimensionality by replacing an equilateral triangle with the correct simplex, and generalizing the formulae (4.108) and (4.109) to the general multidimensional case. However, essential in the construction of the algorithm, a cloning procedure does not exist for $d > 2$. The cloning procedure can be replaced by different partitioning procedures; but partitioning produces different (not only correct) simplices. Therefore the statistical model should be defined to general simplices, and the impact of the deformation of simplices should be evaluated. This chapter aims to investigate the basic features of the multidimensional statistical model, to prove the convergence of a general algorithm, and to investigate experimentally the efficiency of different partitioning methods.

Let us consider the global minimization problem $\min_{x \in A} f(x),\ A \subset \mathbb{R}^d$, where $f(\cdot)$ is a continuous function and A is a compact set. The very general assumptions on $f(\cdot)$ imply that the family of random Gaussian variables ξ_x may be considered as a model of $f(\cdot)$. The mean and variance of ξ_x will be defined generalizing the properties of the Wiener process similarly as to the previous section.

Let A be covered by simplices $S_j,\ j = 1, ..., m,$ and let the observation points coincide with the vertices of S_j; it is assumed that $n > d$. The Markov property (restricting the dependence of the conditional mean and conditional variance of a stochastic process on two neighboring known process values; see

(4.33)) is generalized as the following property of the simple multidimensional statistical model: for $x \in S$ all the weights not corresponding to the vertices of S are defined to be equal to zero:

$$m_n(x|(x_i, y_i), i = 1, \ldots, n) = \sum_{i=1}^{n} w_i^n(x, x_j, j = 1, \ldots, n) \cdot y_i = \\ = \sum_{i=0}^{d} \nu_i(x, \omega_j, j = 0, \ldots, d) \cdot \varphi_i, \quad (4.115)$$

where ω_i , $i = 0, \ldots, d$, denote the vertices of S, and φ_i denote the corresponding function values. However, this is not the only goal of the generalization. The linearity with respect to x of the conditional mean of the Wiener process is very important for the efficient implementation of a one-dimensional algorithm; therefore the linearity of $m_n(x|\cdot)$ with respect to x is also wanted in the multidimensional case: (4.115) is piecewise linear with respect to x if the weights in (4.115) are defined to be equal to the baricentric coordinates of x with respect to ω_i

$$x = \sum_{i=0}^{d} \nu_i(x, \omega_j, j = 0, \ldots, d) \cdot \omega_i.$$

By similarity to the Wiener process, the variance of ξ_x, $x \in S$, for a regular simplex may be defined as a quadratic function with zero values at the vertices and a maximum at the point equidistant from the vertices:

$$s_n^2(x|x_i, i = 1, \ldots, n) = \sigma_0^2 \cdot \left(D^2 - ||C - x||^2\right), \quad (4.116)$$

where D is the distance between C and vertices, and σ_0 is the only parameter of the model which may be estimated from the data collected during the minimization.

In the case of an arbitrary simplex the definition of the average (4.115) remains valid. The variance is obtained by means of the inverse mapping of the considered simplex on the correct simplex with equal perimeter, where the increasingly ordered function values φ_i should correspond to the vertices ω_i, $i = 0, ..., d$.

To start the optimization the feasible region A is covered by simplices with disjoint interiors. The family of Gaussian random variables ξ_x with the characteristics defined above is accepted as a multidimensional statistical model. To the simplex S the criterion $\max_{x \in S} P_n(x)$ is assigned where

$$P_n(x) = \mathbf{P}\{\xi(x) \leq \tilde{y}_{on} - \varepsilon)|\xi(\omega_0) = \varphi_0, ..., \xi(\omega_d) = \varphi_d\}. \quad (4.117)$$

The simplex of maximal criterion value (4.117) is selected for the partitioning.

Let us consider the maximization of $P_n(x)$ over the regular simplex S (4.112). It is assumed that $\varphi_0 = 0 \leq \varphi_1 \leq \ldots \leq \varphi_d$. Such an assumption does not reduce generality, but $\tilde{y}_{on}$ should be correspondingly normalized for each simplex; after the normalization the inequality $\tilde{y}_{on} < 0$ holds.

Theorem 4.22. *$P_n(x)$ has either, a single local maximum in the interior of S at the point $x_{\max}$, or no local maximum in the interior of S at all, where*

$$x_{max} = C - \frac{1}{-\frac{d+1}{d}\tilde{y}_{on} + \frac{1}{d}\sum \varphi_i} \cdot (\varphi - \varphi_* \cdot I), \tag{4.118}$$
$$\varphi = (\varphi_1, \dots, \varphi_d),\ I = (1, ..., 1),$$
$$\varphi_* = b\sum \varphi_i = \frac{\sqrt{d+1}-1}{d\sqrt{d+1}} \sum \varphi_i.$$

Remark 1. If the point defined by (4.118) does not belong to the simplex, then the maximum point of $P_n(x)$ is at the facet defined by the vertices $\omega_0, ..., \omega_{d-1}$. To find the maximum point taking into account the latter constraint, the proof of the theorem should be repeated for the $(d-1)$ dimensional simplex and the projection of the gradient onto the constraint, etc. If not at a previous step, the maximum point will finally be found at the edge of the simplex corresponding to the vertices ω_0, ω_1. However, it is very likely, that the neighbor simplex should be selected if the maximum of $P_n(x)$ is achieved on a facet of the considered simplex. Therefore, a rough estimate of the maximum point may be sufficient, e.g. x_0 obtained by means of a step from the point C in the gradient direction

$$x_0 = C - \frac{\varphi - \varphi_* \cdot I}{||\varphi - \varphi_* \cdot I||\sqrt{d(d+1)}}, \tag{4.119}$$

where the step length is equal to the shortest distance from the point C to the facet of the simplex.

Remark 2. The selection criterion of an arbitrary simplex is defined to be equal to the selection criterion of the regular simplex whose perimeter is equal to the perimeter of the considered simplex. Therefore, $P_n(x_{max})$ for an arbitrary simplex is calculated using formulae (4.106), (4.115), (4.118), (4.114), and substituting (4.116) with

$$s_n^2(x|x_i, i = 1, \dots, n) = \sigma_0^2 \cdot \left(D^2 - ||C - x||^2\right) \cdot h^2/2, \tag{4.120}$$

where $h = \frac{2p}{(d+1)d}$ denotes the average length of the edge of the simplex with the perimeter length equal to p.

Remark 3. If $\varphi_0 = \varphi_1 = \ldots = \varphi_d = 0$, then for the regular simplex there holds the equality $x_{max} = C$, and the maximal probability is equal to

$$P_n(x_{max}) = \Pi\left(\tilde{y}_{on}\sqrt{d+1}/(\sigma_0 h\sqrt{2d}\right).$$

Let A_0 denotes the union of the simplices of an initial covering of A, $A_0 \supseteq A$. The refinement of the cover includes the selection of a simplex and its subdivision into several smaller simplices, which are, loosely speaking, as

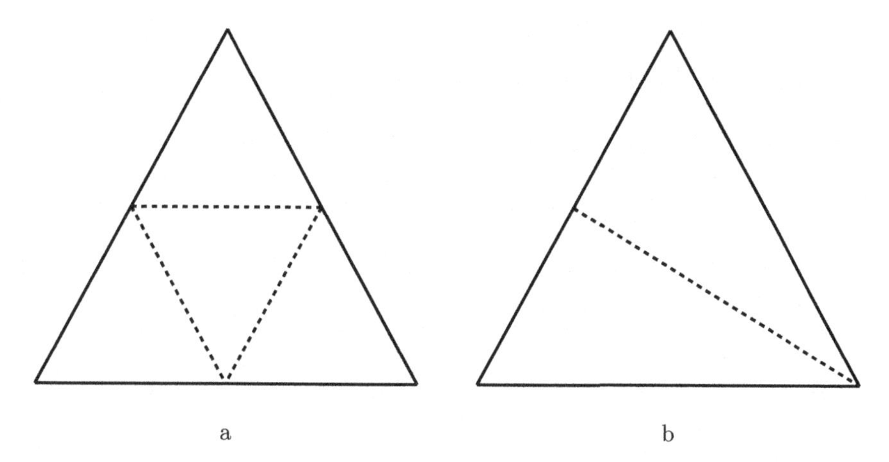

Fig. 4.11. Examples of partitioning of two - dimensional simplices

regular as possible. In the two - dimensional case a special case of the partitioning, i.e. cloning, may be applied producing descendants similar to the parent. Two examples are presented in Fig. 4.11. No method of the partitioning a regular simplex into the regular subsimplices is known for $d \geq 3$. To restrict the analysis with regular simplices the cloning might be generalized to the covering: instead of the partitioning of a parent simplex it might be covered by similar smaller correct simplices. The disadvantage of the latter type of cloning is overcovering of A. If the requirement of similarity of descendants to the parent is relaxed, then a variety of the partitioning procedures might be proposed. For example, from a regular d dimensional simplex $d+1$ descendant regular simplices with half sized edges may be obtained as a result of cutting of the regular parent simplex by means of $d+1$ hyperplanes, which are parallel to the facets of the parent simplex and divide the corresponding edges with ratio 1:1. The vertices of the descendant simplices are: one vertex coincides with the vertex of the parent simplex and the d other vertices coincide with centers of edges, whose intersection produces the corresponding vertex of the parent simplex. The rest of the parent simplex should be partitioned taking into account the new available vertices. A three dimensional version of such a semi regular partitioning is shown in Figures 4.12a, 4.13. A simple procedure of bisect the longest is illustrated by Figures 4.11b, 4.12b. Similar procedures are successfully applied also in Branch and Bound methods based on the deterministic models of the objective function, see e.g. [87], [124], [123]. We will assume that the partitioning produces the bounded number of descendants with the bounded ratio of the longest to the shortest edges: $h_{max}/h_{min} \leq \Gamma$.

To start the optimization the initial simplices should be available. We consider the two versions of initial covering presented in Figures 4.14, 4.15.

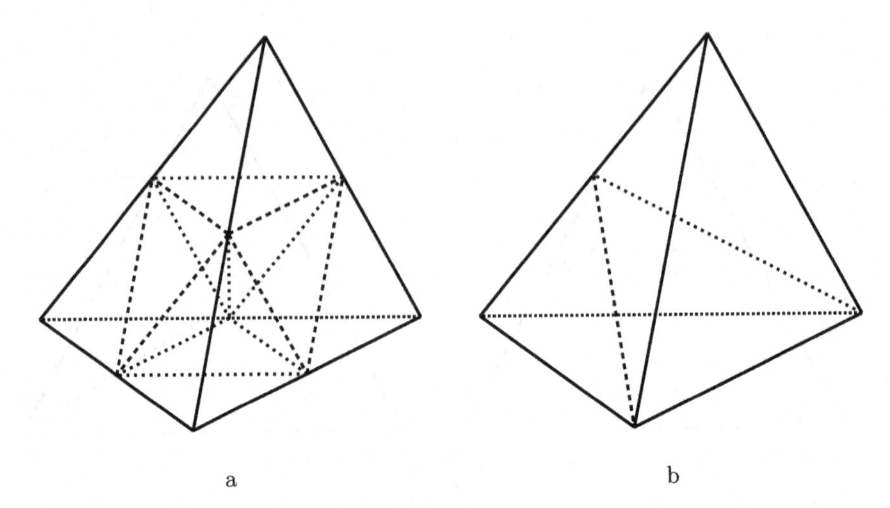

Fig. 4.12. Examples of partitioning of three - dimensional simplices

Let $F(x)$ be a continuous function extending $f(x)$ to A_0 where A_0 denotes an initial simplicial cover of A, $F(x) = f(x)$ for $x \in A$; $F(x) > \min_{x \in A} f(x)$ for $x \in A_0 \backslash A$. The function $F(x)$ is minimized on A_0 by means of the algorithm described in previous chapters.

Theorem 4.23. *The observation points are everywhere dense in* A_0 *implying*

$$\lim_{n \to \infty} y_{on} = \min_{x \in A} f(x), \tag{4.121}$$

where $y_{on} = \min\{y_1, ..., y_n\}$.

For the multidimensional problems the estimates of the global minimum obtained by means of a global technique in practically acceptable time may be rather rough. Therefore it may be reasonable to combine this global technique with a fast converging local algorithm. Such a combination may be considerably faster than a pure global optimization algorithm [39]. Let us suppose that an objective function is sufficiently smooth for the convergence of a descent method with the convergence order $p > 1$. The modified select and partition technique consists of N iterations of the global search technique alternating with one iteration of the local descent method from the best point found.

Theorem 4.24. *Let the above assumptions be satisfied and assume that the objective function* $f(x)$ *is twice continuously differentiable at the unique global minimum point* x_*. *The modified select and partition technique generates a sequence of estimates converging to the global minimum with convergence order better than 1.*

The results of the present section show that the widely used for one-dimensional global optimization statistical model can be generalized to the

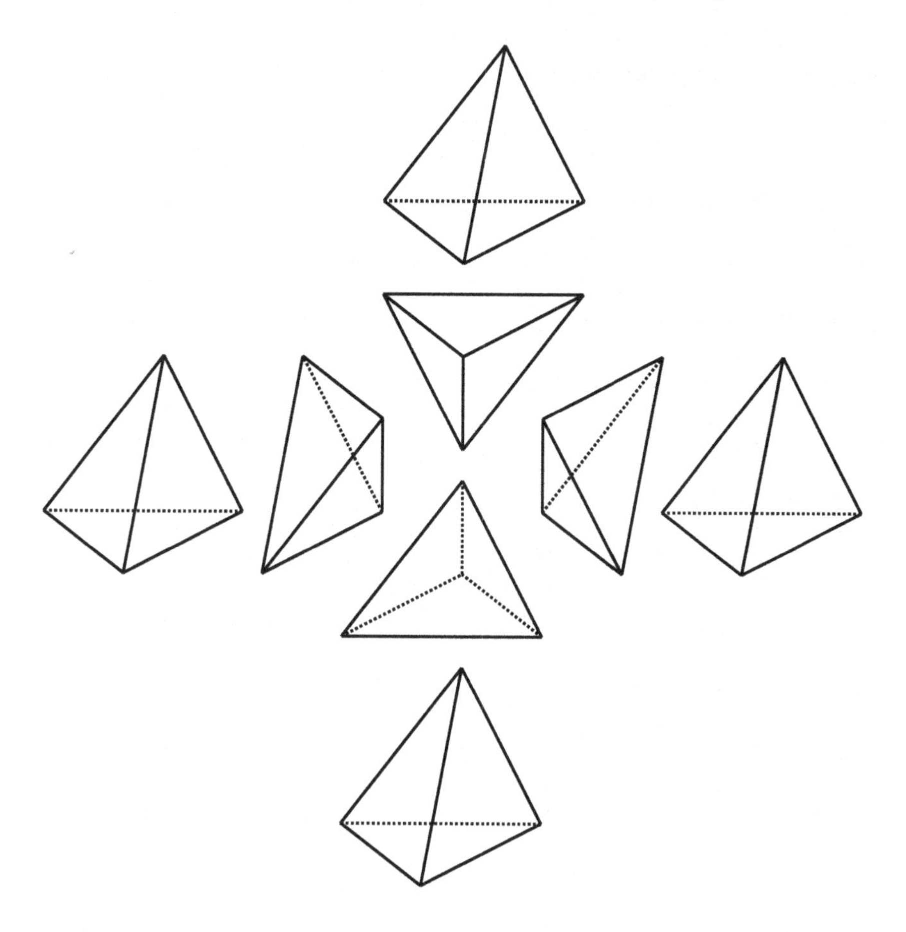

Fig. 4.13. Subsimplices of partitioning of a three - dimensional simplex

multi dimensional case avoiding some crucial computational difficulties characteristic to the earlier used statistical models of multimodal functions. The proposed algorithm inherits many advantages of the one-dimensional prototype. The results of numerical experiments are presented in the next section.

4.4.5 Testing of the P-algorithm with Simplicial Partitioning

The goal of testing was to compare the efficiency of different strategies of the subdivision of simplices, and the initial covering. For such a comparison the test functions representing different properties of difficult optimization problems (proposed for the testing of Lipshitz model based algorithms in [112]) were used. These test functions are presented in Tables 4.1, 4.2. For the criterion of the efficiency we have accepted the number of evaluations of the objective function required to find the known global minimum with the predefined accuracy used in [112].

Table 4.1. Test functions of two variables

Problem number	Test function	Domain	Lipshitz constant
1	$4xy\sin(4\pi y)$	[0,1]× [0,1]	50.2665
2	$\sin(2x+1)+2\sin(3y+2)$	[0,1]× [0,1]	6.3183
3	$-(y-\frac{5x^2}{4\pi^2}+\frac{5x}{\pi}-6)^2-$ $-10(1-\frac{1}{8\pi})\cos x-10$	[-5,10]× [0,15]	112.44
3.1	same as 3	[-5,10]× [-5,15]	142.71
3.2	same as 3	[-5,10]× [0,20]	112.44
3.3	same as 3	[-5,10]× [-5,20]	142.71
4	$-\max(\sqrt{3}x+y,-2y,y-\sqrt{3}x)$	[-1,1]× [-1,1]	2
5	$e^{-x^2}\sin x-\|x\|$	[0,10]× [0,10]	$\sqrt{2}$
6	$-2x^2+1.05x^4-y^2+xy-\frac{1}{6}y^6$	[-2,4]× [-2,4]	1059.59
7	$-100(y-x)^2-(x-1)^2$	[-3,3]× [-1.5,4.5]	12781.7
8	$-(x-2y-7)^2-(2x+y-5)^2$	[-2.5,3.5]× [-1.5,4.5]	86.3134
9	$-[1+(x+y+1)^2(19-14x+\times$ $+3x^2-14y+8xy+3y^2)]$ $\times[30+(2x-3y)^2(18-32x+$ $12x^2+48y-36xy+27y^2)]$	[-2,2]× [-2,2]	2225892
9.1	same as 9	[-4,4]× [-4,4]	96595000
9.2	same as 9	[-3,3]× [-3,3]	11607000
9.3	same as 9	[-1,1]× [-1,1]	277525
10	$-\sin(x+y)-(x-y)^2+$ $+1.5x-2.5y-1$	[-1.5,4]× [-3,3]	17.034
11	$-(x-y)^2-(y-1)^2-$ $-\frac{0.04}{-x^2/4-y^2+1}-\frac{(x-2y+1)^2}{0.2}$	[1,2]× [1,2]	47.426
12	$-0.1\left[12+x^2+\frac{1+x^2}{x^2}+\frac{x^2+y^2+100}{x^4y^4}\right]$	[1,3]× [1,3]	56.862
13	$-\frac{1}{2}\sum_{i=1}^{2}x_i^2+$ $+\prod_{i=1}^{2}\cos(10\ln((i+1)x_i))-1$	[0.01,1]× [0.01,1]	988.82

In Table 4.3 the results of minimization by the proposed algorithm with the accuracy from [112], Tables XI and XIV are given. The test function numbers are presented in the first column; the predefined accuracies used in [112] are presented in the second columns of these tables. The numbers of test function evaluations before the termination are given in the further columns. The results of the version of the algorithm corresponding to the partitioning of the feasible region (like in Fig. 4.14 b for two dimensions, and like in Fig. 4.15 for three dimensions) are given under headings nfe1 and nfe2. The columns under headings nfe3 and nfe4 presents the results of the version of embedding of a feasible region into a regular simplex as shown in Fig. 4.14 a. The results of the version of the algorithm using bisection of the longest edge as shown

Table 4.2. Test functions of three variables

$$f_{30}(X) = -100\left[x_3 - (\frac{x_1+x_2}{2})^2\right]^2 - (1-x_1)^2 - (1-x_2)^2,$$
$$A = [0,1]\times[0,1]\times[0,1],\ L = 244.95;$$

$$f_{31}(X) = \sum_{i=1}^{4} c_i \exp\left(-\sum_{j=1}^{3}\alpha_{ij}(x_j - p_{ij})^2\right),$$
$$\alpha = \begin{pmatrix} 3.0 & 10.0 & 30.0 \\ 0.1 & 10.0 & 35.0 \\ 3.0 & 10.0 & 30.0 \\ 0.1 & 10.0 & 35.0 \end{pmatrix},\ p = \begin{pmatrix} 0.3689 & 0.1170 & 0.2673 \\ 0.4699 & 0.4387 & 0.7470 \\ 0.1091 & 0.8732 & 0.5547 \\ 0.03815 & 0.5743 & 0.8828 \end{pmatrix},$$
$$c^T = (1.0,\ 1.2,\ 3.0,\ 3.2),$$
$$A = [0,1]\times[0,1]\times[0,1],\ L = 42.626;$$

$$f_{32}(X) = -\sum_{i=1}^{10}\left[e^{-x_1 z_i} - x_3 e^{-x_2 z_i} - y_i\right]^2,$$
$$y_i = e^{-z_i} - 5e^{-10z_i},\ z_i = 0.1i,$$
$$A = [0,10]\times[0,10]\times[0,10],\ L = 1163.6;$$

$$f_{33}(X) = -\frac{1}{2}\sum_{i=1}^{2} x_i^2 + \prod_{i=1}^{2}\cos(10\ln((i+1)x_i)) - 1,$$
$$A = [0.01,1]\times[0.01,1]\times[0.01,1],\ L = 971.59;$$

$$f_{34}(X) = \sin x_1 \times \sin(x_1 x_2) \times \sin(x_1 x_2 x_3),$$
$$A = [0,4]\times[0,4]\times[0,4],\ L = 19.39;$$

$$f_{35}(X) = (x_1^2 - 2x_2^2 + x_3^2)\sin x_1 \times \sin x_2 \times \sin x_3,$$
$$A = [-1,1]\times[-1,1]\times[-1,1],\ L = 2.919;$$

$$f_{36}(X) = (x_1 - 1)(x_1 + 2)(x_2 + 1)(x_2 - 2)x_3^2,$$
$$A = [-2,2]\times[-2,2]\times[-2,2],\ L = 130.$$

Table 4.3. Minimization results with original ϵ

a.

function	ϵ	nfe1	nfe2	nfe3	nfe4	nfe5	nfe6
1	0.355	35	29	68	73	643	489
2	0.0446	11	11	14	14	167	137
3	11.9	2	2	2	2	3531	2618
3.1	17.5	5	5	5	5	3953	3245
3.2	13.8	2	2	2	2	3035	2665
3.3	19.6	10	10	12	13	3689	3387
4	0.0141	5	5	5	5	45	41
5	0.1	42	58	26	33	73	53
6	44.9	3	3	3	3	969	629
7	542.0	2	2	2	2	7969	6370
8	3.66	3	3	3	3	301	255
9	62900	1	1	1	1	13953	8759
9.1	$5.47\ 10^6$	1	1	1	1	14559	9531
9.2	$4.93\ 10^5$	1	1	1	1	13281	9002
9.3	$3.93\ 10^3$	1	1	1	1	12295	8917
10	0.691	10	12	18	23	1123	820
11	0.335	6	6	6	6	2677	2222
12	0.804	2	2	2	2	12643	10851
13	6.92	1	1	1	1	15695	10643

b.

function	ϵ	nfe1	nfe2
20	2.12	1	1
21	0.369	166	91
22	101	1	1
23	8.33	1	1
24	0.672	26	26
25	0.0506	1781	1747
26	4.51	2	2

in Fig. 4.11 b and in Fig. 4.12 b are presented in columns nfe1 and nfe3. The results of the version with semi regular partitioning (see Fig. 4.11 a and Fig. 4.12 b) are presented in columns nfe2 and nfe4.

The numbers of function evaluations by the deterministic algorithms guaranteeing prescribed accuracy is much larger as shown in two last columns of the Table 4.3; nfe5 corresponds to the best result of several algorithms considered in [112], and nfe6 represents the results of a Lipshitzian algorithm with simplicial partitioning [305]. These results are presented only for general information; it would be not reasonable to compare them with the results of statistical model based algorithms because of completely different stopping conditions.

The results of the optimization with 100 times higher accuracy are given in Table 4.4. However, for several test functions the global minimum is found with such a predefined accuracy during the stage of constructing the initial

Table 4.4. Minimization results with higher accuracy

a.

function	ϵ	nfe1	nfe2	nfe3	nfe4
1	0.00355	74	119	1035	1149
2	0.000446	84	68	35	36
3	0.119	83	79	199	204
3.1	0.175	29	44	103	70
3.2	0.138	46	42	154	164
3.3	0.196	79	68	41	46
4	0.000141	5	5	5	5
5	0.001	38	58	79	90
6	0.449	6	6	6	6
7	5.420	25	16	16	30
8	0.0366	418	309	56	52
9	629	5	5	5	5
9.1	54700	5	5	5	5
9.2	4930	5	5	5	5
9.3	39.3	2	2	2	2
10	0.00691	365	500	339	532
11	0.00335	81	75	211	204
12	0.00804	41	27	71	30
13	0.0692	230	319	135	76

b.

function	ϵ	nfe1	nfe2
20	0.0212	27	27
21	0.00369	395	680
22	1.01	437	477
23	0.0833	13766	35004
24	0.00672	297	1528
25	0.000506	$> 10^5$	$> 10^5$
26	0.0451	122	363

Table 4.5. Minimization results with high accuracy

function	ϵ	nfe1	nfe2	nfe3	nfe4
6	0.00449	353	326	95	89
7	0.0542	2647	3180	996	919
9	6.29	14	12	17	33
9.1	547	38	24	13	15
9.2	49.3	22	14	30	51
9.3	0.393	2	2	2	2

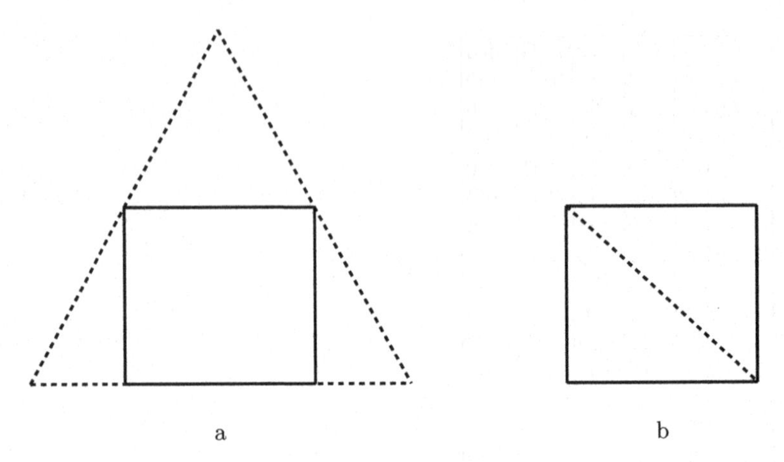

Fig. 4.14. Covering of a square by triangles

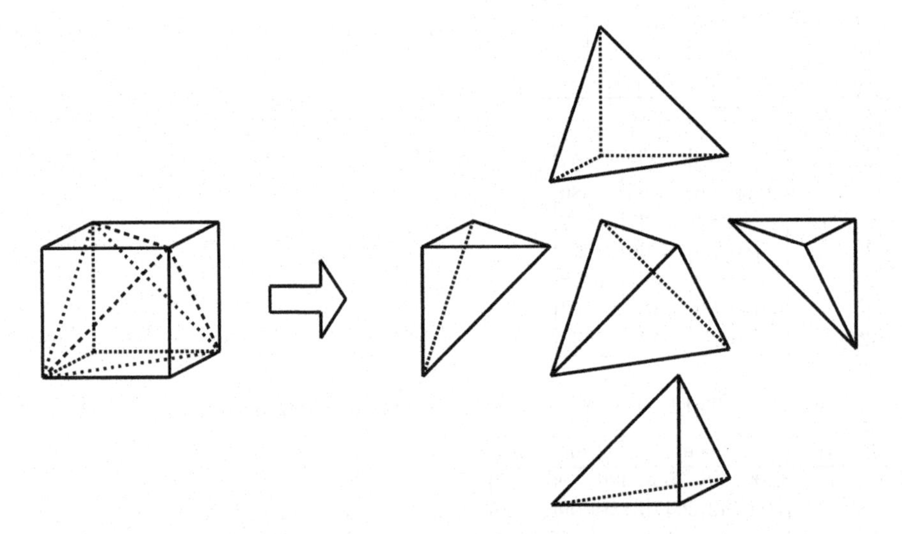

Fig. 4.15. Partitioning of a three - dimensional cube by simplices

cover. These test functions are optimized with 10^4 times higher accuracy than used in [112]. The results are presented in Table 4.5.

The results of Tables 4.3 - 4.5 may be summarized as follows. The initial partitioning of a feasible region, generally speaking, has more advantage than its embedding into a regular simplex. The simply implementable partitioning by 'bisect the longest' is not inferior with respect to the more complicated semi regular partitioning.

Table 4.6. Minimization results for two oscillating functions by the proposed algorithm

Func.	ϵ	N	ϵ	N	ϵ	N
Rastrig.	0.1	668	0.0392	1113	0.01	1140
Shubert	1.37	366	0.1	2664	0.01	2464

The performance of the superior version (partitioning of the feasible region and 'bisect the longest') of the proposed algorithm is illustrated below for two popular test functions. The functions from [112] are frequently used for testing of global optimization algorithms. They represent specific difficulties, especially for algorithms with guaranteed accuracy. Let us consider two more test functions which represent a different type of challenge. The function

$$f(x) = \sum_{i=1}^{2} x_{\cdot i}^2 - cos(18x_{\cdot i}),\ -1 \leq x_{\cdot i} \leq 1,\ i = 1, 2,$$

with global minimum point $x_* = (0, 0)$ and minimum value $f(x_*) = 0$, is known as the Rastrigin test function. It is widely used for testing of global optimization algorithms [248]. The second function is generalization of popular one-dimensional test function by Shubert to two dimensions:

$$f(x) = \sum_{i=1}^{2} \sum_{j=1}^{5} -j \cdot sin((j + 1)x_{\cdot i} + j),\ -10 \leq x_{\cdot i} \leq 10,\ i = 1, 2.$$

where the global minimum of both summands is equal to -12.0312 and it is attained at three points: -6.77458, -0.49139, 5.79179. Both functions represent oscillating objective functions containing neither subregions of the very steep growth nor subregions of flatness. Such functions were considered as the prototypes of the objective functions for the construction of the statistical models of multimodal functions.

The stopping criterion of Lipshitz algorithm guarantees the estimating of the global minimum within tolerance ϵ. The stopping criterion of the proposed algorithm is applied meaning the high probability of estimating of global minimum with accuracy ϵ. To choose the reasonable ϵ the accuracy guaranteed by the quadratic grid 100×100 is estimated using the Lipshitz constant; for the constants (27.7, 96.8), the estimated accuracies are equal to 0.392 and 13.7 correspondingly. For the Rastrigin function the tolerance 0.1, 0.0392, 0.01 was chosen and for the generalized Shubert function the tolerance 1.37, 0.1, and 0.01 was chosen.

The number of function evaluations by the proposed algorithm in the case of different predefined accuracy are presented in Table 4.6. In all cases the value -2.0000 at the point (0.0000, 0.0000) was found for the Rastrigin

Table 4.7. Minimization of two oscillating functions by the algorithm of [47]

Func.	ϵ	N	ϵ	N	ϵ	N
Rastrig., $L = 27.7$	0.1	1015	0.0392	1407	0.01	2057
Shubert, $L = 96.8$	1.37	7776	0.1	11640	0.01	11948

function. The following values were found for the Shubert function: the value -23.9665 at the point (-0.4688, 5.7813) in the case $\epsilon = 1.37$, the value -24.0534 at the point (-0.4883, -6.7676) in the case $\epsilon = 0.1$, the value -24.0612 at the point (-6.7773, 5.7910) in the case $\epsilon = 0.01$. The rare anomaly of the number of function evaluations is observed for the Shubert function. The algorithm stops with a smaller number of function evaluations for ϵ=0.01 than for ϵ=0.1. This anomaly is explained by the influence of ϵ not only to the stopping condition but also to the search strategy. For this particular case the *select* defined by $\epsilon = 0.01$ was more efficient, and important simplices were *cloned* earlier than in the case $\epsilon = 0.1$.

The same test functions were minimized by the Lipshitz algorithm with simplex based covering [47]. The results are presented in Table 4.7. The number of function evaluations of the proposed algorithm is much smaller than the number of function evaluations by algorithm of [47]. On the other hand, the Lipshitz model based algorithm finds the global minimum with the guaranteed accuracy.

4.4.6 Proofs of Theorems

Proof of the Theorem 4.21

The proof of the theorem follows from the fact that any point $z \in A$ is the accumulation point of the sequence x_i. To prove this fact suppose that it is not true, i.e. assume that there exists $z \in A$ and $\rho > 0$ such that $||x_i - z|| > \rho$, $i = 1, 2, \ldots$.

The current point of the evaluation of $f(\cdot)$ by P-algorithm is defined as the maximum point of $P_n(x)$ defined by the following equation

$$P_n(x) = \Phi\left(\frac{\tilde{y}_{on} - m_n(x|(x_i, y_i), i = 1, \ldots, n)}{s_n(x|x_i, y_i, i = 1, \ldots, n)}\right), \tag{4.122}$$

where $\tilde{y}_{on} = y_{on} - \varepsilon$ and $\Phi(z) = \frac{1}{\sqrt{2\pi}} \int_{-\infty}^{z} \exp(-t^2/2)dt$. For the proof of convergence of the algorithm dependence of $P_n(x)$ on different variables of the expression (4.122) is important. To facilitate investigation of these dependencies let us introduce the following notations

$$U(\mu, \sigma) = \Phi\left(\frac{\tilde{y}_{on} - \mu}{\sigma}\right), \tilde{y}_{on} < \mu,$$

$$\phi(x, \tilde{y}_{on}) = \Phi\left(\frac{\tilde{y}_{on} - m_n(x, \cdot)}{s_n(x, \cdot)}\right).$$

It is obvious that $U(\mu, \sigma)$ is the continuous decreasing function of μ and the increasing continuous function of σ.

Let m^+ and m^- denote the upper and lower bounds for values of $f(x)$, $x \in A$ respectively. Since $s_n^2(z|x_i, y_i, i = 1, ..., n) > \tau^- \rho$, then the inequality

$$\phi(z, \tilde{y}_{on}) > U(m^+, \sqrt{\tau^- \rho})$$

holds for any n.

Let z^+ be a cluster point of the sequence x_i. For arbitrary γ, δ' there exists such N that the inequality

$$\sum_{i \in I(z^+, \delta')} w_i^n(z^+, z_j, j = 1, ..., n) > 1 - \gamma,$$

holds for $n > N$. Since the weights are continuous functions of x there exists such $0 < \delta < \delta'$ that for $||x - z^+|| < \delta$ the inequalities

$$\sum_{i \in I(z^+, \delta')} w_i^n(x, z_j, j = 1, ..., n) > 1 - \gamma,$$
$$s_n^2(x|x_i, y_i, i = 1, ..., n) \leq (2\delta' + \gamma G) \cdot \tau^+ = s_-^2,$$

hold where $G = \sup_{v,w \in A} ||v - w||$. Therefore the inequality

$$\phi(x, \tilde{y}_{on}) \leq U(m^-, s_-)$$

is valid for x such that $||x - z^+|| < \delta$.

Since for any finite μ, $U(\mu, 0) = 0$, and $U(\mu, s)$ is the continuous function of s, then the inequality $U(m^-, s) < U(m^+, \sqrt{\tau^- \delta})$ is valid for $s < s_-$ and sufficiently small s_-.

Let for the given τ^+, G the values γ, δ are chosen as follows

$$\gamma = \frac{s_-^2}{2\tau^+ G}, \ \delta = \frac{s_-^2}{4\tau^+}.$$

Then for sufficiently large N and $||x - z^+||$, the inequality

$$\mathbf{P}\{\xi(x) \leq \tilde{y}_{on} | \xi(x_1) = y_1, ..., \xi(x_n)\} = \phi(x, \tilde{y}_{on}) <$$
$$U(m^+, \sqrt{\tau^- \delta}) \leq \mathbf{P}\{\xi(z) \leq \tilde{y}_{on} | \xi(x_1) = y_1, ..., \xi(x_n)\},$$

is valid, showing that the P-algorithm should choose at a current iteration the point z but not a point in the neighborhood of z^+. But this conclusion contradicts to the assumption that z is not an accumulation point of the sequence x_i. The obtained contradiction proves that x_i is everywhere dense in A.

Proof of the Theorem 4.22

By the definition 4.117 there holds the equality

$$P_n(x) = \Phi\left(\frac{\tilde{y}_{on} - m_n(x|(\omega_i, \varphi_i), i = 0, \ldots, d)}{s_n(x|\omega_i, i = 0, \ldots, d)}\right). \tag{4.123}$$

Since the function $\Phi(\cdot)$ is monotonically increasing, then for the constant $s_n(x|\omega_i, i=0,\dots,d)$ the probability $P_n(x)$ increases with increasing numerator of (4.122). Therefore, in the interior of the simplex the maximum of $P_n(x)$ can be achieved only on the line $x = c - t\cdot\nabla_x m_n(x|(\omega_i,\varphi_i), i=0,\dots,d)$. The fact that $m_n(\cdot)$ is a linear function in x with values φ_i at the points ω_i can be written as

$$
\begin{aligned}
m_n(x|\cdot) &= \textstyle\sum_{i=1}^d v_i x_i + u,\\
m_n(\omega_i|\cdot) &= v_i + u = \varphi_i,\ i=1,\dots,d,\\
m_n(\omega_o|\cdot) &= -\tfrac{\sqrt{d+1}-1}{d}\textstyle\sum_{i=1}^d v_i + u = 0.
\end{aligned}
$$

The addition of $m_n(\omega_i|\cdot)$, $i=1,\dots,d$, and subtraction from the obtained sum of $d\cdot m_n(\omega_o|\cdot)$ yields

$$
\begin{aligned}
\textstyle\sum_{i=1}^d v_i + du + (\sqrt{d+1}-1)\sum_{i=1}^d v_i - du &= \textstyle\sum_{i=1}^d \varphi_i,\\
\textstyle\sum_{i=1}^d v_i &= \textstyle\sum_{i=1}^d \varphi_i/\sqrt{d+1},\\
u = \tfrac{\sqrt{d+1}-1}{d\sqrt{d+1}}\textstyle\sum_{i=1}^d \varphi_i &= b\textstyle\sum_{i=1}^d \varphi_i = \varphi_*.
\end{aligned}
\tag{4.124}
$$

The equality

$$v_i = \varphi_i - \varphi_*,$$

easily follows from the equalities above, implying the following expression for the gradient of $m_n(x|\cdot)$:

$$\nabla m_n(x|\cdot) = \varphi - I\varphi_*. \tag{4.125}$$

Since the maximization of

$$\frac{\tilde{y}_{on} - m_n(x|(\omega_i,\varphi_i), i=0,\dots,d)}{s_n(x|\omega_i, i=0,\dots,d)} \tag{4.126}$$

over the simplex is reduced to the maximization over the ray

$$x = c - t\cdot\nabla_x m_n(x|(\omega_i,\varphi_i), i=0,\dots,d) = c - t(\varphi - I\varphi_*),\ t\ge 0, \tag{4.127}$$

let us substitute x in (4.126) with (4.127). After the following substitutions

$$
\begin{aligned}
m_n(x|\cdot) &= \textstyle\sum_{i=1}^d v_i x_i + u =\\
&\textstyle\sum_{i=1}^d(\varphi_i - \varphi_*)(b - t(\varphi_i - \varphi_*)) + \varphi_* =\\
&(b - b^2 d)\textstyle\sum_{i=1}^d \varphi_i + \varphi_* - t||\varphi - I\varphi_*||^2 =\\
&\tfrac{1}{d+1}\textstyle\sum_{i=1}^d \varphi_i - t||\varphi - I\varphi_*||^2,
\end{aligned}
\tag{4.128}
$$

$$s_n^2(x|\cdot) = \sigma_o^2(D^2 - ||c - x||^2) = \sigma_o^2(\tfrac{d}{d+1} - t^2||\varphi - I\varphi_*||^2); \tag{4.129}$$

we have to maximize the expression

$$\frac{\tilde{y}_{on} - \frac{1}{d+1}\sum_{i=1}^{d}\varphi_i + t||\varphi - I\varphi_*||^2}{\sqrt{\frac{d}{d+1} - t^2||\varphi - I\varphi_*||^2}}. \quad (4.130)$$

It is easy to verify that the maximum point of the function

$$\frac{\alpha - \beta t}{\sqrt{\gamma - \kappa t^2}}$$

is $t_* = \frac{\beta\gamma}{\alpha\kappa}$, therefore maximum point of (4.130) in t is

$$\frac{-||\varphi - I\varphi_*||\frac{d}{d+1}}{(\tilde{y}_{on} - \frac{1}{d+1}\sum_{i=1}^{d}\varphi_i)||\varphi - I\varphi_*||},$$

implying

$$x_{max} = c - \frac{1}{-\frac{d+1}{d}\tilde{y}_{on} + \frac{1}{d}\sum\varphi_i} \cdot (\varphi - \varphi_* \cdot I).$$

Proof of the Theorem 4.23

It will be proved that the algorithm generates a sequence of observation points x_i that are everywhere dense in A_0; (4.121) follows immediately from the inequality $F(x) > \min_{x \in A} f(x)$ for $x \in A_0 \backslash A$, and the continuity of $f(\cdot)$. Assume that the theorem is false: there exists a point $x_- \in A_0$ which is not a limit point of x_i, i.e. there exists an ϵ neighborhood of x_- without points x_i. Let us suppose that x_- belongs to the simplices of initial cover S^l, $l \in L$. By means of bread-first search partitioning of S^l after a finite number of steps the descendant simplices S_-^l with the average edge length no longer than h_- will be obtained such that x_- is an inner point of $S_- = \cup S_-^l$, $S_- \subset Sph(x_-, \epsilon)$. Since $P_n(x)$, $x \in S_-^l$, is a monotonic decreasing function of the objective function values at the vertices of S_-^l [288], the inequality

$$\max_{x \in S_-} P_n(x) \geq \Phi\left((\min_{x \in A} f(x) - \max_{x \in A_0} f_0(x) - \epsilon)\sqrt{d+1}/\sigma_0 h_- \sqrt{2d}\right)$$

follows from Remark 3.

Since the set A_0 is bounded, the sequence x_i has at least one limit point, e.g. x_+. The sequence of simplices, whose average edge lengths $h_+^j \to 0$, should be generated by the algorithm in order to generate the points of x_i in any neighborhood of x_+. From the mentioned above monotonicity of $P_n(x)$, and Remark 3, it follows that for the simplex with edge length d_+ the maximal probability is not larger than $\Phi\left(-\epsilon\sqrt{d+1}/\sigma_0 h_+ \sqrt{2d}\right)$.

The assumption that x_- is not a limit point of x_i implies that the simplices in S_- are never selected by the minimization algorithm performing the best first search partitioning. Therefore the simplices with average edge lengths

$$h_+ < h_- \epsilon / (\max_{x \in A_0} f_0(x) - \min_{x \in A} f(x) + \epsilon),$$

whose maximal probability is smaller than maximal probability of simplices in S_-, also can not be selected. Therefore, x_+ can not be a limit point of the sequence x_i. The obtained contradiction proves the theorem.

Proof of the Theorem 4.24

Since the select and partition algorithm generates an everywhere dense sequence of points, then, after a finite number of iterations n_* the best point found will belong to the neighborhood of x_*. Since the objective function is twice continuously differentiable at x_*, then

$$f(x_{o\ i}) - f(x_*) = O(||x_{o\ i} - x_*||^2), \tag{4.131}$$

where $x_{o\ i}$ is the best point found after $i > n_*$ iterations.

The sequence of local descent iterations generates the points $x_{l(t)}$, $l(t) = t \cdot (N+1)$, $t > n_*/(N+1)$, $t = 1, 2, ...$, which converge to x_*, and the average order of convergence is p with respect to t:

$$\limsup ||x_{o\,l(t)} - x_*||^{1/(p^t)} = 1,\ p > 1. \tag{4.132}$$

The equalities (4.131), (4.132) yield

$$\limsup ||x_{o\,i} - x_*||^{1/(r^t i)} = 1, \tag{4.133}$$

$$\limsup\ (f(x_{o\ i}) - f(x_*))^{\frac{1}{2r^i}} = 1,$$

where $r = p^{1/(N+1)} > 1$.

References

1. E. Aarts and J. Korst, *Selected topics in simulated annealing*, Essays and surveys in metaheuristics, Kluwer Acad. Publ., Dordrecht, 2002, pp. 1–37.
2. I. Akrotirianakis and C. Floudas, *Computational experience with a new class of convex underestimators: box constrained nlp problems*, J. Global Optim. **29** (2004), 249–264.
3. H. Al-Mharmah and J. M. Calvin, *Optimal random non-adaptive algorithm for global optimization of brownian motion*, J. Global Optim. **8** (1996), 81–90.
4. M. Ali, C. Storey, and A. Törn, *Application of stochastic global optimization algorithms to practical problems*, J. Optimiz. Theory Appl. **95** (1997), 545–563.
5. F. Archetti and B. Betrò, *A probabilistic algorithm for global optimization*, Calcolo **16** (1979), 335–343.
6. B. C. Arnold, N. Balakrishnan, and H. N. Nagaraja, *Records*, John Wiley & Sons, New York, 1998.
7. N. Baba, *Convergence of a random optimization method for constrained optimization problems*, J. Optim. Theory Appl. **33** (1981), no. 4, 451–461.
8. T. Bäck, *Evolutionary algorithms in theory and paxis*, Oxford University Press, 1996.
9. M. Bakr and et all, *An introduction to the space mapping technique*, Optimiz. Eng. **2** (2002), 369–384.
10. N. Balakrishnan and C. R. Rao (eds.), *Order statistics: applications*, Handbook of Statistics, vol. 17, North-Holland Publishing Co., Amsterdam, 1998.
11. N. Balakrishnan and C. R. Rao (eds.), *Order statistics: Theory and methods*, Handbook of Statistics, vol. 16, North-Holland Publishing Co., Amsterdam, 1998.
12. R. Barr and et all, *Designing and reporting on computational experiments with heuristic methods*, J. Heuristics **1** (1995), 9–32.
13. J. Beirlant, Y. Goegebeur, J. Teugels, and J. Segers, *Statistics of extremes*, John Wiley & Sons, Chichester, 2004.
14. C. Bélisle, *Convergence theorems for a class of simulated annealing algorithms on* $\mathbf{R}^d$, J. Appl. Probab. **29** (1992), no. 4, 885–895.
15. ———, *Slow convergence of the Gibbs sampler*, Canad. J. Statist. **26** (1998), no. 4, 629–641.
16. B. Betrò and F. Schoen, *Sequential stopping rules for the multistart algorithm in global optimisation*, Math. Programming **38** (1987), no. 3, 271–286.

17. ———, *Optimal and sub-optimal stopping rules for the multistart algorithm in global optimization*, Math. Programming **57** (1992), no. 3, Ser. A, 445–458.
18. L. Biegler and I. Grossman, *Retrospective on optimization*, Comp. Chem. Eng. **28** (2004), 1169–1192.
19. N. H. Bingham, C. M. Goldie, and J. L. Teugels, *Regular variation*, Cambridge University Press, Cambridge, 1989.
20. C. G. E. Boender and A. H. G. Rinnooy Kan, *Bayesian stopping rules for multistart global optimization methods*, Math. Programming **37** (1987), no. 1, 59–80.
21. ———, *On when to stop sampling for the maximum*, J. Global Optim. **1** (1991), no. 4, 331–340.
22. C. G. E. Boender, A. H. G. Rinnooy Kan, G. T. Timmer, and L. Stougie, *A stochastic method for global optimization*, Math. Programming **22** (1982), no. 2, 125–140.
23. C. G. E. Boender and H. E. Romeijn, *Stochastic methods*, Handbook of global optimization, Kluwer Acad. Publ., Dordrecht, 1995, pp. 829–869.
24. A. Booker and et all, *A rigorous frameworkfor optimization of expensive functions by surrogates*, Struct. Optimiz. **17** (1999), 1–13.
25. I. Borg and P. Groenen, *Modern multidimensional scaling*, Springer, NY, 1997.
26. R. Brunelli, *Teaching neural nets through stochastic minimization*, Neural Nets **7** (1994), 1405–1412.
27. M. Buhmann, *Radial basis functions*, Cambridge University Press, 2003.
28. D. Bulger, W. P. Baritompa, and G. R. Wood, *Implementing pure adaptive search with Grover's quantum algorithm*, J. Optim. Theory Appl. **116** (2003), no. 3, 517–529.
29. D. W. Bulger, D. Alexander, W. P. Baritompa, G. R. Wood, and Z. B. Zabinsky, *Expected hitting times for backtracking adaptive search*, Optimization **53** (2004), no. 2, 189–202.
30. D. W. Bulger and G. R. Wood, *Hesitant adaptive search for global optimisation*, Math. Programming **81** (1998), no. 1, Ser. A, 89–102.
31. J. Bunge and C.M. Goldie, *Record sequences and their applications*, Stochastic processes: theory and methods, Handbook of Statist., vol. 19, North-Holland, Amsterdam, 2001, pp. 277–308.
32. F. Caeiro and M. I. Gomes, *A class of asymptotically unbiased semi-parametric estimators of the tail index*, Test **11** (2002), no. 2, 345–364.
33. J. M. Calvin, *Average performance of passive algorithms for global optimization*, J. Mat. Anal. Appl. **191** (1995), 608–617.
34. ———, *Average performance of a class of adaptive algorithms for global optimization*, Ann. Appl. Probab. **7** (1997), 711–730.
35. ———, *Convergence rate of the P-algorithm for optimization of continuous functions*, In Approximation and Complexity in Numerical Optimization: Continuous and Discrete Problems, P. Pardalos (Ed.), Kluwer Academic Publishers, Boston (1999), 116–129.
36. ———, *Lower bounds on complexity of optimization of continuous functions*, J. Complexity **20** (2004), 773–795.
37. J. M. Calvin and P. W. Glynn, *Complexity of non-adaptive optimization algorithms for a class of diffusions*, Comm. Stat. Stoch. Models **12** (1996), 343–365.
38. ———, *Average case behavior of random search for the maximum*, J. Appl. Probab. **34** (1997), 631–642.

39. J. M. Calvin and A. Žilinskas, *On convergence of the P-algorithm for one-dimensional global optimization of smooth functions*, J. Optimiz. Theory Appl. **102** (1999), no. 3, 479–495.
40. ______, *On the choice of statistical model for one-dimensional P-algorithm*, Control and Cybernetics **29** (2000), no. 2, 555–565.
41. ______, *A one-dimensional P-algorithm with convergence rate $O(n^{-3+\delta})$ for smooth functions*, J. Optimiz. Theory Appl. **106** (2000), 297–307.
42. ______, *On convergence of a P-algorithm based on a statistical model of continuously differentiable functions*, J. Global Optim. **19** (2001), 229–245.
43. ______, *One-dimensional global optimization for observations with noise*, Comp. Math. Appl. **50** (2005), 157–169.
44. J.M. Calvin, *Polynomial acceleration of Monte–Carlo global search*, Proceedings of the 1999 Winter Simulation Conference (eds Farrington P.A. et al), 1999, pp. 673–677.
45. W. Cheney and D. Kincaid, *Numerical mathematics and computing*, Thomson Learning, 2004.
46. S. Cheng and L. Peng, *Confidence intervals for the tail index*, Bernoulli **7** (2001), no. 5, 751–760.
47. J. Clausen and A. Žilinskas, *Global optimization by means of Branch and Bound with simplex based covering*, Comp. Math. Appl. **44** (2002), 943–955.
48. S. Conte and DeBoor C., *Elementary numerical analysis*, McGraw-Hill, NY, 1980.
49. A. Converse, *The use of uncertainty in a simultaneos search*, Operations Research **10** (1967), 1088–1095.
50. J. H. Conway and N. J. A. Sloane, *Sphere packings, lattices and groups*, third ed., Springer-Verlag, New York, 1999.
51. P. Cooke, *Statistical inference for bounds of random variables*, Biometrika **66** (1979), no. 2, 367–374.
52. ______, *Optimal linear estimation of bounds of random variables*, Biometrika **67** (1980), no. 1, 257–258.
53. T.M. Cover and J.A. Thomas, *Elements of information theory*, John Wiley & Sons Inc., New York, 1991.
54. T. Cox and M. Cox, *Multidimensional scaling*, Chapman and Hall/CRC, Boca Raton, 2001.
55. S. Csörgő and D.M. Mason, *Simple estimators of the endpoint of a distribution*, Extreme value theory (Oberwolfach, 1987), Lecture Notes in Statist., vol. 51, Springer, New York, 1989, pp. 132–147.
56. Gamerman D. and Lopes H.F., *Markov chain Monte Carlo: Stochastic simulation for bayesian inference, second edition*, Chapman & Hall / CRC, Boca Raton, 2005.
57. H. A. David, *Order statistics*, John Wiley & Sons Inc., New York, 1970.
58. H. A. David and H. N. Nagaraja, *Order statistics*, third ed., John Wiley & Sons, Hoboken, NJ, 2003.
59. L. de Haan, *On regular variation and its application to the weak convergence of sample extremes*, Mathematisch Centrum, Amsterdam, 1970.
60. L. de Haan and S. Resnick, *Second-order regular variation and rates of convergence in extreme-value theory*, Ann. Probab. **24** (1996), no. 1, 97–124.
61. B. de Sousa and G. Michailidis, *A diagnostic plot for estimating the tail index of a distribution*, J. Comput. Graph. Statist. **13** (2004), no. 4, 974–995.

62. P. Deheuvels, *Strong bounds for multidimensional spacings*, Z. Wahrsch. Verw. Gebiete **64** (1983), no. 4, 411–424.
63. L. P. Devroye, *Progressive global random search of continuous functions*, Math. Programming **15** (1978), no. 3, 330–342.
64. L.C.W. Dixon and G.P. Szegö, *Towards global optimization 2*, North Holland, Amsterdam, 1978.
65. H. Drees and E. Kaufmann, *Selecting the optimal sample fraction in univariate extreme value estimation*, Stochastic Process. Appl. **75** (1998), no. 2, 149–172.
66. M. Drmota and R.F. Tichy, *Sequences, discrepancies and applications*, Lecture Notes in Mathematics, vol. 1651, Springer-Verlag, Berlin, 1997.
67. V. Drobot, *Uniform partitions of an interval*, Trans. Amer. Math. Soc. **268** (1981), no. 1, 151–160.
68. P. Embrechts, C. Klüppelberg, and T. Mikosch, *Modelling extremal events for insurance and finance*, Springer-Verlag, Berlin, 2003.
69. S. Ermakov, A. Zhigljavsky, and M. Kondratovich, *Reduction of a problem of random estimation of an extremum of a function*, Dokl. Akad. Nauk SSSR **302** (1988), no. 4, 796–798.
70. M. Falk, *Rates of uniform convergence of extreme order statistics*, Ann. Inst. Statist. Math. **38** (1986), no. 2, 245–262.
71. D. Famularo, P. Pugliese, and Y. Sergeyev, *Test problems for lipschitz univariate global optimization with multiextremal constraints*, In Stochastic and Global Optimization, G.Dzemyda, V.Saltenis and A.Žilinskas (Eds.), Kluwer Academic Publishers, Boston (2002), 93–110.
72. L. Farhane, *Espacements multivariés généralisés et recouvrements aléatoires*, Ann. Sci. Univ. Clermont-Ferrand II Probab. Appl. (1991), no. 9, 15–31.
73. V. Fedorov and P. Hackl, *Model-oriented design of experiments*, Lecture Notes in Statistics, vol. 125, Springer-Verlag, New York, 1997.
74. T. Fine, *Optimal search for location of the maximum of a unimodal function*, IEEE Trans. Inform. Theory (1966), no. 2, 103–111.
75. ———, *Extrapolation when very little is known about the source*, Information and Control **16** (1970), 331–359.
76. ———, *Theories of probabilities*, Academy Press, 1973.
77. P. Fishburn, *Utility theory for decision making*, Wiley, 1970.
78. C. Floudas, *Deterministic global optimization: Theory, algorithms and applications*, Kluwer: Dodrecht, 2000.
79. ———, *Research challenges, opportunities and synergism in systems engineering and computational biology*, AICHE Journal **51** (2005), 1872–1884.
80. C. Floudas and P. Pardalos (Eds.), *Optimization in computational chemistry and molecular biology: Local and global approaches*, Kluwer: Dodrecht, 2000.
81. C. Floudas and R. Agraval (Eds.), *Sixt international conference on foundations of computer-aided process design: Discovery through product and process design*, Omnipress, 2004.
82. C. Floudas and P. Pardalos, *A collection of test problems for constrained global optimization algorithms, lncs, vol.455*, Springer, Berlin, 1990.
83. ———, *Handbook of test problems in local and global optimization*, KAP, 1999.
84. ———, *Frontiers in global optimization*, Kluwer: Dodrecht, 2003.
85. E. Fraga and A. Žilinskas, *Evaluation of hybrid optimization methods for the optimal design of heat integrated distillation sequences*, Advances Eng. Softw. **34** (2003), 73–86.

86. J. Galambos, *The asymptotic theory of extreme order statistics*, second ed., Robert E. Krieger Publishing Co. Inc., Melbourne, FL, 1987.
87. E. Galperin, *Global solutions in optimal control and games*, NP Research, Montreal, 1991.
88. M. Gaviano, D. Kvasov, D. Lera, and Y. Sergeev, *Algorithm 829: Sotware for generation of classes of test functions with known local and global minima for global optimization*, ACM Trans. Math. Softw. **29** (2003), 469–480.
89. S.B. Gelfand and S.K. Mitter, *Metropolis-type annealing algorithms for global optimization in* $\mathbf{R}^d$, SIAM J. Control Optim. **31** (1993), no. 1, 111–131.
90. S. Geman and C.-R. Hwang, *Diffusions for global optimization*, SIAM J. Control Optim. **24** (1986), no. 5, 1031–1043.
91. V. Gergel and Y. Sergeev, *Sequential and parallel algorithms for global minimizing functions with Lipshitz derivatives*, Comp. Math. Appl. **37** (1999), 163–179.
92. C.J. Geyer, *Estimation and optimization of functions*, Markov chain Monte Carlo in practice (W. R. Gilks, S. Richardson, and D. J. Spiegelhalter, eds.), Chapman & Hall, London, 1996, pp. 241–258.
93. W. R. Gilks, S. Richardson, and D. J. Spiegelhalter (eds.), *Markov chain Monte Carlo in practice*, Chapman & Hall, London, 1996.
94. M. Glick, A. Rayan, and A. Goldblum, *A stochastic algorithm for global optimization algorithms and for best populations: a test case of side chains in proteins*, Proc. Nat. Acad. Sci. **99** (2002), 702–708.
95. F.G. Glover and M. Laguna, *Tabu search*, Kluwer Acad. Publ., Dordrecht, 1997.
96. C.M. Goldie and L. C. G. Rogers, *The k-record processes are i.i.d*, Z. Wahrsch. Verw. Gebiete **67** (1984), no. 2, 197–211.
97. J. Goldman, *An approach to estimation and extrapolation with possible applications in an incompletly specified environment*, Information and Control **30** (1976), 203–233.
98. E. Gourdin, B. Jaumard, and E. Hansen, *Global optimization of multivariate Lipshitz functions*, Les Cahiers du GERAD **May** (1994).
99. P. Green, *A primer on Markov chain Monte Carlo*, Complex stochastic systems, Monogr. Statist. Appl. Probab., vol. 87, Chapman & Hall/CRC, Boca Raton, FL, 2001, pp. 1–62.
100. ———, *Trans-dimensional Markov chain Monte Carlo*, Highly structured stochastic systems, Oxford Statist. Sci. Ser., vol. 27, Oxford Univ. Press, Oxford, 2003, pp. 179–206.
101. A. Greven, G. Keller, and G. Warnecke (eds.), *Entropy*, Princeton University Press, Princeton, NJ, 2003.
102. A. Groch, L. Vidigal, and S. Director, *A new global optimization method for electronic circuit design*, IEEE Trans. Circ. Sys. **32** (1985), 160–170.
103. P. Groenen, *Majorization approach to multidimensional scaling: Some problems and extensions*, DWO Press, Leiden, 1993.
104. I. Grossman and L. Biegler, *Future perspective on optimization*, Comp. Chem. Eng. **28** (2004), 1193–1218.
105. E. J. Gumbel, *Statistics of extremes*, Columbia University Press, New York, 1958.
106. H.-M. Gutman, *A radial basis function method for global optimization*, J. Global Optim. **19** (2001), 201–227.
107. H. Haario and E. Saksman, *Simulated annealing process in general state space*, Adv. in Appl. Probab. **23** (1991), no. 4, 866–893.

108. B. Hajek, *Cooling schedules for optimal annealing*, Math. Oper. Res. **13** (1988), no. 2, 311–329.
109. P. Hall, *On estimating the endpoint of a distribution*, Ann. Statist. **10** (1982), no. 2, 556–568.
110. Savani V. Hamilton, E. and A. Zhigljavsky, *Estimating the minimal value of a function in global random search: Comparison of estimation procedures*, Global Optimization, Springer, New York, 2007, p. in press.
111. P. Hansen and B. Jaumard, *On Timonov's algorithm for global optimization of univariate Lipshitz functions*, J. Global Optim. **1** (1991), 37–46.
112. ______, *Lipschitz optimization*, Handbook of Global Optimization (R.Horst and P.Pardalos, eds.), KAP, 1995, pp. 404–493.
113. G. H. Hardy, *Orders of infinity. The infinitärcalcül of Paul du Bois-Reymond*, Hafner Publishing Co., New York, 1971, Reprint of the 1910 edition, Cambridge Tracts in Mathematics and Mathematical Physics, No. 12.
114. W. E. Hart, *Sequential stopping rules for random optimization methods with applications to multistart local search*, SIAM J. Optim. **9** (1999), no. 1, 270–290 (electronic).
115. R.L. Haupt and S.E. Haupt, *Practical genetic algorithms*, second ed., Wiley & Sons, NJ, 2004.
116. P. Hellekalek and G. Larcher (eds.), *Random and quasi-random point sets*, Lecture Notes in Statistics, vol. 138, Springer-Verlag, New York, 1998.
117. M. Heyman, *Optimal simultaneos search for the maximum by the principle of statistical information*, Operations Research (1968), 1194–1205.
118. B. M. Hill, *A simple general approach to inference about the tail of a distribution*, Ann. Statist. **3** (1975), no. 5, 1163–1174.
119. J. Holland, *Adaptation in natural and artificial systems*, University of Michigan Press: Ann Arbor, 1975.
120. J. Hooker, *Needed: an empirical science of algorithms*, Operations Research **42** (1994), 201–212.
121. ______, *Testing heuristics: We have it all wrong*, J. Heuristics **1** (1995), 33–42.
122. R. Horst and P. Pardalos (Eds.), *Handbook of global optimization*, Kluwer: Dodrecht, 1995.
123. R. Horst, P. Pardalos, and N. Thoai, *Introduction to global optimization*, KAP, 1995.
124. R. Horst and H. Tuy, *Global optimization: Deterministic approaches, third edition*, Springer: Berlin, 1996.
125. Z. Ignatov, *Point processes generated by order statistics and their applications*, Point processes and queuing problems, Colloq. Math. Soc. János Bolyai, vol. 24, North-Holland, Amsterdam, 1981, pp. 109–116.
126. D. Jamrog, G. Phillips, R. Tapia, and Y. Zhang, *A global optimization method for molecular replacement problem in X-ray crystallography*, Math. Program. **Ser.B 103** (2005), 399–426.
127. S. Janson, *Maximal spacings in several dimensions*, Ann. Probab. **15** (1987), no. 1, 274–280.
128. P. Jizba and T. Arimitsu, *The world according to Rényi: thermodynamics of multifractal systems*, Ann. Physics **312** (2004), no. 1, 17–59.
129. N. L. Johnson, S. Kotz, and N. Balakrishnan, *Discrete multivariate distributions*, John Wiley & Sons Inc., New York, 1997.
130. D. Jones, *A taxonomy of global optimization methods based on response surfaces*, J. Global Optim. **21** (2001), 345–383.

131. D. Jones, C. Perttunen, and B. Stuckman, *Lipshitzian optimization without the Lipshitz constant*, J. Optimiz. Theory Appl. **79** (1993), 157–181.
132. D. Jones, M. Schonlau, and W. Welch, *Efficient global optimization of expensive black-box functions*, J. Global Optim. **13** (1998), 455–492.
133. J. Jurečková, *Statistical tests on tail index of a probability distribution*, Metron **61** (2003), no. 2, 151–175.
134. J. Jurečková and J. Picek, *A class of tests on the tail index*, Extremes **4** (2001), no. 2, 165–183 (2002).
135. S. Karlin, *A first course in stochastic processes*, Academic Press, 1969.
136. A. Kearsley, R. Tapia, and M. Trosset, *The solution of the metric stress and sstress problems in multidimensional scaling using Newton's method*, Comput. Stat. **13** (1998), 369–396.
137. J. Kiefer, *On large deviations of the empiric D. F. of vector chance variables and a law of the iterated logarithm*, Pacific J. Math. **11** (1961), 649–660.
138. M. Kondratovich and A. Zhigljavsky, *Comparison of independent and stratified sampling schemes in problems of global optimization*, Monte Carlo and Quasi-Monte Carlo methods 1996 (Salzburg), Lecture Notes in Statist., vol. 127, Springer, New York, 1998, pp. 292–299.
139. S. Kotz and S. Nadarajah, *Extreme value distributions*, Imperial College Press, London, 2000.
140. H. Kushner, *A versatile stochastic model of a function of unknown and time-varying form*, J. Math. Anal. Appl. **5** (1962), 150–167.
141. ———, *A new method of locating the maximum point of an arbitrary multipeak curve in the presence of noise*, J. Basic Engineering **86** (1964), 97–106.
142. H.J. Kushner and G. Yin, *Stochastic approximation and recursive algorithms and applications*, second ed., vol. 35, Springer-Verlag, New York, 2003.
143. G. Lindgren, *Local maxima of Gaussian fields*, Ark. Math. **10** (1972), 195–218.
144. M. Locatelli, *Baeysian algorithms for one-dimensional global optimization*, J. Global Optim. **10** (1997), 57–76.
145. M. Locatelli, *Convergence of a simulated annealing algorithm for continuous global optimization*, J. Global Optim. **18** (2000), no. 3, 219–234.
146. ———, *Simulated annealing algorithms for continuous global optimization*, Handbook of global optimization, Vol. 2, Kluwer Acad. Publ., Dordrecht, 2002, pp. 179–229.
147. M. Locatelli and F. Schoen, *Random Linkage: a family of acceptance/rejection algorithms for global optimisation*, Math. Program. **85** (1999), no. 2, Ser. A, 379–396.
148. M. Locatelli and F. Schoen, *Efficient algorithms for large scale global optimization: Lennard-Jones clusters*, Comput. Optimiz. Appl. **26** (2003), 173–190.
149. J.-C. Lu and L. Peng, *Likelihood based confidence intervals for the tail index*, Extremes **5** (2002), no. 4, 337–352 (2003).
150. D. Luenberger, *Introduction to linear and nonlinear programming*, Addison-Wesley, 1973.
151. K.F. Man, K.S. Tang, and S. Kwong, *Genetic algorithms: Concepts and designs*, Advanced Textbooks in Control and Signal Processing, Springer-Verlag, London, 1999.
152. M.C. Markót and T. Csendes, *A new verified optimization technique for the "packing circles in a unit square" problems*, SIAM J. Optim. **16** (2005), no. 1, 193–219 (electronic).

153. R. Mathar, *A hybrid global optimization algorithm for multidimensional scaling*, Classification and Knowledge Organization (R. Klar and O. Opitz, eds.), Springer: Berlin, 1996, pp. 63–71.
154. R. Mathar and A. Žilinskas, *On global optimization in two-dimensional scaling*, Acta Appl. Math. **33** (1993), 109–118.
155. ______, *A class of test functions for global optimization*, J. Global Optim. **5** (1994), 195–199.
156. C. McGeoch, *Towards an experimental method for algorithm simulation*, INFORMS J. Comput. **8** (1996), no. 1, 1–15.
157. ______, *Experimental analysis of algorithms*, Handbook of Global Optimization, Volume 2 (P. Pardalos and E. Romeijn, eds.), Kluwer: Dodrecht, 2002, pp. 489–514.
158. K. L. Mengersen and R. L. Tweedie, *Rates of convergence of the Hastings and Metropolis algorithms*, Ann. Statist. **24** (1996), no. 1, 101–121.
159. Rosenbluth M. Teller A. Metroplis N., Rosenbluth A. and Teller E., *Equations of state calculations by fast computing machines*, J. Chem. Phys. **21** (1953), 1087–1091.
160. Z. Michalewich, K. Deb, M. Schmidt, and T. Stidsen, *Test-case generator for nonlinear continuous parameter optimization techniques*, IEEE Trans. Evolut. Comput. **4** (2000), 197–215.
161. Z. Michalewicz, *Genetic algorithms + data structures = evolution programs*, second ed., Springer-Verlag, Berlin, 1994.
162. Romeo F. Mitra, D. and A. Sangiovanni-Vincentelli, *Convergence and finite-time behavior of simulated annealing*, Adv. in Appl. Probab. **18** (1986), no. 3, 747–771.
163. J. Mockus, *On Bayesian methods of search for extremum*, Avtomatika i Vychislitelnaja Technika (1972), no. 3, 53–62, in Russian.
164. ______, *Bayesian approach to global optimization*, KAP, 1988.
165. J. Mockus and et all, *Bayesian heuristic approach to discrete and global optimization*, KAP, 1996.
166. A. Monin and A.Yaglom, *Statistical hydrodinamics, volumes 1 and 2*, Nauka, 1965 and 1967, in Russian.
167. V.V. Nekrutkin and A.S. Tikhomirov, *Some properties of markovian global random search*, AMS transl. from Vestn. Leningr. Univ., ser. I (1989), no. 3, 23–26.
168. ______, *Speed of convergence as a function of given accuracy for random search methods*, Acta Applicandae Mathematicae **33** (1993), 89–108.
169. A. Neumaier, O. Shcherbina, W. Huyer, and T. Vinko, *A comparison of complete global optimization solvers*, Math. Program., Ser.B **103** (2005), 335–356.
170. P. Neuman, *An asymptotically optimal procedure for searching a zero or an extremum of a function*, Proceedings of 2nd. Prague Symp. Asymp. Statist. (1981), 291–302.
171. V. B. Nevzorov, *Records: mathematical theory*, American Mathematical Society, Providence, RI, 2001.
172. J. Neymark and R. Strongin, *Information approach to search for minimum of a function*, Izv. AN SSSR, Eng. Cybern. (1966), no. 1, 17–26, in Russian.
173. H. Niederreiter, *Low-discrepancy and low-dispersion sequences*, J. Number Theory **30** (1988), no. 1, 51–70.
174. ______, *Random number generation and quasi-Monte Carlo methods*, SIAM, Philadelphia, PA, 1992.

175. E. Novak, *Deterministic and stochastic error bounds in numerical analysis, lecture notes in mathematics*, vol. 1349, Springer, 1988.
176. E. Novak and K. Ritter, *Some complexity results for zero finding for univariate functions*, J. Complexity (1993), 15–40.
177. B. Orsier and C. Pellegrini, *Using global line searches for finding global minima of MLP error functions*, International Conference on Neural Networks and their Applications, Marseilles, 1997, pp. 229–235.
178. G. Ostrovski, L. Achenie, and M. Sinka, *A reduced dimension branch-and-bound algorithms for molecular design*, Comp. Chem. Eng. **27** (2003), 551–567.
179. G. Ostrovski and et all, *Flexibility analysis of chemical processes: selected global optimization problems*, Optimiz. Eng. **3** (2002), 31–52.
180. P. Pardalos and E. Romeijn, *Handbook of global optimization, volume 2*, Kluwer: Dodrecht, 2002.
181. P. Pardalos, D. Shalloway, and G. Xue (eds.), *Global minimization of nonconvex energy functions: Molecular conformation and protein folding*, AMS, 1996.
182. N. R. Patel, R. L. Smith, and Z. B. Zabinsky, *Pure adaptive search in Monte Carlo optimization*, Math. Programming **43** (1989), no. 3, (Ser. A), 317–328.
183. V. Paulauskas, *A new estimator for a tail index*, Proceedings of the Eighth Vilnius Conference on Probability Theory and Mathematical Statistics, Part II (2002), vol. 79, 2003, pp. 55–67.
184. P. H. Peskun, *Optimum Monte-Carlo sampling using Markov chains*, Biometrika **60** (1973), 607–612.
185. D. T. Pham and D. Karaboga, *Intelligent optimisation techniques: Genetic algorithms, tabu search, simulated annealing and neural networks*, Springer-Verlag, London, 2000.
186. J. Pickands, *Statistical inference using extreme order statistics*, Ann. Statist. **3** (1975), 119–131.
187. J. Pillard and L. Piela, *Smoothing technique of global optimization: distance scaling method in seaarch for most stable lenard-jones atomic clusters*, J. Comput. Chem. **18** (1997), 2040–2049.
188. J. Pintér, *Convergence properties of stochastic optimization procedures*, Math. Operationsforsch. Statist. Ser. Optim. **15** (1984), no. 3, 405–427.
189. J. Pinter, *Extended univariate algorithms for n-dimensional global optimization*, Computing **36** (1986), 91–103.
190. ———, *Global optimization in action*, Kluwer: Dodrecht, 1996.
191. ———, *Global optimization: software, test problems, and applications*, Handbook of Global Optimization, Volume 2 (P. Pardalos and E. Romeijn, eds.), Kluwer: Dodrecht, 2002, pp. 515–569.
192. L. Plaskota, *Noisy information and computational complexity*, Cambridge University Press, 1996.
193. L. Pronzato, H.P. Wynn, and A. Zhigljavsky, *Dynamical search*, Chapman & Hall/CRC, Boca Raton, FL, 2000.
194. D.J. Reaume, H.E. Romeijn, and R.L. Smith, *Implementing pure adaptive search for global optimization using Markov chain sampling*, J. Global Optim. **20** (2001), no. 1, 33–47.
195. C.R. Reeves and J.E. Rowe, *Genetic algorithms: principles and perspectives*, Kluwer Academic Publishers, Boston, MA, 2003.
196. A. Rényi, *On the extreme elements of observations*, Selected papers of Alfréd Rényi, Vol. III (Pál Turán, ed.), Akadémiai Kiadó, Budapest, 1976, pp. 50–65.

197. S. Resnick, *Extreme values, regular variation, and point processes*, Springer-Verlag, New York, 1987.
198. A. H. G. Rinnooy Kan and G. T. Timmer, *Stochastic global optimization methods. II. Multilevel methods*, Math. Programming **39** (1987a), no. 1, 57–78.
199. ———, *Stochastic global optimization methods. I. Clustering methods*, Math. Programming **39** (1987b), no. 1, 27–56.
200. K. Ritter, *Approximation and optimization on the Wiener space*, J. Complexity **6** (1990), 337–364.
201. K. Ritter, *Average-case analysis of numerical problems, Lecture Notes in Mathematics, vol. 1733*, Springer, 2000.
202. C.P. Robert and G. Casella, *Monte Carlo statistical methods*, Springer-Verlag, New York, 1999.
203. G.O. Roberts and J.S. Rosenthal, *General state space Markov chains and MCMC algorithms*, Probab. Surv. **1** (2004), 20–71 (electronic).
204. C. A. Rogers, *Packing and covering*, Cambridge University Press, New York, 1964.
205. S.K. Sahu and A. Zhigljavsky, *Self-regenerative Markov chain Monte Carlo with adaptation*, Bernoulli **9** (2003), no. 3, 395–422.
206. P. Salamon, P. Sibani, and R. Frost, *Facts, conjectures, and improvements for simulated annealing*, SIAM, Philadelphia, PA, 2002.
207. P. Salamon, P. Sibani, and R.Frost, *Facts, conjectures and improvements for simulated annealing*, SIAM, 20002.
208. L. Savage, *Foundations of statistics*, J.Wiley, 1954.
209. F. Schoen, *A wide class of test functions for global optimization*, J. Global Optim. **3** (1993), 133–138.
210. F. Schoen, *Two-phase methods for global optimization*, Handbook of global optimization, Vol. 2, Kluwer Acad. Publ., Dordrecht, 2002, pp. 151–177.
211. L. Schwartz, *Analyse mathématique. I*, Hermann, Paris, 1967.
212. H.-P. Schwefel, *Evolution and optimum seeking*, J.Wiley, NY, 1995.
213. E. Seneta, *Regularly varying functions*, Springer-Verlag, Berlin, 1976.
214. Y. Sergeyev, *An information global optimization algorithm with local tuning*, SIAM J. Optimiz. **5** (1995), 858–870.
215. ———, *Global one-dimensional optimization using smooth auxiliary functions*, Math. Program. **81** (1998), 127–146.
216. ———, *Parallel information algorithm with local tuning for solving multidimensional global optimization problems*, J. Global Optim. **15** (1999), 157–167.
217. Brooks S.H., *Discussion of random methods for locating surface maxima*, Operations Research **6** (1958), 244–251.
218. ———, *A comparison of maximum-seeking methods*, Operations Research **7** (1959), 430–457.
219. I. Shagen, *stochastic interpolation applied to the optimization of expensive objective functions*, COMPSTAT 1980 (M. Barritt and D. Wishart, eds.), Physika Verlag: Vienna, 1980, pp. 302–307.
220. ———, *Internal modelling of objective functions for global optimization*, J. Optimiz. Theory Appl. **51** (1986), 345–353.
221. V. Shaltenis, *On a method of multiextremal optimization*, Avtomatika i Vychislitelnaja Technika (1971), no. 3, 33–38, in Russian.
222. ———, *Structure analysis of optimization problems*, Vilnius: Mokslas, in Russian, 1989.

223. Y. Shang and B. Wah, *Global optimization for neural network training*, IEEE Computer **29** (1996), 45–54.
224. L. A. Shepp, *The joint density of the maximum and its location for a wiener process with drift*, J. Appl. Probab. **16** (1976), 423–427.
225. H. Sherali, *Thight relaxations from nonconvex optimization problems using the reformulation-linearization/convexification technique (rlt)*, Handbook of Global Optimization (P. Pardalos and H. Romeijn, eds.), Kluwer: Dodrecht, 2002, pp. 1–63.
226. A. N. Shiryaev, *Probability*, second ed., Springer-Verlag, New York, 1996.
227. S.A. Sisson, *Transdimensional Markov chains: a decade of progress and future perspectives*, J. Amer. Statist. Assoc. **100** (2005), no. 471, 1077–1089.
228. R. L. Smith, *Estimating tails of probability distributions*, Ann. Statist. **15** (1987), no. 3, 1174–1207.
229. F.J. Solis and R.J.-B. Wets, *Minimization by random search techniques*, Math. Oper. Res. **6** (1981), no. 1, 19–30.
230. A. J. Stam, *Independent Poisson processes generated by record values and inter-record times*, Stochastic Process. Appl. **19** (1985), no. 2, 315–325.
231. M. Stein, *Interpolation of spatial data*, Springer, 1999.
232. R. Strongin, *Numerical methods in multiextrtemal optimization*, Nauka, 1978, in Russian.
233. R. Strongin and Y. Sergeyev, *Global optimization with non-convex constraints*, Kluwer, 2000.
234. A. Sukharev, *Minimax algorithms in problems of numerical analysis*, Nauka: Moskow, 1989, in Russian.
235. A.G. Sukharev, *Optimal strategies of the search for an extremum*, Zh. Vychislit. Math. and Math. Phys. (Russian) **1** (1971).
236. W. Szpanowski, *Average case analysis of algorithms*, J.Wiley: New York, 2001.
237. M. Tawarmalani and N. Sahinidis, *Convexification and global optimization in mixed-integer nonlinear programming*, Kluwer Academic Publishers: Dordrecht, 2002.
238. L. Tierney, *Markov chains for exploring posterior distributions*, Ann. Statist. **22** (1994), no. 4, 1701–1762, With discussion and a rejoinder by the author.
239. A.S. Tikhomirov, *Markov sequences as optimization algorithms*, Model-Oriented Data Analysis. Proceedings of the 3rd International Workshop in Petrodvorets. Russia. 25-30 May 1992., Physica-Verlag, Heidelberg, 1993, pp. 249–256.
240. ———, *On the rate of convergence of Markov random search methods*, Vestnik of Novgorod State University (1996), no. 3, 90–92 (in Russian).
241. ———, *On the rate of convergence of Markov monotonous random search methods*, Vestnik of Novgorod State University (1997), no. 5, 66–68 (in Russian).
242. ———, *Optimal markov monotonic symmetric random search*, Computational Mathematics and Mathematical Physics **38** (1998), no. 12, 1894–1902.
243. ———, *On the Markov homogeneous optimization method*, Computational Mathematics and Mathematical Physics **46** (2006), no. 3, 361 – 375.
244. A.S. Tikhomirov, T.Yu. Stojunina, and V.V. Nekrutkin, *Monotonous random search on a torus: Integral upper bounds of the complexity*, Journal of Statistical Planning and Inference (2007), (submitted).

245. A.S. Tikhomirov and Nekrutkin V.V., *Markov monotonous random search of an extremum. Review of theoretical results*, Mathematical models. Theory and applications, vol. 4, University of St. Petersburg, St.Petersburg (in Russian), 2004, pp. 3–47.
246. L. Timonov, *An algorithm for search of a global extremum*, Izv. Acad. Nauk SSSR, Eng. Cybern. **15** (1977), 38–44, in Russian.
247. H. Tjelmeland and J. Eidsvik, *On the use of local optimizations within Metropolis-Hastings updates*, J. R. Stat. Soc. Ser. B Stat. Methodol. **66** (2004), no. 2, 411–427.
248. A. Törn and A.Žilinskas, *Global optimization*, Springer, 1989.
249. A. Törn and A. Žilinskas, *Parallel global optimization algorithms in optimal design*, Lecture Notes in Control and Information Scences, vol.143, Springer, 1990, pp. 951–960.
250. A. Törn and S. Viitanen, *Topographical global optimization using pre-sampled points*, J. Global Optim. **5** (1994), 267–276.
251. ______, *Iterative topographical global optimization*, State of the Art in Global Optimization (C. Floudas and P. Pardalos, eds.), Princeton University Press: Princeton, 1996, pp. 353–363.
252. A. Törn, S. Viitanen, and M. Ali, *Stochastic global optimization: problem classes and solution techniques*, J. Global Optim. **13** (1999), 437–444.
253. J. Traub, G. Wasilkowski, and H. Wozniakowski, *Information, uncertainty, complexity*, Addison-Wesley: NY, 1983.
254. J. Traub and H. Wozniakowski, *A general theory of optimal algorithms*, Academic Press: NY, 1980.
255. M. Trosset and R. Mathar, *On existence of nonglobal minimizers of the stress criterioin for metric multidimensional scaling*, Proceedings of the Statistical Computing Section, ASA (1997), 195–199.
256. H. Tuy, *Convex analysis and global optimization*, KAP: Dodrecht, 1998.
257. P. van der Watt, *A note on estimation of bounds of random variables*, Biometrika **67** (1980), no. 3, 712–714.
258. P.J.M. van Laarhoven and E.H.L. Aarts, *Simulated annealing: theory and applications*, D. Reidel Publishing Co., Dordrecht, 1987.
259. ______, *Simulated annealing: theory of the past, practice of the future?*, Proceedings of the Third European Conference on Mathematics in Industry, European Consort. Math. Indust., vol. 5, Teubner, Stuttgart, 1990, pp. 45–57.
260. E. M. Vaysbord and D. B. Yudin, *Multiextremal stochastic approximation*, Engineering Cybernetics. English Edition of Tekhnicheskaya Kibernetika (1968), no. 5, 1–11 (1969).
261. P. K. Venkatesh, M. H. Cohen, R. W. Carr, and A. M. Dean, *Bayesian methods for global optimization*, Physical Review E **55** (1997), no. 5, 6219–6232.
262. L. Weiss, *Asymptotic inference about a density function at an end of its range.*, Naval Res. Logist. Quart. **18** (1971), 111–114.
263. I. Weissman, *Confidence intervals for the threshold parameter. II. Unknown shape parameter*, Comm. Statist. A—Theory Methods **11** (1982), no. 21, 2451–2474.
264. D. Wingo, *Fitting three parameter lognormal model by numerical global optimization-an improved algorithm*, Comput. Stat. Data Anal. (1984), no. 2, 13–25.
265. G. R. Wood and Z. B. Zabinsky, *Stochastic adaptive search*, Handbook of global optimization, Vol. 2, Kluwer Acad. Publ., Dordrecht, 2002, pp. 231–249.

266. M. Wright, *Direct search methods: once scorned, now respectable*, Numerical Analysis 1995 (D. Griffits and G. Watson, eds.), Addison-Wesley: Reading, MA, 1996, pp. 191–208.
267. Z. B. Zabinsky, *Stochastic adaptive search for global optimization*, Kluwer Acad. Publ., Dordrecht, 2003.
268. Z. B. Zabinsky and R. L. Smith, *Pure adaptive search in global optimization*, Math. Programming **53** (1992), no. 3, Ser. A, 323–338.
269. Z. B. Zabinsky and G. R. Wood, *Implementation of stochastic adaptive search with hit-and-run as a generator*, Handbook of global optimization, Vol. 2, Kluwer Acad. Publ., Dordrecht, 2002, pp. 251–273.
270. J. Zamora and I. Grossman, *Continuous global optimization of structured process systems models*, Comp. Chem. Eng. **22** (1998), 1749–1770.
271. A. Zhigljavsky, *Mathematical theory of global random search*, Leningrad University Press, Leningrad, 1985, in Russian.
272. ———, *Branch and probability bound methods for global optimization*, Informatica **1** (1990), no. 1, 125–140.
273. ———, *Theory of global random search*, Kluwer Acad. Publ., Dordrecht, 1991.
274. ———, *Semiparametric statistical inference in global random search*, Acta Appl. Math. **33** (1993), no. 1, 69–88.
275. A. Zhigljavsky and M. V. Chekmasov, *Comparison of independent, stratified and random covering sample schemes in optimization problems*, Math. Comput. Modelling **23** (1996), no. 8-9, 97–110.
276. A. Zhigljavsky and A. Žilinskas, *Methods of search for global extremum*, Nauka: Moscow, 1991, in Russian.
277. R. Zieliński, *Global stochastic approximation: a review of results and some open problems*, Numerical techniques for stochastic systems (Conf., Gargnano, 1979), North-Holland, Amsterdam, 1980, pp. 379–386.
278. ———, *A statistical estimate of the structure of multi-extremal problems*, Math. Programming **21** (1981), no. 3, 348–356.
279. A. Žilinskas, *One-step Bayesian method for the search of the optimium of one-variable functions*, Cybernetics (1975), no. 1, 139–144, in Russian.
280. ———, *On global one-dimensional optimization*, Izv. Acad. Nauk USSR, Eng. Cybern., **4** (1976), 71–74, in Russian.
281. ———, *On one-dimensional multimodal optimization*, Trans. of Eighth Prague Conf. on Inform. Theory, Stat. Dec. Functions, Random Processes, vol. B, Reidel: Dodrecht, 1978, pp. 393–402.
282. ———, *On statistical models for multimodal optimization*, Math. Operat. Stat., ser. Statistics **9** (1978), 255–266.
283. ———, *Optimization of one-dimensional multimodal functions, algorithm 133*, Appl. Stat. **23** (1978), 367–385.
284. ———, *Axiomatic approach to extrapolation problem under uncertainty*, Automatics and Remote Control (1979), no. 12, 66–70, in Russian.
285. ———, *MIMUN-optimization of one-dimensional multimodal functions in the presence of noise, algoritmus 44*, Aplikace Matematiky **25** (1980), 392–402.
286. ———, *Two algorithms for one-dimensional multimodal minimization*, Math. Operat. Stat., ser. Optimization **12** (1981), 53–63.
287. ———, *Axiomatic approach to statistical models and their use in multimodal optimization theory*, Math. Program. **22** (1982), 104–116.
288. ———, *Axiomatic characterization of a global optimization algorithm and investigation of its search strategies*, Operat. Res. Letters **4** (1985), 35–39.

289. ______, *Global optimization: Axiomatics of statistical models, algorithms and their application*, Mokslas: Vilnius, 1986, (in Russian).
290. ______, *Note on Pinter's paper*, Optimization **19** (1988), 195.
291. ______, *A note on "Extended Univariate Algorithms" by J.Pinter*, Computing **41** (1989), 275–276.
292. ______, *On convergence of algorithms for broad classes of objective functions*, Informatica **3** (1992), no. 2, 275–279.
293. ______, *A review of statistical models for global optimization*, J. Global Optim. **2** (1992), 145–153.
294. ______, *A quadratically converging algorithm of multidimensional scaling*, Informatica **7** (1996), 268–274.
295. ______, *Statistical models for global optimization by means of select and clone*, Optimization **48** (2000), 117–135.
296. ______, *Hybrid search for optimum in a small implicitly defined region*, Control and Cybernetics **33** (2004), no. 4, 599–609.
297. A. Žilinskas and et all, *On intelligent optimization in bio-medical data analysis and visualization*, IDAMAP, Workshop at MEDINFO 2001, see also http://magix.fri.uni-lj.si/idamap/idamap2001/papers/zilinskas.pdf (2001), 52–55.
298. ______, *Adaptive search for optimum in a problem of oil stabilization process design*, Adaptive Computing in Design and Manufacture **VI** (2004), 87–98.
299. A. Žilinskas, E. Fraga, and A. Mackute, *Data analysis and visualization for robust multi-criteria process optimization*, Comp. Chem. Eng. **30** (2006), 1061–1071.
300. A. Žilinskas and A. Katkauskaite, *On existence of a random function compatible with the relation of conditional likelihood*, Cybernetics (1982), no. 4, 80–83, in Russian.
301. A. Žilinskas and J. Mockus, *On a Bayesian method of search of the minimum*, Avtomatika i Vychislitelnaja Technika (1972), no. 4, 42–44, in Russian.
302. A. Žilinskas and A. Podlipskyte, *On multimodality of the sstress criterion for metric multidimensional scaling*, Informatica **14** (2003), 121–130.
303. A. Žilinskas and E. Senkene, *On estimation of the parameter of a Wiener process*, Lith. Math. J. (1978), no. 3, 59–62, in Russian.
304. A. Žilinskas and E. Senkiene, *On estimation of the parameter of a Wiener random field using observations at random dependent points*, Cybernetics (1979), no. 6, 107–109, in Russian.
305. A. Žilinskas and J. Žilinskas, *Global optimization based on a statistical model and simplicial partitioning*, Comp. Math. Appl. **44** (2002), 957–967.

Index